Preface

The term tribology is scarcely twenty-five years old and yet there can be few university or college courses in mechanical engineering which do not now include material under this heading. Of course, the problem of producing bearings, slides, seals, and other tribological systems to give smooth machine running and long component lives is one which has faced practitioners for generations, and consideration of their design has always played a part in the education of mechanical engineers. What has become increasingly obvious in recent years is the inherently interdisciplinary nature of the tribologist's task; as well as involving practising and academic engineers, advances in the subject have drawn upon the ingenuity and expertise of physicists, chemists, metallurgists, and material scientists. Consequently, although envisaged principally for use by final year undergraduates and post-graduate students in mechanical engineering, I hope that this volume may be of interest to students and specialists in these other related areas. Tribology is still very much an area of active research and the published literature in the fields of lubrication, friction and wear — already dauntingly voluminous — continues to grow at an alarming rate. I have made no attempt to produce a research monograph but rather to provide a framework of fundamental analytical tools which can be used in a wide variety of different physical situations. Each chapter is concluded with a short list of suggestions for further reading which provide access to the more specialised literature.

As well as introducing some of the more important aspects of material behaviour referred to throughout the text, Chapter 1 includes a short historical survey of the subject. The analytical aspects of engineering tribology can be said to have their foundation just over a century ago when two important landmarks stand out. In 1882 Heinrich Hertz published his classic work on the dry contact of solid bodies and, four years later, Osborne Reynolds, stimulated by the then surprising experimental observations on lubricated journal bearing performance made by Beauchamp Tower, produced his seminal paper on which all subsequent work on hydrodynamics depends. Progress over the succeeding century in our understanding of how a load is transmitted from one solid body to another can be seen as the gradual relaxation of the restrictions imposed by the initial Hertzian treatment, while the corresponding development in the analysis of bearings and lubricated contacts has involved broadening the range of effects incorporated into Reynolds' model. These two strands merge elegantly in the relatively new, and still developing, area of elasto-hydrodynamics.

Tribology is essentially concerned with surfaces and Chapter 2 contains brief descriptions of the principal methods of their qualitative observation and quantitative measurement. The following chapter, a precursor to those covering the topics of friction and wear, describes those elements of the discipline of contact mechanics which the tribologist is likely to encounter. Wherever possible, explicit equations for the induced stress states have been used with the aim of retaining the physical insight provided by analytical expressions which can be so easily lost in particular numerical solutions. Practical problems associated with friction are as old as the design of any sort of mechanical device and have been the subject of experimental investigation for more than three hundred years; the phenomenon of wear — the inevitable companion of friction and often of rather greater economic significance — has been the recipient of similar attention for only about fifty years. Our understanding of these important physical effects has developed extremely rapidly over the last twenty years or so, aided particularly by the widespread use of electron microscopy and other instrumental methods of surface micro-analysis; nevertheless there are many questions on the fundamental mechanisms of friction and wear that remain less then fully answered and so Chapters 4 and 5, which deal with these topics, can be considered only as statements of our current understanding which will certainly be modified by future advances.

The majority of engineering bearings operate in the presence of a lubricant film generated between the solid surfaces which is sufficiently thick to physically separate them and thus overcome the adverse effects of friction and wear associated with solid-to-solid contact. The pressure within the film which forces the component surfaces apart is either provided by some external pressure source, as in hydrostatic bearings, or generated within the confines of the bearing itself through the mechanisms of hydrodynamics. These two regimes are analysed in Chapters 6 and 7 and a number of design guidelines for satisfactory bearing operation presented. Stresses within such concentrated contacts, particularly those in which the geometry is nonconformal, as occur in rolling-element bearings or between loaded gear teeth, may be sufficiently large to modify both the shape of the solid surfaces and the rheological behaviour of the fluid: Chapter 8 describes some of these effects which play such an important role in the area of elasto-hydrodynamics.

The various regimes of lubricant action can be graphically displayed on a curve of friction coefficient versus the severity of the loading — the well-known Stribeck curve. As the intensity of the loading is increased so the operating point of the contact moves from hydrodynamic conditions through those of elasto-hydrodynamics into the so-called 'mixed' regime of operation where the surface chemistry of the lubricant starts to be of comparable significance to its bulk rheological properties. Under still more arduous conditions, chemical effects become dominant and the state of affairs becomes one of boundary lubrication, the subject of Chapter 9: here it is more difficult to produce the neat, quantitative analyses preferred by engineers. Chapter 10 describes a number of dry running contacts, of both low and high friction, as well as so-called 'marginally lubricated' bearings. Finally, Chapter 11 is concerned specifically with mechanics of rolling which finds such important applications in all the various forms of rolling-element bearings. The appendices at the end of the body of the text contain a resumé of some of the 'essential' design formulae as well as some characteristic or typical numerical material data which are useful when carrying out initial or trial calculations.

The greater part of the material contained in this book is based upon a lecture course which has for some years been available to final year engineering students at Cambridge University and I am grateful to colleagues on the teaching staff for their advice and comments in its preparation: in particular I should like to thank Professor K. L. Johnson and Dr J. A. Greenwood. Several chapters are provided with a number of numerical problems and where appropriate I am grateful to the University for permission for their inclusion. I should also like to take this opportunity to thank Professor D. Tabor for his continuing support and encouragement over a number of years as well as Professor T. Farris at Purdue and Dr N. A. Fleck and Dr A. J. Kapoor in Cambridge for a number of very helpful comments on the manuscript.

Cambridge
November 1993   J. A. W.

Contents

xii Contents

Nomenclature

Except where otherwise stated symbols have the following meanings.

General

A	area
a	half-width of contact strip; radius of contact circle
α	pressure viscosity index
$\mathcal{A}, \mathcal{B}, \mathcal{C}, \mathcal{D}$	constants
B	bearing dimension
b	half-width of region of sticking friction
c	radial clearance
c	specific heat
D	diameter
Δ	mutual approach of two solid bodies
δ	displacement of centre of curvature of solid body or surface asperity
De	Deborrah number
E	elastic modulus
E'	reduced elastic modulus
e	bearing eccentricity
E^*	contact modulus
f	friction angle
F	friction or tangential force
G	shear modulus
γ	shear strain
$\mathcal{H}$	hardness
$\bar{h}$	value of h at which $dp/dx = 0$
$\dot{H}$	rate of generation, or flow, of thermal energy
$\dot{h}$	steady rate of thermal energy input per unit area
H	quantity of thermal energy
H	ratio h_i/h_o
h	film or gap thickness
$\hslash$	sampling interval

η	dynamic viscosity
ϑ	angle of asperity or surface slope
K	Archard or dimensionless wear coefficient
k	dimensional wear coefficient $K/\mathcal{H}$
K	thermal conductivity
$\hbar$	material flow stress in pure shear
k	stiffness
κ	thermal diffusivity
λ	damping coefficient or wavelength
Λ	ratio of film thickness to surface roughness
L	contact pattern or bearing pad dimension
M	torque or moment
$\dot{m}$	mass flow rate
μ	coefficient of friction
μ_R	coefficient of rolling friction
N	number of points, asperities, pads, etc.
$\mathfrak{N}$	rotational speed, revolutions per second
ν	Poisson's ratio
p	pressure above ambient
p^*	non-dimensional pressure
p_a	atmospheric pressure
Pe	Peclet number
p_o	peak pressure
Θ	temperature
Q	volumetric flow rate
θ	polar coordinate, or angle
q	volumetric flow rate per unit width
Q^*	non-dimensional flow
R	radius of curvature
R	reaction force
ρ	material density
r	radius or radial polar coordinate
R_a	centre-line average roughness
Re	Reynolds number
R_q	root mean square roughness
R_t	peak to valley height
R_z	average value of R_t (DIN), or ten-point height (ISO)
S	slide: roll ratio
S	Sommerfeld number $\eta\omega LD/W(R/c)^2$
Σ	squeeze number
σ	direct stress
S'	US Sommerfeld number $\eta\mathfrak{N}LD/W(R/c)^2$
Sk	skewness

$\mathfrak{I}$	time constant
τ	shear or traction stress
t	time
τ', τ''	surface tractions
τ_1	principal shear stress
Ta	Taylor number
τ_i	shear stress to initiate flow
U	velocity component in Ox-direction
$\bar{U}$	entraining velocity in Ox-direction
u	general displacement or velocity within bulk in Ox-direction
V	velocity component in Oy-direction
$\bar{V}$	entraining velocity in Oy-direction
$\mathcal{W}$	component of velocity in Oz-direction
W	normal load on a surface
Ω	rotational speed, rad s^{-1}
ω	rotational speed, rad s^{-1}
w	surface displacement
$\mathfrak{w}$	wear rate
W^*	non-dimensional load
x, y, z	Cartesian coordinates
$\bar{x}$	value of x at which $dp/dx = 0$
Y	material yield stress in simple tension
Ψ	plasticity index
ξ	non-dimensional position coordinate $|x/a|$
ψ	attitude angle
R	molar gas constant

Subscripts

1, 2	surface of component 1 and 2
i, o	conditions at inlet (or entry) and outlet (or exit)
m	mean or average value
r, θ, z	cylindrical polar axes
x, y, z	Cartesian coordinate axes
c	central or critical
min	minimum
max	maximum
nom	nominal
eff, e	effective
op	optimum
mp	melting-point

Chapter 1

α	pressure viscosity index
β	temperature viscosity parameter
L, U, H	values of viscosity in determination of viscosity index (VI)
$\mathcal{H}_B$	Brinell hardness
$\mathcal{H}_K$	Knoop hardness
$\mathcal{R}_A, \mathcal{R}_B, \mathcal{R}_C$	Rockwell hardness numbers

Chapter 2

β	correlation length
β^*	correlation length at which $\rho(\tau) = 0.1$
D	fractal dimension
δ	angle of taper section
$\phi(z)$	probability density function
K	kurtosis
κ	curvature
L	sampling length
m_i	ith moment of area of probability density function
$\bar{m}, \bar{\kappa}$	root mean square values of m and κ
n	number of sampled points
n_p, n_s	number of peaks and summits
$\rho(\tau)$	autocorrelation function
σ^2	variance of a statistical distribution
σ	standard deviation of a statistical distribution
τ	spatial delay
V_a	accelerating voltage
z	height of surface above mean

Chapter 3

$\bar{p}$	hydrostatic stress in core beneath indenter
b	half-width of region of sticking friction
c	limit of elastic zone under indenter
χ	non-dimensional roughness parameter $\sigma R/a^2$
Δu_i	magnitudes of tangential velocity discontinuities in upper bound calculations
ε	direct strain
$\phi(x, z), \phi(r, \theta)$	Airy stress functions

$\phi(z_s)$	surface roughness probability density function
η	proportion of thermal energy
κ	thermal diffusivity
λ	surface roughness constant
r_0	constant, defining shape of deformed surface
R_s	characteristic summit radius
s	dummy variable or surface slip
s_i	lengths of interfaces in upper bound calculations
σ_s	standard deviation of summit heights
T	dimensional time parameter
W^S	shakedown load
W^Y, p_0^Y, p_m^Y	values of W, p_0, and p_m at yield
X	dimensionless group $Ux/2\kappa$
z_s	surface summit height

Chapter 4

α_1, α_2	coefficients in material yield criterion
α_B	contact patch radius at pull-off
γ	surface energy
μ_s, μ_k	static and kinetic coefficients of friction
p'	ploughing pressure
Θ_g	glass transition temperature
w	width of wear track
W_B	pull-off load

Chapter 5

ϕ	asperity dihedral half-angle
f_A^*	area fraction of inclusions for shear failure
$\tilde{p}$	non-dimensional pressure
$\tilde{w}$	non-dimensional wear rate
$\tilde{U}$	non-dimensional speed
f_m	fraction of molten oxide lost from surface
f_v	volume fraction of inclusions
γ_0	plastic shear strain accumulated per cycle
Q_0	activation energy
ψ	angle of attack
ε	specific weight loss in erosion
V	abrasive particle velocity

Chapter 6

C_d	discharge coefficient
H_f	friction power
h_o, Q_o, k_o	conditions at design or operating point
H_p	bearing pumping power
H_{tot}	total power
k	bearing stiffness
λ_b, λ_s	damping coefficients of bearing and structure
p_r	recess pressure
p_s	supply pressure
c	bearing land width
w, l	width and length of rectangular pad

Chapter 7

B	length of pad bearing
β	angle of bearing arc
χ	value of x/B at centre of pressure
Δh	$(h_i - h_o)$
f	friction factor
P	ratio B_i/B_o
p_s	pressure at position of step
R	total contact force
u_c	velocity of contact patch
$\xi = \arctan \gamma$	Sommerfeld variable $x/\sqrt{2Rh_o}$
ψ	attitude angle

Chapter 8

B	non-dimensional bearing group
Δt	event duration
g_1	elasto-hydrodynamic viscosity parameter
g_2	elasto-hydrodynamic pressure parameter
g_3	elasto-hydrodynamic elasticity parameter
g_4	elasto-hydrodynamic 'character'
H	ratio h/h_o
h^*	elasto-hydrodynamic film thickness parameter
k	fluid consistency
Λ	bearing or compressibility number
P	ratio p/p_a

T	relaxation time
X	non-dimensional position x/B
ζ	non-dimensional group $\mathsf{B}\xi$

Chapter 9

r_a	fractional reduction in cross-sectional area
α	die semi-angle
m	friction factor

Chapter 10

$\dot{H}$	rate of frictional work
B	thickness of porous layer
B_f	seal balance factor
Φ	material permeability factor
P	total spring force
R	reaction force
s	brake sensitivity
T, T'	belt tensions
Ψ	non-dimensional design parameter $B\Phi/c^3$

Chapter 11

α	fractional hysteretic energy loss
β	elastic material parameter
d	offset of sticking zone in tractive rolling contact
$\mathcal{F}$	probability of failure
Γ	conformity of rolling contact
γ	contact angle in angular contact ball bearing
L_{10}	fatigue life for 10 per cent probability of failure
L_s	fatigue life for probability of failure s
ξ	creep ratio
$\bar{U}$	rolling velocity
$\dot{s}$	velocity of microslip
N	number of rolling elements
p_0^S	maximum Hertz shakedown pressure

1
Introduction

1.1 Tribology

The word *tribology* was coined only just over twenty years ago and appears in only the most up to date of dictionaries; however, the topics with which tribologists are concerned have been of vital interest to scientists, engineers, and those who design or operate machinery, for as long as mechanical devices have existed. Formally, tribology is defined as the *science and technology of interacting surfaces in relative motion and of related subjects and practices*; it deals with every aspect of friction, lubrication, and wear. The word is derived from the Greek τριβοσ (TRIBOS) meaning rubbing, although the subject embraces a great deal more than just the study of rubbing surfaces.

Perhaps as much as one third of our global energy consumption is consumed wastefully in friction: at a time when energy resources are at a premium the contribution that can be made to their efficient utilization, as well as to the reduction of pollution, by making use of the best tribological practices is obvious. In addition to this primary saving of energy there are very significant additional economies to be made by reductions in the cost involved in the manufacture and replacement of prematurely worn components. An important landmark in the development of the subject was the publication in 1966 in Great Britain of the report of the government committee, chaired by Mr Peter Jost, which had been formed to report on the position of industrial lubrication in the United Kingdom: it was asked specifically to identify those areas of industrial practice where significant improvements could be made. The committee's report (HMSO 1966) concluded that, by utilizing the best techniques in machine design and operation, annual savings of more than £500 million sterling (at 1966 levels) could be made: the sources of these economies are illustrated in Fig. 1.1.

Subsequent studies in many other countries, including the United States, Germany, Australia, and China, have come to similar conclusions about the potential impact of tribology on their national economies. For the engineering designer a significant outcome of this early initiative was the publication of the first edition of the *Tribology Handbook* (Neale 1973) which contains a wealth of practical advice and data on the design and application of

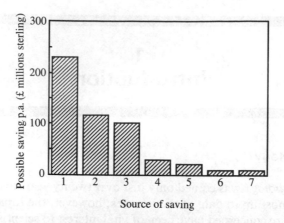

Fig. 1.1 Savings indicated by the Jost report by improved tribological practices in the UK (1966 levels). 1: reduced maintenance and replacement; 2: fewer breakdowns; 3: longer machine life; 4: reduced frictional dissipation; 5: savings in investment; 6: savings in lubricants; and 7: savings in manpower (from HMSO 1966).

tribological elements in an easily accessible form, as does the more recently published *Wear Control Handbook* (Winer and Peterson 1978) and *Handbook of Tribology* (Bhushan and Gupta 1991). All these volumes remain important and useful sources of information.

For most engineers friction is really rather a nuisance. Of course, there are obvious exceptions to this — devices such as clutches and brakes, friction and traction drives, and so on, in which friction plays a vital role in machine operation. However, the aim of the engineering designer is virtually always to bring about the transmission of mechanical power with the lowest possible frictional losses. When loads are to be transmitted between surfaces which move relative to one another some form of *bearing* is required, and in practice there are a limited number of distinct ways of satisfying this requirement. A bearing can be formally defined as a device which permits two components in a mechanism to move relative to one another in either one or two dimensions while constraining their movement in the remaining dimensions. The simplest arrangement which, while supporting substantial normal loads, allows motion in one direction only is the linear bearing, illustrated in Fig. 1.2(a). In even more common engineering use than this is the journal bearing which permits a cylindrical shaft, the *journal*, to rotate while carrying a radial load, Fig. 1.2(b). A thrust bearing also allows rotation of the shaft but now sustains a load parallel with the axis of rotation, Fig. 1.2(c); some designs of bearing can carry both radial and thrust loads.

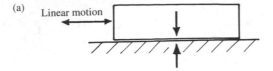

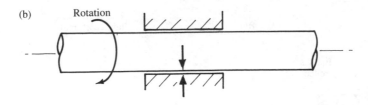

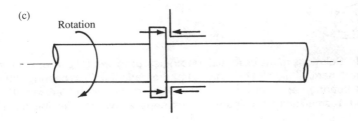

Fig. 1.2 Three types of important engineering bearing illustrated schematically: (a) linear bearing carrying a normal load; (b) journal bearing sustaining a radial load; and (c) thrust bearing carrying an axial load.

There are a limited number of ways of satisfying both the kinematic and load-carrying requirements illustrated in Fig. 1.2. The general classes of possible design solutions are illustrated diagrammatically in Fig. 1.3; note that in these diagrams the sizes of the bearing clearances and thickness of the fluid films are shown greatly exaggerated.

In *dry bearings*, case (a), the loaded surfaces are permitted to rub together without the provision of a lubricant. The surfaces should be chosen to have intrinsically low friction and/or wear characteristics and this usually means that one of them will be non-metallic. The surfaces may be coated or treated to have different mechanical or chemical properties from the bulk so as to improve their tribological performance: these are examples of *surface engineering*. A variant on this design is one in which the bearing sleeve is

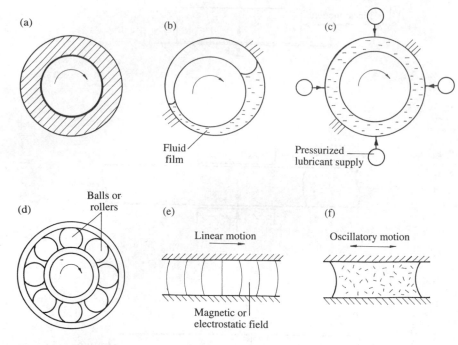

Fig. 1.3   Solutions to the essential tribological problem: (a) dry rubbing bearing; (b) hydrodynamic fluid film bearing; (c) hydrostatic fluid film bearing; (d) rolling-element bearing; (e) linear motion allowed by magnetic or electrostatic levitation; and (f) small-scale oscillation accommodated within a flexible elastomeric block.

made of a porous metal which has been impregnated with a liquid lubricant or a low-friction solid. These are examples of *marginally lubricated bearings*.

In *fluid film bearings*, cases (b) and (c), the loaded surfaces are separated by a continuous lubricating film of either a liquid or a vapour or gas. Relative sliding motion takes place within the film, so that energy losses are associated with viscous shearing rather than solid sliding. The film must be adequately pressurized to carry the normal load applied to the surfaces without their coming into direct physical contact. This pressure can be produced within the film itself, as a result of the combination of surface geometry and relative motion, to give *hydrodynamic* (or, if the film is air, *aerodynamic*) support, or be generated by an external pump as is the case in *hydrostatic* (or *aerostatic*) bearings.

Rolling-element bearings, case (d), are designed in such a way that one surface is allowed to roll on the other. This clearly imposes a geometric or kinematic limitation on the type of motion that is possible and to overcome this restriction either there must be a combination of rolling and sliding or, as is more usual, a third body (or bodies) such as a ball or roller is placed between the bearing surfaces. The wide range of rolling-element bearings commercially available, in terms of physical size and load-carrying capacity, is evidence of the value of this form of solution.

In bearings depending on *magnetic* or *electrostatic levitation*, case (e), a positive load-carrying capacity is generated by the mechanisms of either magnetic or electrostatic repulsion. Although a very specialized solution to the basic tribological design problem, examples can be found both of very lightly loaded instrument bearings as well as bearings operating on the same principle but carrying massively greater loads — for example, those in magnetically levitated vehicles.

In the special case of small-amplitude oscillatory motion the two surfaces may be joined together either by an elastomeric block or *flexible bushing* case (f), or by flexible metal strips. Examples of the former can be found in vehicle suspension bushes while metallic crossed-strip flexible pivots can be attractive is some instances (for example, in some instrumentation applications) because of their very low friction losses.

The dry and fluid film bearings illustrated above as cases (a)–(c) are varieties of *lower* kinematic pairs. They depend on either sliding or turning, and use a geometry in which the load is transferred from one element to the other over a comparatively large surface area. They are examples of what are known as *plain* bearings and involve surfaces in matching or *conformal* contact. The geometry of rolling-element bearings leads to contacts which are *non-conformal* (or *counter-conformal*) and produce, at least in principle, contacts at either single points (in the case of ball bearings) or along straight lines (as occur in roller bearings). In such *higher* kinematic pairs there may be a combination of rolling and sliding. Other familiar engineering contacts of this sort are those that occur between a wheel and a rail or roadway, or between two meshing gear teeth.

For the simple case involving the support of a loaded, rotating shaft (or journal) the two most immediately obvious imposed service conditions are the radial load and the speed of rotation. Figure 1.4 can be used to give a swift indication of the limits of the performance of three types of bearing most appropriate for this situation, that is, types (a), (b), or (d) of Fig. 1.3. For the cases of the plain rubbing bearing and the hydrodynamic fluid film bearing the axial length has been assumed to be equal to the bearing diameter D; the bearing is said to be 'square'. At low speeds dry, or marginally lubricated, bearings perform as well as those involving more complex arrangements of rolling elements but, as the speed is increased, rolling-element bearings offer progressively greater load-carrying capacities.

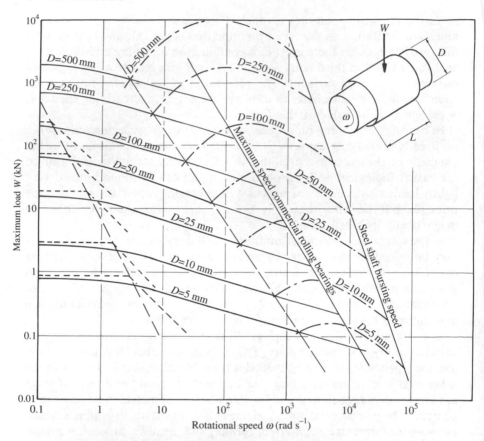

Fig. 1.4 Selection of journal bearings for continuous rotation (from Neale 1973): dry and marginally lubricated bearings (– – – – – – –); fluid film bearings (– – – – –); rolling-element bearings (——————).

Once the rotational speed is sufficiently high to maintain a hydrodynamic oil film, plain hydrodynamic journal bearings will invariably carry greater loads than rolling-element bearings of a similar size. Their speed is ultimately, in theory at least, limited by the bursting strength of the shaft. In many practical cases the particular operating environment or circumstances will impose more stringent requirements than the load-carrying capacity alone; some of these limitations are indicated in Table 1.1. In the case of oscillatory motion, hydrodynamic bearings would be inappropriate whereas it may be possible to use crossed-strip flexures.

Any bearing of the types illustrated in Fig. 1.3 has a finite practical life. In the case of elastomeric bushes or flexible metal strips, the bearing may

Table 1.1 Continuously rotating journal bearing—qualitative comparisons.

Type of bearing	Dry rubbing	Hydrodynamic fluid film	Hydrostatic fluid film	Rolling element
Accuracy of radial location	Fair	Fair	Excellent	Good
Axial load capacity	Some	None	None	Some
Low starting torque	Good	Fair	Excellent	Excellent
High-temperature performance	Limited by materials	Limited by lubricant	Limited by lubricant	Good, limited by materials
Lubricating system required	Simple	Usually requires circulation	Complex, high-pressure system	Relatively simple if sealed
Tolerance of dirt	Good if sealed	Needs sealing and filtering	Sealing and filtering essential	Good with sealing

fail suddenly because of some form of fatigue failure of the deformable element. Fluid film and rolling-element bearings can also suffer sudden, and sometimes catastrophic, failure because of material fatigue but are more likely to display a progressive loss of performance as their surfaces deteriorate with use—the bearing gradually wears out. Although this has long been recognized as a most important, if regrettable, feature of mechanical devices, the scientific study of the phenomena that are actually involved in wear is relatively recent. This is understandable when we remember that earlier generations of engineers, with limited sources of power at their disposal, were principally concerned to devise and design machines that would operate at all in spite of the effects of friction between their moving parts; it is only with our increasing confidence to do this successfully that greater attention can be focussed on designing for adequate life and low maintenance.

1.2 Bearing materials

In common with other engineering components the elements of both plain and rolling bearings must be capable of sustaining the loads applied to them without undergoing such dimensional or material changes that would constitute failure, or at least they must resist such changes for an acceptable lifetime. The mechanisms that can lead to the failure of tribological components are many and various; failure may be initiated on the actual load-bearing surface (perhaps by material fatigue associated with the periodic nature of the loading pattern), by corrosion, or perhaps by the intervention of some unforeseen event such as a bearing overload, the failure of the lubricant supply, or the action of abrasive contaminant particles swept into the bearing gap by the lubricant. In other circumstances the site of the initiation may be less obvious and lie below the component surface; it might again be associated with an excessive bearing load or arise from subsurface material imperfections or inclusions.

1.2.1 *Mechanical properties*

To the engineer, metallic materials are characterized by a number of relatively simple physical properties: the values of Young's modulus E and Poisson's ratio v are needed to evaluate the elastic strains (and thus recoverable changes in shape) from a knowledge of the component geometry and the applied loads. In a metallurgical alloy these elastic constants are governed by the principal constituent element and are little affected by relatively small changes in composition or by any heat treatments or thermal processing undergone by the component. With increasing loads, metals show a transition from elastic to plastic behaviour and this change often takes place at a characteristic stress. Unlike E and v, the value of this *yield stress* is extremely sensitive to small changes in composition, or prior mechanical working and heat treatment.

The 'strength' of a structural material is very commonly assessed by measuring its *hardness* and this property is of particular significance to the tribologist. The hardness of a solid can be loosely defined as the resistance of its surface to permanent indentation; it is usually expressed as the ratio of the applied load to the area of the resulting impression. However, this ratio is not a single distinct property but depends on a combination of more fundamental attributes (such as the yield stress in shear and the bulk modulus); it is also significantly affected by the shape of the hardness indenter used. A number of different tests have been developed: they range from the simple Mohs scratch test (proposed in 1822 and still used as a basic practical test by field mineralogists) to highly sophisticated microhardness tests that can distinguish between the individual metallurgical phases in a metallic alloy. The impression may be in the form of a scratch, or an indentation made by a steel or diamond tool under a static force, or it may be the depression remaining after a dynamic rebound test.

In the *Brinell* test, introduced in the early years of this century, a small (usually steel) sphere of diameter D (the standard value is 10 mm) is forced by a load W (measured in kgf and usually equal to 3000 kgf) into the surface of the specimen; the diameter d of the resultant circular dent is then measured by a low-powered microscope. The surface area of the dent is calculated on the assumption that it has the same shape as the undeformed ball: the Brinell hardness number $\mathcal{H}_B$ is given by the formula

$$\mathcal{H}_B = \frac{2W}{\pi D\{D - (D^2 - d^2)^{1/2}\}}. \tag{1.1}$$

As this has the dimensions of stress, or pressure, one might at first sight expect to obtain the same result in tests made at different loads on the same specimen; however, this is not the case because, as a result of the shape of the indenter, impressions made at differing loads do not have the same proportions—they are not *geometrically similar*. Despite this drawback the Brinell test is still widely used, particularly on large castings and forgings as the test equipment is rugged and well suited to use in open workshop conditions.

The *Rockwell* hardness test was developed about seventy years ago specifically to test hardened steels and similar materials used in rolling-element bearings. The test machine essentially measures the penetration of an indenter (conventionally either a diamond cone with an included angle of 120° or a steel ball) into the specimen surface. An initial *minor* load, of 10 kgf is first applied and the penetration depth indicator of the machine is set to zero. The load on the indenter is then progressively increased to a higher value, the *major* load, and then reduced again to the minor value. The depth-measuring device then shows the increased depth of penetration caused by the application of the higher load and the scale of the instrument is calibrated to read the Rockwell hardness number directly. A variety of major loads can be used and it is necessary to indicate which has been used by prefixing the hardness number by a letter; the most common combinations are shown in Table 1.2. The *Rockwell superficial hardness* test is similar in principle, but in order to keep the depth of the impression small the minor load is restricted to 3 kgf. Hard sheet materials as thin as 0.125 mm can be successfully tested and the test can also be used for coated or surface-treated specimens.

Geometric similarity between hardness impressions of different sizes can be achieved by using a *pyramidal* indenter. Two such geometries are now regularly in use. The *Vickers* hardness test (sometimes also known as the *diamond pyramid* hardness test) uses a square-based diamond pyramid as the indenter. To allow comparison with the established Brinell test an included angle between the faces of the pyramid of 136° was chosen (this corresponds to the mid-point of the recommended Brinell range). In the test the indenter is forced into the surface of the specimen and the resulting square impression is viewed in an optical microscope fitted with a form of linear measuring

Table 1.2   Rockwell hardness scales.

Scale	Indenter	Loads (kgf)		
		Minor	Major	Total
$\mathcal{R}_A$	Diamond Cone	10	50	60
$\mathcal{R}_C$	Diamond cone	10	140	150
$\mathcal{R}_B$	$\frac{1}{16}$-inch steel ball	10	90	100
$\mathcal{R}_{45\text{-}N}$	Diamond cone	3	45	48
$\mathcal{R}_{45\text{-}T}$	$\frac{1}{16}$-inch steel ball	3	45	48

device. A hardness number $\mathcal{H}$, defined as the indenting force divided by the surface area of the impression, is calculated from the formula

$$\mathcal{H} = \frac{1.854\,W}{D^2}.$$ (1.2)

This value, called the Vickers hardness number (VHN or HV) or sometimes the diamond pyramid number (DPN) is increasingly quoted in SI units, usually GPa, in place of the previously conventional units of kgf mm^{-2}; (1 GPa = 102 kgf mm^{-2}). The test is widely used in scientific circles and in Europe, but less so in the US. One particular advantage of the Vickers hardness test is that there is a useful effectively linear correlation between the value of $\mathcal{H}$ and the material yield stress—see section 3.5.1. The test is also capable of resolving small differences of hardness and is especially sensitive when carried out on the microscale, in other words, when the indentation is less than 0.1 mm in size. It is possible to test extremely small components or to identify differences in hardness between separate metallurgical phases although careful specimen preparation and alignment of the instrument is essential. The *Knoop* indenter was developed at the National Bureau of Standards in the US specifically to enable hardness tests to be made on frangible materials such as glass, gemstones, and ceramics. It has the geometry of a four-sided diamond pyramid of elongated shape in which the angle between the two opposite edges is 172° 30′ and that between the other two edges is 130°. The impression left by the indenter is a rhombus with the longer diagonal D about seven times the length the shorter and about thirty times the depth of the indentation. The Knoop hardness $\mathcal{H}_K$ is defined once again as the load divided by the projected area of the impression, in this case this means that

$$\mathcal{H}_K = \frac{14.23\,W}{D^2}.$$ (1.3)

A great many other devices have been developed and are available for the assessment of 'hardness'. The *Scleroscope* test differs from those described above in that it relies on dynamic effects rather than plastic indentation; a diamond-tipped rod, the hammer, is dropped from a fixed elevation on to the test surface and the height of the rebound measured. The difference between these two values in height is a measure of the energy lost in the deformation of the surface during the impact and therefore, in some way, of its mechanical strength properties. Devices of this sort offer the particular advantage of portability and are very convenient for on-site investigations.

Because hardness is so widely used in material specifications there is an inevitable demand for a universal conversion table between the various numerical scales on which it is measured. However, because each test samples material properties in a different way, no such universal relation is possible. Conversion tables have been established for some specific materials (Table 1.3 is such a table that can be used for carbon steels) but these are only approximations and values should not be extrapolated beyond the ranges specified.

Many tribological components are subject to fatigue, either because of rotating or reciprocating relative motion, or because of cyclical changes in

Table 1.3 Hardness conversion factors for steel.

HV		Brinell	Rockwell hardness numbers		
(kfg mm^{-2})	(GPa)	$\mathcal{H}_B$ (kgf mm^{-2})	$\mathcal{R}_C$	$\mathcal{R}_A$	$\mathcal{R}_{45\text{-}N}$
900	8.83		67.0	85.0	83.6
860	8.44	–	65.9	84.4	82.7
820	8.04	733	64.7	83.8	81.7
780	7.65	710	63.3	83.0	80.4
740	7.26	684	61.8	82.2	79.1
700	6.87	656	60.1	81.3	77.6
660	6.47	620	58.3	80.3	75.9
620	6.08	582	56.3	79.2	74.2
580	5.69	545	54.1	78.0	72.1
540	5.30	507	51.7	76.7	70.0
500	4.91	471	49.1	75.3	67.7
460	4.51	433	46.1	73.6	64.9
420	4.12	397	42.7	71.8	61.9
380	3.73	360	38.8	60.8	58.4
340	3.36	328	34.4	67.6	54.4
300	2.94	295	29.8	65.2	50.2
260	2.55	255	24.0	62.4	45.0
220	2.16	235	20.3	60.7	41.7
180	1.77	175	–	56	–
140	1.37	135	–	48	–

load magnitude (or possibly as a result of both effects). Ferritic steels with hardnesses below about 3 GPa exhibit a fatigue limit, that is, an amplitude of reciprocating stress below which their fatigue lives become effectively infinite. On the other hand, high-strength steels, aluminium alloys, and non-ferrous metals in general, do not in general show such behaviour. The absense of a fatigue limit in aluminium alloys is the reason for the mandatory inspection of highly stressed aluminium structures carried out in order to locate fatigue cracks and replace such damaged parts before the defects become dangerously large. In tribological components, failure by fatigue can be an especially important limitation to the lives of rolling-element bearings; see Section 11.2.3.

1.2.2 *Materials for plain bearings*

The greatest spur to the development of the plain bearing over the last fifty years has been the internal combustion engine: its importance can be judged from the fact that the annual world-wide production of plain crankshaft bearings is currently in excess of 1 billion. The geometry of typical hydrodynamic bearings and the conditions under which they operate are chosen to encourage the formation of lubricant films between the bearing surfaces: these films can be comparatively thick, that is, of a thickness dimension which is significantly greater than the sum of the expected asperities on the two opposing surfaces, so that no solid contact would be expected. In theory, therefore, there should be minimal wear — eventual loss of material depending on surface or subsurface fatigue. However, in practice such lubricated contacts do suffer degradation and often this is as a result of hard particulate contaminants being swept through the bearing gap. Such contaminants can arise from the external environment or may be wear debris from other friction pairs lubricated by the same service fluid. For long component lives lubricant cleanliness and filtration is of paramount importance: in addition, it is usual practice to protect the more important, or expensive, component (invariably the steel shaft or journal) by making it the harder element of the pair and using a much softer, and to some extent, sacrificial bearing or *sleeve*.

Imperfections in manufacture, deflections under load, or the collapse of the film under stop–start conditions can all lead to some degree of contact between the journal and the bearing. Rubbing at high loads and speeds brings with it the increased danger of material seizure or welding. To minimize this risk, bearing alloys are formulated for good compatibility, that is, a minimum tendency to weld to the shaft in areas of dry rubbing. Most bearing alloys contain a soft low-melting-point phase which smears over the bearing surface whenever high temperatures are generated thus delaying the onset of gross failure. In most cases this bearing alloy is applied as a thin lining (0.025–0.25 mm in thickness) over a steel backing or shell and it is this form of construction, in some cases with multiple layers, which enables modern machine bearings to carry such high specific surface loadings.

1.2.3 *Materials for rolling-element bearings*

The great majority of rolling-element bearings produced throughout the world are manufactured from a 1 per cent C, 1.5 per cent Cr steel known variously as SAE 52100, EN31, or 535A99. The material is spheroidized-annealed for ease of formability prior to machining to form both the races and the rolling elements, that is, the balls or rollers. Subsequent heat treatment, usually oil quenching and air tempering, produces a uniform martensitic structure throughout the section so providing both a high resistance to deformation and a long fatigue life. The final hardness or HV value is usually in excess of 7 GPa. Standard-quality bearings are made from steel melted in electric arc furnaces and then refined to high levels of purity in secondary steel making processes. These are designed to remove non-metallic impurities since these are liable to act as sites of eventual fatigue crack initiation. Oxide inclusions are particularly deleterious because, as well as being brittle, they are also surrounded by residual tensile stresses (arising from differences between their coefficients of thermal expansion and that of the matrix). Consequently, for applications where the load and fatigue requirements are particularly severe, the additional costs of vacuum arc refining, which produces extremely low inclusion levels, may be justified.

The high hardness and excellent temperature stability of ceramics make them attractive candidate materials for rolling-element bearing applications. In addition, their low density (compared to that of steel) leads to reduced centrifugal forces and so less danger of skidding of the balls or rollers at high rotational speeds. Silicon nitride (Si_3N_4) bearings have been shown to operate successfully, running dry, at temperatures above 500°C though it is necessary to provide some boundary lubrication, perhaps in the form of graphite or molybdenum disulphide; see Section 9.6.3.

1.3 Lubricants and lubricant properties

1.3.1 *Classification of hydrocarbons*

Prior to the middle of the nineteenth century lubricants for the moving parts of machinery had been almost exclusively derived from animal or vegetable sources. In the latter half of the century the rise of the oil drilling and refining industries made large quantities of mineral oils available and, over a period of little more than twenty years, these assumed the dominant position in the market place that they have retained ever since.

Mineral oils are obtained during the distillation of crude petroleum. In the early days of the industry there was little difference between the fractions used for lubrication and those, of a similar viscosity, used as heavy fuel oils. However, over the past few decades there have been many improvements in the production of lubricating oils: these have involved both the removal of unwanted or undesirable naturally occurring elements within the oil (such as excessive sulphur compounds, and some waxy elements) as well as the introduction of additives to improve natural lubricity or oiliness. Mineral oils are made up predominantly of hydrocarbons, compounds of carbon and hydrogen, which fall into one of three classes; paraffinic, naphthenic, and aromatic. The atomic arrangements of each of these is illustrated diagrammatically in Fig. 1.5. Oils are usually classified on the basis of their major hydrocarbon constituent, which is often characteristic of the oil field from which they were extracted.

Paraffinic oils

A paraffin is the generic name for the arrangement in which the carbon atoms are linked to form long chains, see Fig. 1.5(a). These may be branched

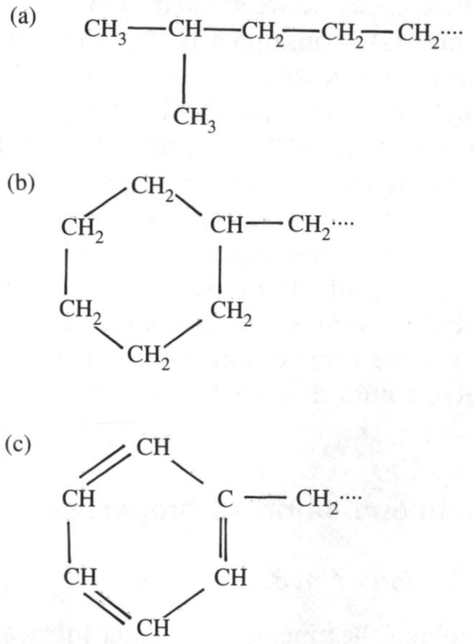

Fig. 1.5 Principal types of hydrocarbons in mineral oils: (a) paraffinic; (b) naphthene; and (c) aromatic. In each case the dashed lines indicate a chain of further methyl (CH_3) groups.

but do not contain any ring structures. Straight and branched chains of *fully saturated* paraffins (that is, those containing as many hydrogen atoms as possible) have the general formula C_nH_{2n+2}. In practice n is a very large number, and a single oil may contain many hundreds of different members of the same general family.

Naphthenic oils

A naphthene is a saturated hydrocarbon in which some of the carbon atoms have formed into ring structures such as that illustrated in Fig. 1.5 (b); these may link together to form *poly*-naphthenes.

Aromatic oils

Aromatics and poly-aromatics have similar ring structures to the naphthenic oils but they are not fully saturated and the carbon rings contain alternate single and double bonds as shown in Fig. 1.5 (c); this gives them some particular and distinctive features. They usually occur naturally in much smaller quantities than the paraffinic and naphthenic constituents. The compositions of some typical paraffinic and naphthenic oils are shown in Table 1.4.

Mineral oils, in common with many other chemicals, do not remain unchanged over long periods of time: the main source of their degradation is the gradual oxidation of some of the hydrocarbon chains. As a rule, aromatic compounds are more reactive and so degrade more rapidly than the naphthenes which in turn are more susceptible to oxidation than the paraffins. Oxidation is always detrimental to the performance of an oil as a lubricant: the process can result in the formation of sludgy deposits which can hinder lubricant supply by blocking oilways and filters, and it can also encourage the formation of organic acids which can be corrosive to some machine components. Consequently, all commercial lubricants contain some small amounts of *antioxidants*: these react preferentially with gaseous oxygen and so slow down the rate of lubricant deterioration. Ultimately, however, they become exhausted and once this happens the rate of oxidation

Table 1.4 Compositions of typical paraffinic and naphthenic oils.

Constituents	Percentage present in	
	paraffinic oils	naphthenic oil
Carbon atoms in paraffin chains	65	50
Carbon atoms in naphthene rings	31	45
Carbon atoms in aromatic rings	2	2
Sulphur (by weight)	0.5	1
Other compounds	1.5	2

increases rapidly. Ideally, mineral oils should be renewed just before this increase takes place. Otber additions, known as *detergents* and *dispersants*, may be included in commercial oils to help reduce, or prevent, the formation of sludges and deposits during operation.

In recent years there have been increasing demands for lubricants which can operate at temperatures greater than can be tolerated by mineral oils (these are limited to about 150°C); such non-hydrocarbons are generically known as *synthetic lubricants*. However, all these have various disadvantages; they may be flammable, toxic, or corrosive towards some metallic, polymeric, or elastomeric materials, or they behave inadequately as boundary lubricants. They can also be very expensive. Organic esters were first developed for aircraft jet engine lubrication and can operate up to about 250°C. Phosphate esters, formed by reaction between phosphoric acid and various alcohols are used principally for their excellent fire resistance; they are, however, liable to thermal deterioration at temperatures much above 120°C. Up to temperatures of about 250°C various types of silicone oils can be used. They have the advantages of being chemically inert and stable, non-toxic, and non-corrosive but they are not inherently good boundary lubricants for ferrous materials. At higher temperatures still polyphenyl ethers are available. Some of the relevant properties of these fluids, including their approximate cost, are gathered together for comparison with those of a typical mineral oil in Table 1.5.

1.3.2   *Viscosity*

Dynamic viscosity—scientific units

To the tribologist, the most important rheological property of a lubricant is its viscosity. The *dynamic* (or *absolute*) viscosity η of a fluid is a measure of the resistance it offers to relative shearing motion; η is defined as the shearing force in the direction of flow between two parallel planes of unit size per unit velocity gradient between them. Since the shear force per unit area is the shear stress τ, and the velocity gradient is equivalent to the local shear strain rate $\dot{\gamma}$, it follows that

$$\eta = \frac{\tau}{\dot{\gamma}}. \tag{1.4}$$

The importance of this fluid property was suggested by Newton in the seventeenth century; he also suggested a means of measuring its value by using rotating concentric cylinders and such a method is utilized in some modern viscometers. Fluids which, under conditions of uniform temperature and pressure, can be characterized by eqn (1.4) are called *Newtonian*, and many common fluids, especially those with relatively simple molecular

Table 1.5 Characteristic properties of some synthetic lubricants.

Fluid property	Typical organic ester	Typical phosphate ester	Typical silicone oil	Polyphenyl ether	Mineral oil
Maximum operating temperature in air °C	240	150	250	350	150
Specific gravity	1.01	1.12	1.06	1.19	0.88
Viscosity index	140	0	175	−60	0–120
Flash point °C*	255	200	290	275	150–200
Boundary lubrication on steels	Good	Good	Poor	Fair	Good
Rubber seals	Use silicone rubbers	Use butyl rubbers	Use neoprene rubbers	Attacks rubber	Use nitrile rubbers
Corrosion	Can present problems	Can present problems	Non-corrosive	Non-corrosive	Depends on impurities
Particular advantage	High temperature stability	Fire resistance	High temperature stability	High temperature stability	
Approximate cost	10	10	25	250	1

* The flash point is a measure of the flammability of a liquid; it is the lowest liquid temperature at which the vapour above the surface ignites in the presence of a flame.

structures, fall into this category. The various classes of *non-Newtonian* behaviour exhibited by tribological fluids are discussed in Section 8.3.

Within the Systeme Internationale of units, force per unit area or shear stress is measured in units of N m^{-2} (or *Pascal*, Pa) and velocity gradient in ms^{-1} ÷ m, that is, s^{-1}, so that the units of dynamic viscosity η are those of

$$\frac{\text{shear stress}}{\text{velocity gradient}} \quad \text{that is, Pa s or Ns m}^{-2} \text{ (or kg m}^{-1}\text{s}^{-1}\text{).} \quad (1.5)$$

No specific name has been allocated to this unit. Within the c.g.s. system the corresponding quantity has the units of dyne s cm^{-2} (or gm cm^{-1} s^{-1}) and this unit is given the name *poise*, after the French doctor Poiseuille who, as part of a study into the mechanics of the flow of blood, investigated the movement of water through very-small-diameter glass tubing. Water at around 20°C has a dynamic viscosity of about one hundredth of a poise, that is, 1 centipoise (cP), and this is the unit in which values of viscosity of most liquids are quoted. By definition, it follows that

$$1 \text{ cP} = 1 \times 10^{-2} \text{ poise} \equiv 1 \times 10^{-3} \text{ Pa s} = 1 \text{ mPa s.} \quad (1.6)$$

Lubricating oils have typical viscosities of a hundred times that of water, that is, about 100 cP; on the other hand, air is about one hundred times less viscous than water.

Within the Imperial (or conventional US) system, viscosity is measured in lbf s in^{-2}. This unit is known, in honour of Reynolds, as the *reyn*. Converting the units of force and area produces the relation that

$$1 \text{ reyn} = 68.95 \times 10^5 \text{ cP,} \quad (1.7)$$

or, perhaps more conveniently, that 1 *microreyn* (that is, 1×10^{-6} reyn) is equal to 6.895 cP.

Kinematic viscosity—industrial and commercial units
The *kinematic* viscosity of a fluid is defined from the relation

$$\text{kinematic viscosity} = \frac{\text{absolute viscosity}}{\text{density}}, \quad (1.8)$$

and is the fluid property which appears in situations where flow is due to self weight or gravity; in the SI system this property has the units m^2 s^{-1}. In the c.g.s. system the corresponding unit (cm^2 s^{-2}) is called the *stoke*. As with the poise this is really too large for practical use and numerical values of kinematic viscosity are usually quoted in centistokes (cS); thus 1 cS $= 1 \times 10^{-2}$ stoke $= 1$ mm^2 s^{-1}.

One method of determining viscosity is to measure the time taken for a given volume of fluid to flow out of a container through a narrow capillary tube and various commercial instruments using this principle have been developed. Since most of these depend on gravity-driven flow, they measure the *kinematic* rather than the *absolute* viscosity. Figure 1.6 shows the essential features of the Ostwald viscometer which is widely used for this sort of determination.

The two glass bulbs, one higher than the other, are connected by a glass capillary whose size is chosen to make inertia and turbulence effects negligibly small. The time of flow for a certain volume of fluid is measured and this is then multiplied by a calibration factor to give the kinematic viscosity. As viscosity is very temperature dependent the apparatus must be immersed in a constant-temperature bath to give accurate results. Most tribology analyses require the value of the absolute rather than the kinematic viscosity so that if data from viscometers of this type are to be used we also need to know the density of the fluid. Most lubricating oils have densities in the range 800–1000 kg m^{-3} and the chart of Fig. 1.7 can be used to make the conversion from cS to cP.

The viscosity of commercial lubricants is still sometimes quoted in terms of Saybolt, or Redwood seconds, or Engler degrees. These figures refer to the times that particular volumes of the fluid take to flow through the short length of capillary tubing that is incorporated in each of the corresponding instruments. Because of turbulence effects there is no simple linear correlation between viscosities measured on these scales and kinematic viscosity.

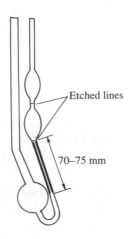

Etched lines

70–75 mm

Fig. 1.6 The Ostwald viscometer. The viscometer is filled through the left-hand tube and then oil drawn up by suction into the right-hand limb. The time taken for the meniscus to fall from the upper to the lower etched line is then measured.

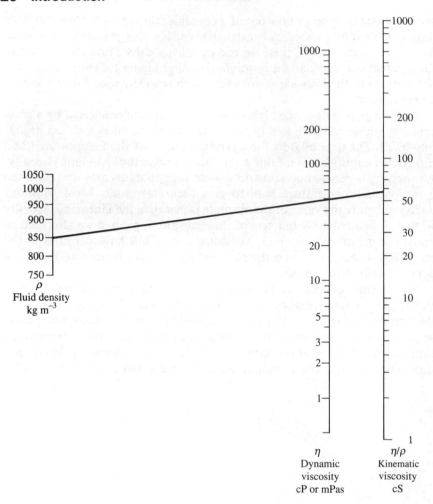

Fig. 1.7 Viscosity conversion chart. In the example shown an oil with a density 850 kg m^{-3} and kinematic viscosity 58 cS is shown to have an absolute viscosity of 49 cP.

Full conversion tables are available in various engineering handbooks, but for most purposes the values in Table 1.6 will suffice. Flow times of less than about 40 seconds in these instruments are unreliable and should be avoided.

In yet another design of viscometer the test fluid is introduced into the narrow gap between two concentric cylinders. When one of these is rotated at a steady speed the torque transmitted to the stationary cylinder depends on the absolute, rather than the kinematic, viscosity of the fluid between

Table 1.6 Approximate viscosity conversion factors.

Kinematic viscosity (cS)	Saybolt Universal (seconds)	Redwood No. 1 (seconds)	Engler (degrees)
6	46	41	1.5
8	52	46	1.7
10	59	52	1.8
15	77	68	2.3
20	98	86	2.9
30	141	125	4.1
50	232	205	6.6
> 50	cS × 4.6	cS × 4.1	cS × 0.132

them. Cylinders with different clearances enable liquids with widely varying viscosities to be handled; once again it is important to keep the temperature constant at a known value.

1.3.3 Temperature and pressure effects

The viscosity of a fluid depends on both the local values of temperature Θ and pressure p and both these effects can be very important in tribological devices. It would be very convenient to have an analytical expression relating η to these parameters and over the years a great deal of effort has been expended on this problem. Reynolds himself suggested an expression connecting viscosity η to temperature Θ, namely

$$\eta = R \exp(-\beta\Theta), \tag{1.9}$$

where R and β are constants. However the range of temperatures over which this is usefully accurate for hydrocarbons is very limited. Over those temperatures commonly encountered in engineering situations most lubricating mineral oils can be described by the empirical equation

$$\log\log(\eta/\rho + \mathcal{C}) = \text{constant} - c\log\Theta. \tag{1.10}$$

where $\mathcal{C}$ and c are constants. Experience has shown that if the kinematic viscosity, η/ρ, is measured in cS then the constant $\mathcal{C}$ has the value 0.6. The equation

$$\log\log(\text{cS} + 0.6) = \text{constant} - c\log\Theta \tag{1.11}$$

forms the basis of the ASTM (American Society for Testing Materials) chart on which the value of the term on the left-hand side of eqn (1.11) is plotted against the log of temperature to give a linear display. A similar basis is used for the classification of lubricants in BS 4231.

SAE and ISO viscosity grade classifications and the viscosity index

Figure 1.8 shows a plot of absolute viscosity versus temperature for a range of mineral oils differentiated by their SAE (Society of Automobile Engineers) designations. Note that the vertical scale in this figure is not logarithmic (as might be suggested by taking logarithms of eqn (1.6)) and that the horizontal scale is not strictly linear but that both have been distorted so that the data plot as straight lines. This rather masks the fact that small changes in temperature can bring about very large changes in

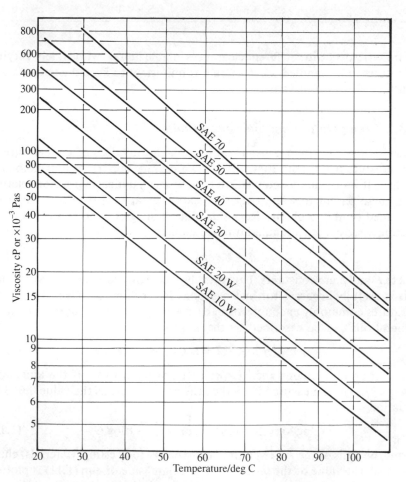

Fig. 1.8 Viscosity versus temperature for a range of mineral oils classified by SAE grades.

viscous resistance; for example, reducing the temperature of an SAE 10 oil from 60°C to 40°C more than doubles its viscosity. In general the more viscous a lubricant then the greater is its susceptibility to temperature (and pressure) effects.

The SAE classification of engine oils is based on viscosities determined at 100°C and −18°C and specifies a permissible range (see Table 1.7(a)) rather than a specific value. It is based on viscosity only and considers no other properties; it therefore does not constitute a full measure of the 'quality' of the oil. SAE grades bearing the suffix W ('winter' grade) are recommended for low ambient temperatures. A *multigrade* oil is one which at −18°C falls within the specification of a W oil but at 100°C satisfies a non-W specification. Thus, an SAE 20W–50 oil is within the specified limits for a 20W oil at low temperatures but has the kinematic viscosity of an SAE 50 oil at higher ranges. The ISO viscosity grade of a lubricant is designated as its kinematic viscosity, measured in cS or $mm^2\ s^{-1}$ at 40°C. There is thus only an approximate correlation between SAE and ISO grades; these are indicated in Table 1.7(b).

An alternative means of indicating the susceptibility of a lubricant to changes in viscosity with temperature is the use of the *viscosity index* or VI. This relates the change in viscosity of the sample lubricant at two temperatures, 40°C and 100°C, to two arbitrary oils. At the time of its introduction, the natural mineral oils which showed the least variation of viscosity with temperature came from the Pennsylvania oil fields in the United States. These have a high proportion of paraffins, and were given a VI of 100. The oils which suffered the greatest decrease of η with temperature came from the coast line of the Gulf of Mexico and any sample of these which has the *same* viscosity as a Pennsylvania oil at the higher temperature was assigned a VI of 0. The viscosity of the test sample is first measured at the higher temperature and the Pennsylvania and Gulf oils with the same viscosity at this temperature looked up in tables. Then the viscosity of the test sample at the lower temperature is measured, say U. The viscosities at this temperature of the chosen datum oils are found from tabulated values and if, for the Pennsylvania and Gulf oils these are respectively H and L, then the VI of the test sample is given as

$$VI = 100 \times \left\{\frac{L - U}{L - H}\right\}. \tag{1.12}$$

In general paraffinic hydrocarbons have high VIs, the aromatic components have low values, while those of the naphthenic fractions are intermediate. A lubricant with a high VI is likely to perform better in a high-temperature situation than one with a low value of VI. Modern methods or refining have improved the values of the VI that can be achieved from poorer quality oils although the highest VI that can be obtained from a simple

Table 1.7 (a) SAE viscosity grades for engine oils.

SAE viscosity grade	Viscosity range		
	cP at −18° max	cS at 100°C	
		min	max
5W	1250	3.8	–
10W	2500	4.1	–
20W	10,000	5.6	–
20	–	5.6	<9.3
30	–	9.3	<12.5
40	–	12.5	<16.3
50	–	16.3	<21.9

Table 1.7 (b) SAE and ISO viscosity grade equivalence.

ISO viscosity grade	Equivalent SAE grade(s)
46	15W
68	5W-30
100	30, 10W-40
150	40, 20W-50
220	50
320	60

mineral oil is only a little in excess of 100. To obtain significant improvements requires the use of polymeric *viscosity index improvers* and it is these that have facilitated the development of multigrade oils and hydraulic fluids capable of working over a wide range of temperatures. Some synthetic fluids have been developed with VIs in excess of 200 and these may be required where the operating temperature range is particularly severe. Further details of these topics can be found in the various tribology handbooks referred to above and in Cameron (1966), Briant *et al.* (1991), Lansdown (1982), Jones and Scott (1983), and Fuller (1984).

Many lubricants, including hydrocarbon oils, undergo a considerable increase in viscosity when subject to high pressures, said to be they are *piezoviscous*. The importance of this effect in the lubrication of machine components was suspected in the 1920s although numerical and analytical treatments which demonstrated this were not available until the early 1950s. The measurement of viscosity under local hydrostatic pressures of perhaps more than a thousand atmospheres is not straightforward and demands considerable experimental skill and ingenuity. Lubricant behaviour is often modelled by an exponential equation of the form

$$\eta = \eta_0 \exp \alpha p, \qquad (1.13)$$

and this relation is known as the *Barus* equation. η_0 is the base viscosity, that is, that at low pressure, and the constant α is called the *pressure viscosity index*; values of η_0 and α for a number of lubricants are shown in Table 1.8.

The implications of this effect in the lubrication of heavily loaded non-conformal contacts, such as those between gear teeth or a cam and its follower, are explored in Chapter 8. Although very high pressures and significant temperatures can be generated in such heavily loaded contacts, comparatively few analyses also take fully into account either the effects of the compressibility or the thermal expansion of the lubricant. These effects can be quite significant. For example, under very high pressures the density of hydrocarbon oils can increase by as much as 30 per cent. This effect can be described by the empirical formula

$$\frac{\rho(p)}{\rho_0} = \frac{1 + 2.27p}{1 + 1.68p} \qquad (1.14)$$

in which $\rho(p)$ is the density under pressure p (measured in GPa) and ρ_0 is the value at low pressures. Further details can be found in Cameron (1966), Dowson and Higginson (1977), Lansdown (1982), and Briant *et al.* (1991).

1.3.4 *Viscosity of gases*

Aerostatic and aerodynamic bearings depend on the load-carrying capacity of a gas or vapour film. Viscosity is still an important fluid property although there is a much narrower spread in the values of the viscosities of gases compared to the range that is observed with liquids, as can be seen

Table 1.8 Base viscosities and pressure viscosity indices for various lubricants.

Lubricant	Temperature (°C)	η_0 base viscosity (cP)	α pressure viscosity index (Pa^{-1})
Kerosene	27	1.9	1.1×10^{-8}
Castor oil	38	266	1.46×10^{-8}
SAE 10	38	40	2.51×10^{-8}
	82	7	2.10×10^{-8}
SAE 30	38	105	3.19×10^{-8}

Table 1.9 Dynamic viscosities ($Pa s \times 10^{-6}$) of some gases and liquids at normal pressures.

	Temperature			
Gases	0°C	20°C	100°C	400°C
Air	17.3	18.2	22.0	33.2
Carbon dioxide	13.6	14.7	18.6	30.7
Hydrogen	8.4	8.8	10.4	15.3
Methane	10.3	11.0	13.5	21.1
Neon	29.8	31.3	37.0	54.3
Nitrogen	16.6	17.6	21.2	31.9
Oxygen	19.5	20.4	24.4	37.6
Liquids				
Acetone	397	324	–	–
Ethanol	1767	1197	–	–
Castor oil	$> 2500 \times 10^3$	986×10^3	17×10^3	–
n-pentane	278	234	130	–
Water	1787	1002	283	–

from Table 1.9. The viscosity of one of the most viscous gases, neon, is only about four times that of the least, hydrogen. On the basis of the kinetic theory of gases we should expect their viscosity to be independent of pressure but to vary with the square root of temperature: the first of these expectations is substantially true but the second requires correction. Although the absolute viscosity of gases increases with temperature, as a result of falling density, their kinematic viscosity is more or less constant over a wide temperature range.

A special challenge is posed to tribologists by the design of bearings and other components to operate in high-vacuum environments. These conditions may be met in the operating columns of electron microscopes and other similar instruments or be experienced by mechanical devices operating on satellites or space stations. No aerodynamic support is of course available and liquid lubricants face problems of lubricant loss by vaporization; this effect can lead to contamination of the vacuum system in electron beam instruments and changes of tribological behaviour with time in space environments. Consequently, bearings for both sorts of applications are often designed to run unlubricated either as plain journals or as rolling elements. Components in space have also to contend with severe thermal conditions; because of the electromagnetic radiation from the sun, the lack of convection, and the eclipsing of the spacecraft by the earth's shadow and changes in its orientation, temperature extremes from −100°C to +150°C

are not uncommon. Spacecraft are subject to proton and electron bombardment and tribological materials must be capable of sustaining such attacks:
PTFE, for example, is one otherwise attractive material which can rapidly
deteriorate with relatively small adsorbed doses of such radiation. At
launch, the severe degree of vibration can lead to short-term accelerations
of the payload which can be as great as 40 g, and materials must be chosen
to minimize problems of fretting or adhesion (see Chapter 5).

Space mechanisms can be broadly classified into two groups: those
required to perform only a very few operations or perform intermittently,
and those which operate continually throughout the spacecraft lifetime; in
practice both face tribological problems. For unmanned spacecraft, a
reliability factor of more than 99 per cent is demanded and in addition all
components must be designed to achieve minimum overall mass. The
majority of components are manufactured from aluminium or titanium
alloys or from beryllium (because of its high strength:weight ratio); none
of these are 'good' tribological materials so that additional and sophisticated
surface treatments are often needed to meet the stringent design requirement; see Section 2.4.

1.4 A brief history of tribology

The history of technology is a fascinating subject in its own right, through
which are woven developments in the understanding of friction and wear,
and the invention of devices to accommodate or ameliorate them. We can
reasonably surmise that ever since our ancestors began to drag loads over
the ground they have sought ways to reduce the magnitude of the frictional
forces they were obliged to overcome. The oldest civilization of which we
have a clear historical record developed in Mesopotamia about five thousand
years ago and there is clear evidence that they had developed, or knew of,
a number of quite sophisticated tribological devices. They possessed, for
example, both wheeled carriages and the potters wheel and each of these
requires some rudimentary form of both journal and thrust bearings. There
is evidence, from wall paintings in both Mesopotamia and Egypt which show
the transport of large stone blocks and carvings, that craftsmen in these
societies appreciated the role of lubrication in reducing friction in both
sliding and rolling contacts; see Fig. 1.9.

Chinese pictographs dating back to the middle of the second millenium
BC have also been found which show wheeled chariots. Potters wheels using
fired porcelain cups as bearings appeared in China about 1500 BC, and by
400 BC. Chinese bearing technology had developed sophisticated plain
lubricated bronze bearings for use on war chariots.

During the Graeco-Roman period, while there are few extant writings

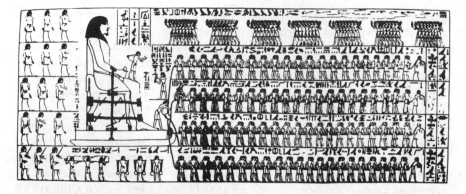

Fig. 1.9 Transport of an Egyptian stone colossus, *c.* 1880 BC. Perhaps an early tribologist can be seen at the front of the sled lubricating its passage from a jug of liquid.

concerned specifically with tribological designs, it is clear, from either an examination of the surviving artefacts from this period or the descriptions of contemporary devices, that such problems exercised the minds of practical men. The Greeks were active in the development of mechanics and mechanical devices including some using water or pneumatic power, and, even if these were only actually built as toys or models, it is clear that to function satisfactorily careful attention had to be paid to the design of the bearings and pivots. It is known that the lathe was developed at this time and there are records of the use of levers, gears, and pulleys in various applications. The construction of large-scale stone buildings required cranes and hoists and there was also a clear appreciation of the importance of good tribological design in such humble items as corn and olive mills. Some of the most illuminating accounts of the implementation of tribological ideas in the technology of this period can be found in the writings of Marcus Vitruvius Polio, the Roman architect and engineer, who lived in the first century AD. His writings were lost during the decline of the Roman Empire but a copy was located in the fifteenth century and is thought to have had a considerable influence on the development of Renaissance architecture. One of the most interesting Roman archaeological finds, from a tribological point of view, was made in the late 1920s in Lake Nemi which is situated about 30 km from Rome. The remains of two large ships were recovered, dating from about AD 50, and on one of these was found a wooden platform, about a metre in diameter, supported on what is clearly a rolling-element bearing. There was also some evidence of a similar device using tapered rollers.

Historians are still somewhat divided on the importance that was placed in Western Europe on technological developments during the middle ages (the period which runs from about AD 400 to 1450) although there appears to be no doubt that steady development occurred in China during this period. Many ideas and inventions seem to have originated in eastern Asia before being introduced into Europe. Much of the machinery used during this medieval period had its origins in earlier times although there were developments which placed greater demands on tribological design and practice. The mechanical clock was probably the outstanding engineering achievement of this era and although several examples are still in existence it is not known whether the brass bushes, in which the majority of their iron shafts are supported, are original or were added very much later. However, there can be little doubt that the medieval craftsmen who developed the various escapement mechanisms required by these magnificent devices must have given some thought to matters of friction and wear.

Large-scale mechanical power during the middle ages was derived principally either from water or wind sources. By the time of the Norman conquest water wheels with power outputs of the order of 30 kW had been built, while the development of windmills, for both grinding corn and particularly pumping water, went on continuously into the late nineteenth century. The journal and thrust bearings in these machines usually involved wooden shafts running against wooden or stone bearing blocks, although there are some notable uses of metallic bearings very early on. Bearings were lubricated principally by animal fats such as tallow and lard.

The end of the fifteenth and early sixteenth century was the great era of maritime exploration, and this activity is intimately related to developments in navigation, surveying instruments, and improved chronometers, as well as to the business of ship building and armament manufacture. That the problems of friction and wear were of concern can be judged from the effort expended upon them by the greatest mind of that, and perhaps any other, age Leonardo da Vinci (1452–1519). His notebooks contain more than 5000 pages of notes and sketches, a good proportion of which are concerned with the scientific investigation of friction and the study of wear and bearing materials (Fig. 1.10), as well as the design of ingenious rolling-element bearings. After his death these documents were effectively lost for two or more centuries as the collection was broken up and kept in a number of private libraries and so, regrettably, his extraordinary insight had comparatively little immediate practical effect on the history of technology. The fifteenth century also saw the invention of movable type printing and it is from some of the books of this period that we can learn something of the contemporary mechanical arts and industrial techniques. Plain bearings increasingly employed metals (rather than wood and stone) and there was growing interest in low-friction bearing alloys. The benefits of rolling-element bearings

(a)

(b)

Fig. 1.10 Some of Leonardo's studies of friction showing his experiments to investigate this phenomenon: (a) the influence of contact area on the force of friction; and (b) the frictional torque on a journal and half bearing (from the *Codex Atlanticus* and *Codex Arundel*).

began to be exploited in not only scientific instruments but also some larger-scale machinery.

The middle part of the seventeenth century is often known as 'the age of reason' and it was during this period that what we would understand as the scientific method of investigation was developed. The importance of science in national life can be judged from the fact that great scientific institutions such as the Royal Society in England and the Académie Royale des Sciences in France date from this time (1660 and 1666 respectively). By the close of the century the 'laws of friction' which we still use had been stated and some important investigations on the nature of solid friction undertaken; see Section 4.1.1. Sir Isaac Newton (1642–1727) had set mechanics on a rational footing and his conjectures on the resistance to the flow of fluids form one of the keystones in the science of fluid mechanics and hence lubrication theory. Although the term *viscosity* was in use it had no precise scientific meaning, and 150 years were to pass after Newton's hypothesis that fluid resistance depended on velocity gradient before the introduction, by the French mathematician Claude Navier, of a coefficient of viscosity in the equations of fluid motion. The invention of the calculus by Newton and Liebnitz provided the conceptual tool essential for the analysis of many scientific and engineering situations. With the development of various comparatively large-scale industrial processes, especially in mining and minerals processing, there were increasing demands for large-scale machines and tribological practice developed with other of the mechanical arts.

It is usually accepted that the industrial revolution started in the final decades of the eighteenth century and its development is intimately involved with that of steam technology. Although stationary atmospheric steam engines had been is use since about 1700 (principally for driving water pumps in mining) it was almost a hundred years later that the development of high-pressure engines, with efficiencies of perhaps twenty times their antecedents, provided the plentiful supply of power which drove subsequent industrial development. As well as the design of the bearings and other tribological elements of the engines themselves, friction, lubrication, and wear were of great importance in many of the contributory technologies — not least the evolution of the machine tools required for their manufacture. In addition to these influences, the design of road and rail vehicles provided a major impetus for improvements in the efficiency of both plain and rolling-element bearings. The value of the use of dissimilar metals, particularly brass against iron or steel, in plain bearings was widely recognized and special-purpose bearing alloys based on tin and lead were developed.

One of the most comprehensive studies of friction during the early part of this period was that undertaken by Charles Coulomb in 1785. This work, submitted in response to the offer of a prize by the French Academy of Sciences, was very much concerned with practical tribological problems in naval and military matters. Coulomb set out to investigate the effects of load and both the area and time of contact on the dry frictional resistance of a wide range of material combinations. He fitted various empirical equations to his observations and these can be interpreted as, for the first time, distinguishing between the effect of *adhesion* and that of *deformation*. He concluded that in the majority of cases frictional resistance arose from the mechanical interlocking of surface asperities and his conclusions formed the basis of the understanding of the genesis of solid friction for many years. Although today we might criticize his model of surface interaction for its lack of emphasis on the actual physical mechanism of energy dissipation, his name is honoured in our use of the phrase Coulomb friction. The detailed analysis that a fuller explanation of surface asperity interactions requires was not initiated for another hundred years and today is still very much a matter of development.

Prior to the middle of the nineteenth century lubricants had been almost exclusively derived from animal or vegetable sources; they included sperm and tallow oils, olive oil, castor oil, and so on, and were often thickened by the addition of solids such as graphite or talc to form a lubricating paste or grease. Mineral oil, extracted from shales or coal, was used although it was not until large quantities were found in various parts of the world, and its value as a superior lubricant confirmed, that it became predominant. In the 1860s and 1870s it was clear that current engineering understanding of the role of lubricants in many commercially important applications was inadequate. In 1883 Nikolai Petrov carried out a series of experiments in St Petersburg to examine how the frictional losses in railway vehicle axle boxes depended on the nature of the lubricant; he showed that the frictional force was proportional to the product of the bearing area, the sliding speed, and the lubricant viscosity, but inversely proportional to the bearing clearance; he was one of the first investigators in the field to make viscosity, rather than density, the crucial lubricant property. However, he did not go on to investigate the influence of these parameters on the load-carrying capacity of the bearing. Within the United Kingdom one area of concern identified by the Institution of Mechanical Engineers for investigation was that of *Friction at High Velocities* and the work carried out into this topic by Beauchamp Tower and presented to the Institution in 1885 represents one of the most important milestones in the history of the development of the subject. These experiments used a 180° plain journal bearing which was loaded against a rotating shaft. In normal use lubrication was provided by the shaft simply dipping into a fluid bath but in one set of experiments Tower decided to supply the bearing directly from a small reservoir, or lubricator, and for this purpose drilled a $\frac{1}{2}$-inch hole at the top centre of the bearing. What was observed is described graphically in his own account:

On the machine being put together again and started with the oil in the bath, oil was observed to rise in the hole which had been drilled for the lubricator. The oil flowing over the top of the cap made a mess, and an attempt was made to plug up the hole, first with a cork and then with a wooden plug. When the machine was started the plug was slowly forced out by the oil which showed that it was acted upon by a considerable pressure. A pressure gauge was screwed into the hole, and on the machine being started the pressure, as indicated by the gauge, gradually rose to above 200 lbs per sq. in. The gauge was only graduated up to 200 lbs per sq. in., and the pointer went beyond the highest graduation. The mean load on the horizontal section of the bearing was only 100 lbs per sq. in. This experiment showed conclusively that the brass was actually floating on a film of oil, subject to a pressure due to the load. The pressure in the middle of the brass was thus more than double the mean pressure.

These observation are illustrated in Fig. 1.11: the concept of *hydrodynamic* lubrication had been born. What was needed now was a sound theoretical basis for these observations, and this was provided almost immediately in the classic work of Osborne Reynolds who at that time held the chair of civil and mechanical engineering at Owens College, Manchester.

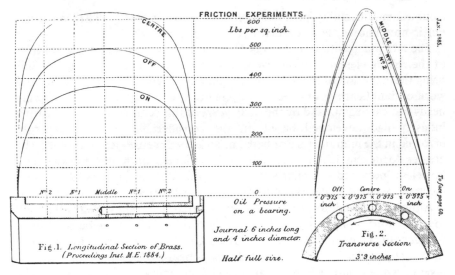

Fig. 1.11 Pressure distributions from Beauchamp Tower's work on a lubricated journal bearing (from Tower 1885).

His paper of 1886, read to the Royal Society, contains the derivation of the governing differential equation which bears his name and is the starting point of all subsequent pieces of lubricated bearing analysis.

At almost the same time as these developments were taking place in the analysis of lubricated contacts an equally important paper had been published on dry contact by Heinrich Hertz in 1882. Hertz was concerned with the effects on interference phenomena that might arise from any geometric changes in the shapes of the glass lenses used for their production when they were pressed together. He proceeded to solve the problem of the contact stresses and deformation between elastic solids of arbitrary shapes, and the results of his work form an important element of all modern studies on non-conforming contacts such as rolling-element bearings and gears. Thus it was that within a few years in the mid-1880s the foundations of much of the modern subject of tribology were laid. Over the next half-century much effort went into deriving solutions of Reynolds' equation for particular cases that were of engineering and commercial interest. Important contributions were made in the case of journal bearings by Sommerfeld and in the case of thrust bearings by Michell, Lord Rayleigh, and Kingsbury who was also instrumental in demonstrating that the fluid film does not have to be a liquid but can also be a vapour or gas. The early years of the twentieth century saw the extension of bearing analyses to include the effects of stiffness and stability, and in particular, there was growing concern about the troublesome instability phenomenon which was originally called *oil whip*, but is now known, less colourfully, as *half-speed whirl*.

The late nineteenth century was also the period when the modern forms of the rolling-element bearing were developed, one major impetus for this being the evolution of the design of the bicycle. A great deal of experimental work on ball bearings applied to bicycle hubs, crank spindles, and steering heads was carried out and by the early years of this century a sizeable bearing industry had developed based on the considerable technical expertise involved in the manufacture of hard, accurately dimensioned steel balls. The greater loads carried by the axles of carriages, both horse-drawn and horseless, provided a similarly important stimulus to the parallel development of cylindrical and taper rolling-element bearings.

Although the axle bearings of vehicles often use rolling elements, plain journals utilizing hydrodynamically derived fluid films, have been retained for most of the engine bearings. By the start of this century it had become the usual practice to support steel shafts in bearings lined with a generous layer of white metal or Babbitt (an alloy of tin, copper, lead, and zinc). The soft low-melting-point metal could be cast in place and then hand scraped to final dimensions. Today it is still the case that similar alloys account for the majority of simple plain bearings, although their mode of manufacture is much changed, and their conditions of operation are a good deal more severe. An early development, introduced in the 1920s, which successfully retained the desirable properties of the white metal, was its incorporation in a harder metal matrix — copper–lead alloys were typical of this two-phase system. With time, the thickness of the white metal layer reduced from perhaps 3 mm at the start of the century to its current value of often less than 50 microns. This change coincided with the introduction of thin-walled bearings in which the layer of bearing metal is mounted on a stiff steel shell. These components can be rapidly mass produced and are now in virtual universal use.

The thickness of the fluid films in such hydrodynamic bearings, although not large, are still many times greater than the molecular size of the lubricant. By the 1920s it was clear that the presence on a pair of dry surfaces of even the smallest amount of a naturally oily or greasy substance could reduce the force of friction between them by a great margin. The term *boundary lubrication* was coined by Sir William Bate Hardy in 1922 to describe this effect; he demonstrated, in simple but elegant experiments, the importance of such layers and showed that they can still be present under normal loads which are sufficiently large to deform the surface of the underlying solids. In turn this led to a reappraisal of the fundamental mechanisms of friction and this is much associated with the work carried out at the Cavendish Laboratories in Cambridge under the direction of Professors Philip Bowden and David Tabor (Bowden and Tabor 1950 and 1964).

Once Reynolds' equation had been solved and the solutions applied so successfully to conformal contacts such as journal bearings it was entirely reasonable to apply a similar analysis to the geometry of non-conforming contacts such as those that occur between gear teeth or in rolling bearings. Unfortunately, the resultant calculations suggested film thicknesses far below the scale of the surface roughnesses on the components themselves, indeed, in some cases of less than the dimensions of the oil molecules making up the lubricant. The solution to the dilemma posed by the fact that, despite the results of such calculations, components of this sort operated successfully and without excessive surface distress or damage, was not clear, although there was some evidence that under the very high local pressures the viscosity of the lubricant could be very much greater than that measured under ambient conditions. The problem remained unresolved for another thirty years. A notable breakthrough occurred in 1949 when two Russian scientists, Ertel and Grubin, successfully incorporated the effects of pressure on both the viscosity of the lubricant and the elastic distortion of the solid surfaces in the entry region of a non-conforming contact. This produced estimates of fluid film thickness close to two orders of magnitude greater than the simpler, constant-viscosity, rigid-solid analysis. Through the next two decades the investigation of this new subject, *elasto-hydrodynamic lubrication*, proved a fruitful area of research and development; later analyses produced more details of pressure distributions and distorted shapes which have largely been verified by experiment. The essential features of both film shape and thickness for both line and point contacts can now be determined with some confidence and this has been of great value in the design of heavily loaded lubricated elements such as cams and followers, roller and ball bearings, gears, and traction drives.

As a result of both a better understanding of the physics of lubricant film formation and improvements in materials and manufacturing technology the specific loads on bearings have risen greatly over the years of this century. Nevertheless, bearings still deteriorate with use, and it seems probable that further advances in tribology will depend on a greater understanding of the mechanisms of surface damage, degradation, and wear. Although the results of wear, that is the progressive loss of substance from an operating surface, can be manifest on a comparatively macroscopic scale, the individual events that go to generate each wear particle are very tiny — perhaps ultimately reducing to the atomic scale. There is still much work to be done to come to a satisfactory understanding of the contact of two surfaces and how the force of friction and the associated effects of wear are generated between them.

1.5 Tribology literature

Many of the wide variety of textbooks on machine analysis contain some guidance on the design of tribological devices. In addition there are a number of more specialized texts on which the student of the subject can call, as well as the tribology and wear handbooks already referred to: a selection of these are listed in Section 1.6. In the field of friction, lubrication and wear, as in many other areas of technology, the number of research papers and publications has grown over the past few years — specialist journals are available and a number of these are likewise listed below.

1.6 Further reading

American Society for Testing and Materials (1984) *Annual Book of ASTM Standards, part 3*. ASTM, Philadelphia.

Arnell, R. D., Davies, P. B., Halling, J., and Whomes, T. L. (1991) *Tribology: principles and design applications*. Macmillan, London.

Barwell, F. T. (1970) *Lubrication of bearings*. Butterworths, London.

Barwell, F. T. (1979) *Bearing systems*. Oxford University Press.

Bhushan, B. and Gupta, B. K. (1991) *Handbook of tribology*. McGraw-Hill, New York.

Billett, M. G. (1979) *Industrial lubrication*. Pergamon Press, Oxford.

Bowden, F. P. and Tabor, D. *The friction and lubrication of solids*: Part I (1950); Part II (1964), Oxford University Press.

Braithwaite, E. R. (ed.) (1967) *Lubrication and lubricants.*, Elsevier, Amsterdam.

Briant, J., Denis, J., and Parc, G. (1991) *Rheological properties of lubricants*. Editions Technip, Paris.

Cameron, A. (1966) *The principles of lubrication*. Longmans London.

Czichos, H. (1978) *Tribology — a systems approach*. Elsevier, Amsterdam.

Derry, T. K. and Williams, T. I. (1960) *A short history of technology*. Oxford University Press.

Dorinson, A. and Ludema, K. C. (1985) *Mechanics and chemistry in lubrication*. Elsevier, Amsterdam.

Dowson, D. (1979) *History of tribology*. Longman, London.

Dowson, D. and Higginson, G. R. (1977) *Elasto-hydrodynamic lubrication*, Pergamon Press, Oxford.

Freeman, P. (1962) *Lubrication and friction*. Pitman, London.

Fuller, D. D. (1984) *Theory and practice of lubrication for engineers*. John Wiley, New York.

Halling, J. (1976) *Introduction to tribology*. Wykeham, London.

Hamrock, B. J. (1991) *Fundamentals of fluid lubrication*. NASA References Publication 1255.

Hersey, M. D. (1966) *Theory and Research in lubrication*. John Wiley, New York.

HMSO (1966) *Lubrication (tribology) education and research*, DES Report, London.

Hutchings, I. M. (1992) *Tribology*. Edward Arnold, London.
Jones, M. H. and Scott, D. (eds.) (1983) *Industrial tribology*. Elsevier, Amsterdam.
Kajdas, C. *et al.* (1990) *Encyclopedia of tribology*. Elsevier, Amsterdam.
Lansdown, A. R. (1982) *Lubrication: a practical guide to lubricant selection*. Pergamon Press, Oxford.
Moore, D. F. (1975) *Principles and applications of tribology*. Pergamon Press, Oxford.
Morner, R. M. and Orsulik, S. T. (eds.) (1992) *Chemistry and technology of lubricants*. Blackie, London.
Neale, M. J. (ed.) (1973) *Tribology handbook*. Newnes-Butterworth, London.
Needham, J. (1965) *Science and civilisation in China. Vol. 4: physics and physical technology*. Cambridge University Press.
Rabinowicz, E. (1965) *Friction and wear of materials*. Wiley, New York.
Sibley, L. B. and Kennedy, F. E. (eds.) (1990) *Achievements in tribology*. ASME.
Stolarski, T. A. (1990) *Tribology in machine design*. Heinemann-Newnes, London.
Summers-Smith, D. (1969) *An introduction to tribology in industry*. Machinery Publishing Co., Brighton,UK.
Tower, B. (1885) First report on friction experiments. *Proceedings of the Institution of Mechanical Engineers*, **39**, 58–73.
Winer, W. O. and Peterson, M. B. (1978) *Wear control handbook*. ASME, New York.
BS 4231 (1992) Classification for viscosity grades of industrial liquid lubricants. British Standards Institution, London.

Journals of tribological interest

Industrial Lubrication and Tribology
International Journal of Mechanical Sciences
Japanese Journal of Tribology
Journal of Applied Mechanics
Journal of Engineering Science
Journal of Materials Science
Journal of the Mechanics and Physics of Solids
Journal of Tribology; Transactions of the American Society of Mechanical Engineers, Series F
Lubrication Engineering
Lubrication Science
Proceedings of the Institution of Mechanical Engineers, Series C: Journal of Mechanical Engineering Science
Proceedings of the Institution of Mechanical Engineers, Series J: Journal of Engineering Tribology
Proceedings of the Royal Society
Soviet Engineering Research
Tribologie & Schmierungstechnik
Tribology and Corrosion Abstracts
Tribology International
Tribology Transactions; American Society of Lubrication Engineers
Wear

2
Engineering surfaces

2.1 The nature of engineering surfaces

No real engineering surface, no matter how carefully, or indeed expensively, prepared can possess perfect geometry. As well as errors in the *form* or *shape* of the component there will always be a roughness on the surface which is apparent when this is examined at a sufficiently high magnification. When two such surfaces are loaded together it is the tips of the surface roughnesses or asperities that must first carry the applied load: the geometry of individual contact spots and the way in which these islands of *real* contact are distributed throughout the *nominal* or *apparent* contact area is clearly of interest to tribologists in attempting to predict the overall performance, or likely life history, of the contact.

The geometric texture of an engineering surface reflects both its production route and the nature of the underlying material. It is possible to produce a truly smooth surface (for example, cleaving specimens of mica can produce a surface with roughness only on the atomic scale) and if two such surfaces are loaded together real and apparent areas are very nearly equal. The asperities on the surface of a very compliant surface, such as a soft rubber, may, if sufficiently small, be squeezed flat by quite modest contact loads, and in this way there can again be equality between real and apparent areas of contact. However, these are special cases; in general, useful metal surfaces exhibit a range of surface fluctuations which, although large compared to molecular dimensions, are small compared to the dimensions of most engineering components. These surface, as opposed to shape, variations are somewhat arbitrarily differentiated on the basis of wavelength into *waviness* (undulations with a relatively long wavelength, perhaps on the scale of millimetres) and *roughness* (variations which have a much shorter wavelength or correspondingly higher frequency); see Fig. 2.1. The distribution of both waviness and roughness may have a distinct directional component or *lay*, particularly if the final manufacturing process has been one such as turning or milling which possesses a strong directionality. Surfaces which have been finished by a non-directional method, such as electropolishing or lapping (provided this has removed all the traces of previous processes) may be much more homogeneous or isotropic in nature.

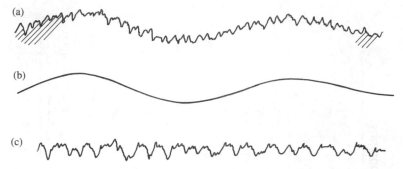

Fig. 2.1   The components of surface topography: the total profile (a) represents the combined effects of waviness (b) and roughness (c) superimposed on the geometric shape or form of the component surface.

2.2   Techniques of surface examination

The roughness of a surface affects many aspects of its behaviour and a variety of methods have been employed, each with considerable ingenuity, to assess and measure it. We shall see that it is not a straightforward matter to quantify roughness, especially if the aim is to use this quantitative assessment in predicting subsequent tribological behaviour. However, it would seem reasonable to suppose that roughness must be characterized by some information on heights *normal* to the mean plane of the surface together with some knowledge of the spatial distribution, or wavelengths, *within* the surface. We can thus classify the various methods available for topographical examination according the range of vertical heights and spatial wavelengths they can each differentiate. This information is summarized in Table 2.1.

2.2.1   *Optical and electron microscopy*

Optical microscopy

The most obvious way to examine an engineering surface is to view it optically. The surface is illuminated by a beam of visible light which is focused, that is, brought to as small a cross-section as possible, on the area of interest. The reflected light rays are then collected by the objective lens and an image of the surface produced by a suitable optical system. This can be formed on a viewing screen, a photographic plate or film, or directly on the retina of the eye. The image is a two-dimensional representation of the relative scattering powers of the various features of the surface and thus can only be an exact representation of the surface if this is absolutely flat or

Table 2.1 Techniques used in the measurement of surface topography.

Method	Magnification range		Resolution at maximum magnification (microns)		Depth of field (microns)		Comments
	at lowest	at highest	horizontal	vertical	lowest	highest	
Optical microscopy	×20	×1000	0.5	0.01* 0.03†	5	0.5	*Using interference contrast. †Using taper section
Scanning electron microscopy SEM	×20	×100,000	0.015	1*	1000	0.2	Operates in vacuo; limits on specimen size. *Using stereoscopic imaging.
Transmission electron microscopy TEM	×2000	×200,000	0.0005	0.0005	5	0.1	Operates in vacuo; requires replication of surface
Surface profilometry	×50	×100,000	5*	0.05	1	0.5	Operates along linear track. *Limited by stylus radius

plane—which is certainly not the case for conventional engineering surfaces. Even if the surface were plane there still remains an important limitation on the detail that can be resolved; as the magnification is increased it is found that small features on the surface appear as a collection of concentric ring patterns. These are the diffraction patterns formed as light is obscured by the aperture of the objective lens. This phenomenon is a fundamental limitation of any form of microscopy and means that detail with linear dimensions anything less than approximately the wavelength of the light being used cannot be resolved. This places an upper limit of roughly × 1000 on conventional optical microscopy which corresponds to a wavelength of about 0.5 micron. Even at magnifications somewhat below this, the very limited depth of field (which means that parts of the specimen only a very small vertical distance away from the mean surface level are out of focus) is a severe restriction on the use of straightforward optical microscopy in obtaining a general view of the topography of a rough or contoured surface.

The profile of a surface along a line can be established by optical examination if first a section is made perpendicular to the surface to be examined. This is metallographically mounted, ground, polished, and perhaps etched, using the conventional techniques of physical metallurgy. To avoid the loss of detail at the very edge (which is the area of most tribological interest) by smearing and local distortion, it is usually necessary to coat the specimen with a hard, adherent, protective layer prior to preparation. The effective resolution of this sectioning technique can be made much greater by making a *taper section*. As the name implies, in this method the section is cut through the surface to be examined at a shallow angle δ so effectively magnifying the height variations by a factor of cot δ; see Fig. 2.2. Protection by a hard adherent layer is essential: for steel a nickel layer 400–500 microns in

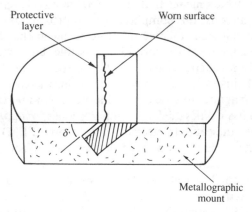

Fig. 2.2 Use of taper sectioning to examine specimen surface topography.

thickness is suitable, and if an angle δ of 5.7° is used then there is an increase in magnification of × 10.

The microtopography of highly reflective surfaces can be investigated to very high rates of resolution by optical interferometry. Using a microscope with appropriate optics surface steps as small as 1/20 of the wavelength of the illuminating light can be made (in the case of monochromatic sodium light this corresponds to about 0.015 microns). Surface roughness measurements of great accuracy can also be made using diffraction techniques, and much recent work in this field has been stimulated by the need to fabricate laser mirrors of great geometric accuracy for fusion research: the roughness requirements for these components are well below the limiting resolution of any other measurement technique. However, the relation between optical scattering and surface topography is complex and certainly beyond the scope of this discussion. For success with either of these examination techniques, it is necessary for the surfaces to be highly reflective which makes them generally inappropriate for the majority of worn tribologically surfaces.

Electron microscopy

The onset of significant and troublesome diffraction phenomena can be postponed to much higher magnifications by using a source of radiation with an effective wavelength very much less than that of visible light. For example, the effective wavelength λ of a beam of electrons is given by

$$\lambda = \sqrt{\frac{150}{V_a}} \times 10^{-4} \, \text{micron} \tag{2.1}$$

where V_a is the electrical potential, in volts, through which the beam is accelerated. Electron microscopes typically use accelerating voltages of the order of 100 kV, which means that the effective wavelength of the beam is about 4×10^{-6} microns; this compares with approximately 0.5 microns for visible light. At first sight this would seem to mean that an improvement in resolution by a factor of 10^5 should be possible. Unfortunately, inherent limitations in the magnetic lenses used in the focusing column prevent an improvement of this order, but nevertheless, an increase in resolution of three orders of magnitude is readily achieved in commercial instruments.

There are two basic types of electron microscopes; *scanning* and *transmission* instruments. In scanning electron microscopy (SEM) a fine beam of high-energy electrons is focused to a point on the surface of the specimen. This causes the emission of secondary electrons which are collected and amplified to give an electrical signal: this, in turn, is used to modulate the intensity of an electron beam in a cathode ray tube (CRT). The intensity of the spot on the CRT screen thus mimics the intensity of electron emission from the specimen. To build up a complete picture the electron beam in the microscope is scanned (or *rastered*) over an area of the surface while the CRT beam is scanned over a geometrically similar raster. The image on the CRT is thus a map of the intensities of electron emission from the specimen in the same way that the image in an optical instrument is a map of light reflected from the surface. As the most important (though not the only) variable influencing the extent of electron emission and collection is surface topography, the CRT screen effectively displays a topographical image of the surface under examination. As well as emitting electrons the bombardment by the primary electron beam also leads to the production of X-rays with an energy spectrum which is characteristic of the atomic species present in the uppermost micron or so of the surface. These can be separately analysed and so it is thus possible to combine 'visual' observation of the surface with chemical analysis of selected areas. Some quantitative topographical information can be obtained, if somewhat laboriously, from the SEM by taking stereo pairs of photographs, differing by a small angle of relative tilt, and using photogrammetric techniques of examination. Various potentially much more rapid methods of extracting topographical information directly from the image signal have been proposed and demonstrated experimentally: when these are commercially exploited scanning electron microscopy will be an even more useful examination technique than it is now.

In many practical cases the major advantage that SEM has over optical examination is not the greater ultimate magnification but rather the much improved depth of field at magnifications in the × 100 to × 1000 range. A comparison of the depths of field at various magnifications (and therefore resolutions) for good-quality optical and electron instruments is shown in Table 2.2.

In transmission electron microscopy (TEM) a focused beam of high-energy or fast electrons is incident on a very thin specimen which deflects and scatters them as they actually pass through it. Subsequent electron lenses

Table 2.2 Depth of field and resolution of the SEM and the optical microscope.

Magnification	Resolution (microns)	Depth of field (microns)	
		SEM	optical
× 20	5	1000	5
× 100	1	200	2
× 200	0.5	100	0.7
× 1000	0.1	20	–
× 5000	0.02	4	–
× 10,000	0.01	2	–

magnify the image, finally focusing it on a fluorescent screen for viewing or recording photographically. Successful examination of bulk material requires the preparation of a specimen that is thin enough to transmit the beam of electrons without too great a loss of intensity, but retaining the microstructural features of interest. In order to examine the *surface* of a component using TEM, it is usual practice to fabricate a replica of the surface which retains its essential topography but is to some extent electron transparent: this means that it must be no more than about 0.5 micron thick. The most common method of replica production involves two stages. First, a plastic replica of the surface is made by pressing a softened plastic film into intimate contact with it. This is then stripped from the surface, coated obliquely with a layer of evaporated carbon, and supported on a fine metal grid. The plastic is then dissolved away in a suitable solvent leaving the carbon film, which retains the surface features of the original, for TEM examination.

Scanning electron microscopy has become one of the standard methods of qualitative surface examination in tribological investigations. Provided the material is electrically conducting it has the great advantage that little specimen preparation is required: materials that are non-conducting can be made so on their surface by coating with a thin evaporated film of a heavy metal. The disadvantage of SEM lies mainly in the fact that the specimen must be placed within the vacuum column of the instrument and there is therefore a restriction on the size of the component that can be examined. Large components may have to be destroyed to yield specimens suitable for examination. This drawback can be overcome by preparing a replica of the surface in the same way as is necessary for TEM, but with the advantage that it does not have to be electron transparent. Rubber, plastic, or acetate surface replicas can all be made conductive by coating with a very thin film of an evaporated metal (such as gold) prior to examination. Specimens must be clean, that is, the surfaces free of volatile fluids or lubricants which might contaminate the vacuum system of the column, and capable of mechanically withstanding the process of being placed *in vacuo*.

The late 1980s have seen the development of a rather different form of electron microscopy known as STM or *scanning tunnelling electron microscopy*: this has great potential for future surface studies. This technique depends on the phenomenon of electron tunnelling which occurs between two conductors separated by an *extremely thin* (less than 0.01 micron) insulating film or layer. When even a modest potential difference is applied across the film a small current will flow due to the overlap of the electron clouds in the outermost layers of atoms in the conducting substrates. In STM one electrode is formed by the sharp, pointed tip of a probe and the other by the surface of the specimen to be examined. The instrument must possess a very precise positional control mechanism for the vertical location of the probe as this must be continuously adjusted as the tip is rastered across the surface so as to maintain the tunnelling current at a constant value. This current is extremely sensitive to tunnelling distance and may decrease by as much as an order of magnitude for each 0.1 nm increase in separation. By monitoring the position of the probe an image of the surface extending down to the atomic scale can be gradually built up. One particularly exciting feature of this technique, in contrast to SEM and TEM, is that since the electron path is so short it is not essential to operate in a vacuum. The presence of gas-phase molecules in the region of the gap leads to only a slight reduction in resolution. The central problems in the use of STM are associated with the design and manufacture of an operating stage capable of the extraordinary precision required and in disentangling the richness of information on the image into its topographic, chemical, and electronic features.

2.2.2 *Surface profilometry*

While both optical, and more especially scanning electron, microscopy can provide enormously valuable *qualitative* information on surface conditions the principle source of *quantitative* data on surface texture is obtained from surface profilometry. The essential features of a modern stylus instrument are shown in Fig. 2.3.

A sharp, very lightly loaded, stylus is drawn at a constant speed over the surface to be examined. The transducer, in modern instruments usually a linear variable differential transformer, produces an electrical signal that is

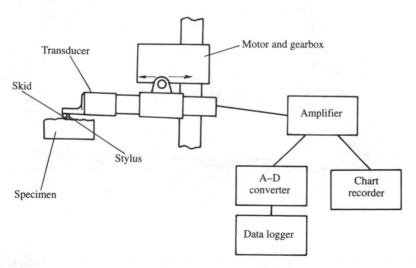

Fig. 2.3 Component parts of a typical stylus surface measuring instrument or profilometer.

proportional to stylus displacement. In its simplest form this analogue signal is fed, after suitable amplification, to a chart recorder so providing a magnified view of the original profile. In order to provide data that are representative of the surface as a whole the horizontal travel of the stylus needs to be many times the maximum amplitude of its vertical motion; thus to contain this information in a manageable length of chart it is usual to use a much greater magnification in the vertical direction than in the horizontal – typically greater by a factor of more than 20. This distortion in the chart record must always be remembered when examining such traces. The dramatic slopes and valleys suggested in the output, see Fig. 2.4(a), are really artefacts of this compression. In reality, surface asperities have much more gentle and undulating slopes, as indicated in Fig. 2.4(b).

The stylus and transducer assembly bear a strong similarity to the components of an audio pick-up: the stylus is tipped by a diamond with an effective radius of a few microns, and although this dimension is very small the effect it has on the quantitative data recorded may not be negligible (see Section 2.3.3). It is important that the stylus traverses the surface at a constant speed – this can usually be set in the range between a few millimetres per second and a maximum of about 100 mm per second. In most instruments the stylus carriage effectively moves along a reference surface, for example an optical flat, which is incorporated in the drive arrangement. The length of this flat, and thus the stroke of the device, can vary between a few millimetres and as much as 10 cm. The quantity actually measured by the transducer is the vertical separation between this flat and the tip of the stylus.

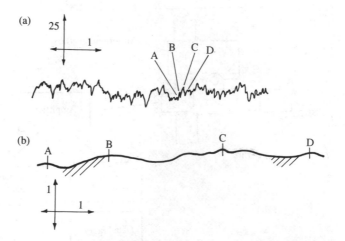

Fig. 2.4 Typical analogue output from a surface profilometer: (a) much compressed horizontal scale; and (b) equal horizontal and vertical magnifications. A, B, C and D are corresponding points in each trace.

If the mean level of the surface to be measured is not accurately set up to be parallel to the reference flat then at all but the very lowest of vertical magnifications the range of linearity of the transducer, amplifier, and chart recorder will very soon be exceeded. It is thus common practice to use a *skid*, effectively a mechanical averaging device, which is attached to the pick-up and rests on the surface either beside or behind the stylus. The transducer then senses the *difference* in level between the stylus and the skid and the setting up procedure is much simplified. These two arrangements are illustrated in Fig. 2.5.

Information from a linear track across the surface can also be obtained from non-contacting optical profilometers. In these devices the conventional stylus assembly is replaced by an optical head. The surface under test is illuminated by an intensive bundle of infrared radiation some of which is scattered back to the instrument. This reflected radiation, whose angular distribution is determined by the surface topography of the specimen, impinges on an array of photodiodes so that the intensity distribution can be measured as a function of the scattering angle. From this information a microcomputer computes the relevant roughness data.

2.3   The statistical nature of surfaces

2.3.1   *Simple statistical parameters*

From the earliest studies of surface topography it has been appreciated that no single numerical parameter can adequately describe surface geometry. The two simplest, and still most widely used, roughness parameters are the R_a value or *centre-line average* (CLA) and R_q value or *root mean square* (RMS) figure. The first of these is defined by the equation

$$R_a = \frac{1}{L} \int_0^L |z| \, dx, \tag{2.2}$$

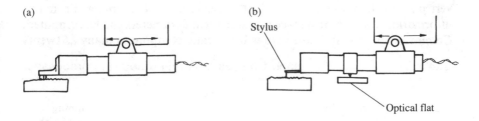

(a)

(b)

Stylus

Optical flat

Fig. 2.5   Operation of a stylus instrument, (a) with skid, and (b) skidless, using an optical flat as the reference surface.

where z is the height of the surface measured above the mean level, that is, the line drawn such that the area of the metal above is equal to that of the voids below, x is the coordinate in the surface, and L the measurement length, as shown in Fig. 2.6(a).

Typical R_a values resulting from a range of widely used production processes are given in Table 2.3.

One immediate difficulty in using R_a values is that they may fail to distinguish between a relatively gently undulating surface and one with a much spikier profile. The surfaces of Figs 2.6 (b) and (c) both have R_a values of $0.64a$ yet might be expected, perhaps under most circumstances, to exhibit significantly different tribological properties.

The RMS roughness parameter R_q goes some way to overcoming this difficulty since, from its definition, the squared term gives greater significance to surface variations some way from the mean. R_q is defined by the equation

$$R_q = \sqrt{\frac{1}{L} \int_0^L z^2 \, dx}. \tag{2.3}$$

A frequent requirement is to be able to estimate the proportion of the nominal contact area between two opposing surfaces that are in true contact: this is sometimes referred to as the *bearing area*, and the way in which this proportion of the nominal contact area increases as the mean levels of the surfaces are driven closer together can be displayed on the *Abbott* (or *Abbott and Firestone*) *bearing area* curve.

A typical point on this curve is established by drawing a line parallel to the mean level at some height d and then expressing the sum of all the metal intercepts along this line as a proportion of the total measurement length, that is, as $(a_1 + a_2 + a_3 + \cdots)/L$ in Fig. 2.7. Strictly this procedure gives a bearing *length* along a line, rather than an area, but if the surface is isotropic (i.e. has no very pronounced lay) then the bearing length and the bearing area are numerically equal.

In modern stylus instruments the analogue output of the transducer is converted into a digital signal which is then fed to a microcomputer. From this information a wealth of surface and statistical parameters can be computed. Commercially available software will typically compute as many as twenty

Table 2.3 Typical R_a values from various manufacturing processes.

	Process				
	Rough casting	Coarse machining	Fine machining	Grinding/ polishing	Lapping/ superfinishing
R_a (μm)	10	3–10	1–3	0.2–1	0.02–0.4

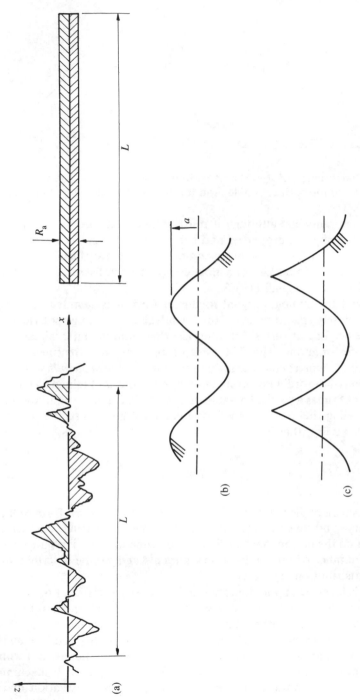

Fig. 2.6 (a) The centre-line average or R_a value of a surface over sampling length L. The shaded areas are equal. Both profiles (b) and (c) have the same R_a value of 0.64a.

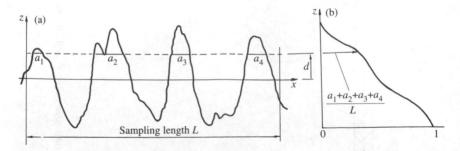

Fig. 2.7 Construction of the Abbott bearing area curve from the topography of a surface: (a) the surface profile; and (b) the associated bearing area curve.

or thirty such numbers, although it must be said that their significance in describing the physical properties and behaviour of the surface is not at all well established. A few of the more commonly quoted surface height parameters are listed in Table 2.4; for a more comprehensive list see, for example, Thomas (1982) or Dagnall (1986).

To convert the analogue signal to digital form it is sampled at a large number of closely spaced points. This is equivalent to taking spot values of the surface height z at very short intervals: this sampling interval can be as small as a few microns. The first stage of any statistical treatment of the surface profile involves consideration of the form of the *probability distribution function* of the ordinate z designated by $\phi(z)$; this is obtained by plotting the number of times a particular value of z occurs in the data versus the value of z and scaling the most closely fitting curve to the data so that the area it encloses is unity, that is,

$$\int_{-\infty}^{+\infty} \phi(z)\, dz = 1. \tag{2.4}$$

For real surfaces the outcome of this process is always some form of 'bell-shaped' curve and much of the analytical treatment of surface contact has been based on the notion that the distribution function $\phi(z)$ is characteristic of a large number of randomly occuring events and can be described by a *Gaussian* distribution (Fig. 2.8).

While surfaces with an approximately Gaussian distribution can be generated (for example by shot or bead-blasting, or spark erosion) they are actually the exception rather than the rule in general engineering. Most machine tools leave their imprint in the surface texture they produce on the workpiece; this pattern or lay may be very apparent as in the case of turning on a lathe but can also be present, though less obvious, in other machining processes. It is not always easy to judge from direct observation of the

Table 2.4 Definitions of some surface roughness parameters.

Symbol	Name	Definition
R_a	Average roughness	$\dfrac{1}{L}\displaystyle\int_0^L |z|\,\mathrm{d}x$ taken over 2–20 samples
R_{max}	Maximum peak-to-valley	Largest peak-to-valley height in five adjoining sampling lengths
R_q	RMS roughness	$\sqrt{\dfrac{1}{L}\displaystyle\int_0^L z^2\,\mathrm{d}x}$
R_t	Peak-to-valley height	Separation of highest peak and lowest valley
R_z (DIN)	Average peak-to-valley height	Average of single R_t values over five adjoining sampling lengths
R_z (ISO)	Ten-point height	Separation of average of five highest peaks and five lowest valleys within single sampling length

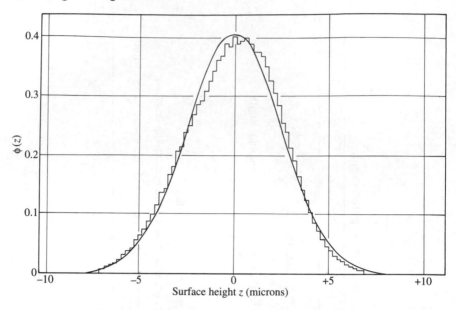

Fig. 2.8 Histograms of heights compared to a Gaussian distribution with the same standard deviation based on 282,309 spot heights from a spark eroded surface (from Sayles and Thomas 1979).

frequency distribution curve the extent to which any given surface deviates from Gaussian; a somewhat more graphic indication can be obtained if the same data on surface heights are plotted on normal probability axes on which a truly random or Gaussian distribution will fall on a straight line (whose gradient provides a measure of the standard deviation of the data). This effect is illustrated for some real surfaces in Fig. 2.9.

The shape of the probability density distribution is associated with a certain amount of useful information on the nature of the surface. Following the conventions of statistics, one method of attempting to quantify the shape of this curve is by means of the moments of the distribution. The nth moment of the distribution m_n is defined as

$$m_n = \int_{-\infty}^{+\infty} z^n \phi(z) \, dz. \tag{2.5}$$

The first moment m_1 represents the *mean* level of the signal; the level from which z is measured is usually chosen so that m_1 is equal to zero. The second moment of the distribution is defined from equation (2.5) as

$$m_2 = \int_{-\infty}^{+\infty} z^2 \phi(z) \, dz. \tag{2.6}$$

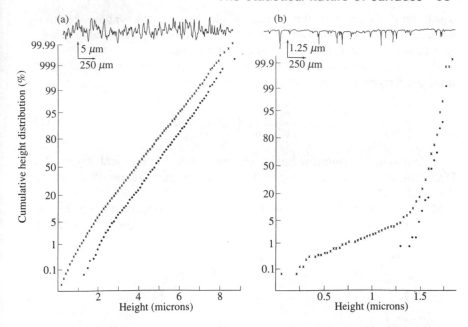

Fig. 2.9 Cumulative height distributions plotted on normal probability axes: circles denote peak heights; crosses denote all heights. (a) Bead blasted aluminium; (b) mild steels abraded and polished (from Williamson 1967–8).

Provided m_1 is set at zero then $\sqrt{m_2}$ is equal to σ, the *standard deviation* of the distribution, but as this is also effectively the definition of R_q it follows that $R_q = \sigma$. A good deal of statistical analysis is in terms of σ^2 rather than σ; this quantity is known as the *variance*. In the special case of a Gaussian distribution it can also be noted that

$$\frac{R_q}{R_a} = \sqrt{\frac{\pi}{2}} \approx 1.25. \tag{2.7}$$

In practice, even for surfaces with non-Gaussian distributions this ratio is usually numerically close to 1.25.

The third moment of area is commonly normalized by the standard deviation to give the *skewness* Sk,

$$\mathrm{Sk} = \frac{m_3}{\sigma^3} = \frac{1}{\sigma^3} \int_{-\infty}^{+\infty} z^3 \phi(z) \, \mathrm{d}z. \tag{2.8}$$

This parameter gives some measure of the degree of asymmetry of the distribution since a symmetrical distribution must have Sk = 0. If in the course of its preparation or service a surface undergoes a series of processes which tend to remove the peaks, while leaving the valleys relatively unscathed, then the distribution can move from a skewness value close to zero to one in which Sk < 0, as indicated in Fig. 2.10.

The fourth moment of the distribution, once again normalized by the standard deviation, is known as the *kurtosis K*

$$K = \frac{m_4}{\sigma^4} = \frac{1}{\sigma^4} \int_{-\infty}^{+\infty} z^4 \phi(z) \, dz. \tag{2.9}$$

A Gaussian distribution has a kurtosis with a numerical value of 3. Distributions with a peak sharper than Gaussian have values of K less than this, and vice versa.

Many other statistical parameters can be readily computed from the trace of the surface once this has been stored in digital form. However, these must

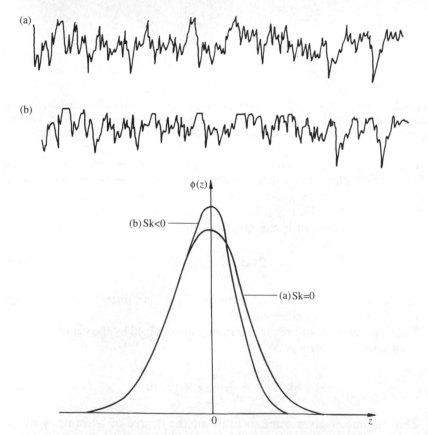

Fig. 2.10 Effect of removal of surface peaks on a surface with an originally Gaussian distribution, curve (a) leading to a probability density distribution with a negative value of skewness.

all be treated with caution especially if any attempt is to be made to use them as predictors of subsequent surface performance or indicators of likely surface failure. One important feature of these statistical parameters is that since they are based on random or pseudorandom data they themselves are subject to random statistical variations. Thus it may not be possible to tell immediately whether a given value of some derived parameter, such as skewness or kurtosis, represents a true functional property of the surface in question or has been unduly influenced by the sampling process itself.

We have seen in previous sections that the simpler measures of surface heights and their distributions may fail to distinguish between surfaces with vary different profiles. None of the statistical parameters deduced from the probability distribution contain information on the *horizontal* or *spatial* distribution. On the basis of height distribution alone the two surfaces of Fig. 2.11 would give identical roughness parameters. However, they may well exhibit very different tribological properties as a result of the 'open' texture of (a) compared to the closer 'closed' texture of (b).

There are a number of possible parameters that could serve as 'horizontal descriptors' of the surface. In essence they reduce to either the number of peaks per unit length of the profile or the number of crossing points per unit length with the mean level that the profile exhibits. Unfortunately, neither of these parameters, or those such as slope or curvature which depend on them, are intrinsic properties of the surface. Many real surfaces appear to exhibit a form of self similarity, that is, the shape of the profile remains more or less the same as the degree of magnification is increased: coarse scales of roughness having finer roughnesses of a similar geometry imposed upon them. Thus the *peak density*, or *zero-crossing density*, *mean slopes*, or

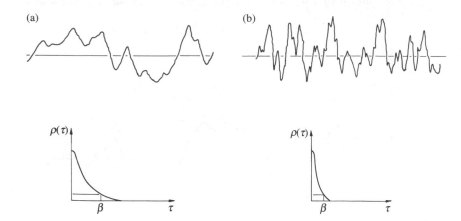

Fig. 2.11 (a) *Open* and (b) *closed* surface textures and their autocorrelation functions.

curvatures, are all highly dependent on the sampling interval at which the signal is digitized.

Two avenues of approach have been explored in attempts to incorporate information on spatial variations and these are briefly described in the following two sections.

2.3.2 *Use of autocorrelation functions*

The autocorrelation function (ACF), $\rho(\tau)$, of a single signal or profile $z(x)$ is defined as

$$\rho(\tau) = \frac{1}{\sigma^2} \left\{ \lim_{L \to \infty} \frac{1}{L} \int_0^L z(x) \times z(x + \tau) \, dx \right\} \tag{2.10}$$

where L is the sampling length and σ the standard deviation of z. It is thus effectively formed by spatially delaying the original profile by some distance τ and then averaging the product of the delayed signal and the original over some suitable representative length L; finally, it is made non-dimensional by the variance. This process is illustrated in Fig. 2.12.

If $\tau = 0$ then clearly

$$\rho(0) = \frac{1}{\sigma^2} \left\{ \frac{1}{L} \int z^2 \, dx \right\} = 1 \tag{2.11}$$

and as τ tends to infinity the extent of correlation decreases and so $\rho(\tau)$ tends to zero; thus if the variation of $\rho(\tau)$ is plotted against τ then the curve must decay from a value of unity at $\tau = 0$ to become asymptotic to zero at large values of τ. The form of this decay can provide some information on the spatial distribution of roughness. If, for example, the surface contains an

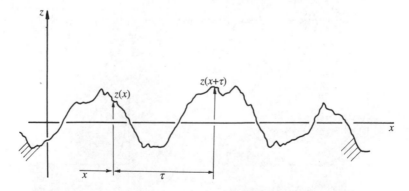

Fig. 2.12 Graphical representation of the autocorrelation function.

inherent periodicity of wavelength λ, as might be introduced by particular machining processes, then the ACF will display a series of maxima when τ takes on values which are integral multiples of λ. This is illustrated in the data displayed in Fig. 2.13.

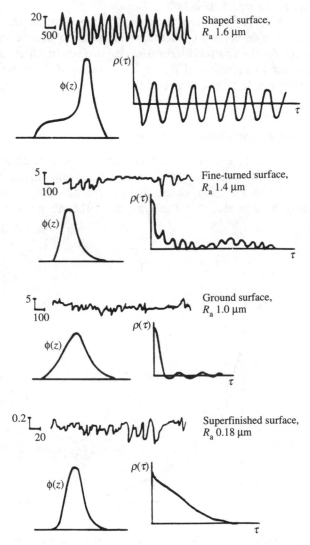

Fig. 2.13 Examples of engineering surfaces, their height distributions, and autocorrelation functions (from Arnell *et al*. 1991).

It has been suggested that the simple exponential decay expression

$$\rho(\tau) = \exp(-\tau/\beta^*) \qquad (2.12)$$

is a sufficiently good fit for many surfaces that approach randomness, for example the ground and superfinished surfaces of Fig. 2.13; $1/\beta^*$ is the decay rate, or slope at $\tau = 0$, see Fig. 2.14. The *correlation length* β is defined as that value of τ for which $\rho(\tau) = 0.1$ and for the simple exponential decay is thus numerically equal to $2.3\beta^*$.

Even if the ACF is not strictly exponential the value of the correlation length β can provide some rudimentary information on the shape of the curve and thus on the spatial distribution of surface heights; for example, in comparing the two profiles of Fig. 2.11 the value of β for the closed surface is significantly less than that for the open.

2.3.3   *Use of slopes and curvatures*

An alternative procedure by which spatial variations can be described makes use of the root mean square values of surface slope and curvature. Suppose that the representative length L of the surface is sampled at discrete intervals each of length $\hbar$. The slope of the surface at the ith point m_i might be defined by

$$m_i = \frac{z_{i+1} - z_i}{\hbar}, \qquad (2.13)$$

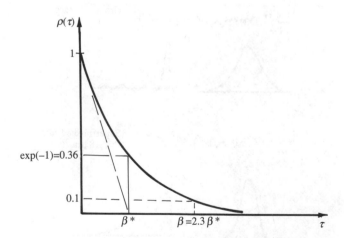

Fig. 2.14   The exponentially decaying autocorrelation function (ACF). Distance β is known as the correlation length.

and likewise the curvature κ_i, that is, the *inverse* of the local radius of curvature R_i, by

$$\kappa_i = \frac{1}{R_i} = \frac{-z_{i+1} + 2z_i - z_{i-1}}{\hbar^2}. \qquad (2.14)$$

The root mean square values of these quantities can then be found from the relationships

$$\bar{m}^2 = \frac{1}{n} \sum m_i^2 \qquad (2.15)$$

when $n = L/\hbar$ is the number of data points, and

$$\bar{\kappa}^2 = \frac{1}{n} \sum \kappa_i^2. \qquad (2.16)$$

Unfortunately, as has been pointed out in the previous section, the numerical values of these quantities must be treated with caution as they are not truly intrinsic properties of the surface but depend on the choice of L and $\hbar$; both tend to increase without limit as $\hbar$ is made smaller, and so shorter and shorter wavelength components are included.

Work since the late 1980s has suggested that for a profile having a near Gaussian distribution of heights the distribution of surface peaks is also near Gaussian; here a peak is defined as a point having a greater z-coordinate than the two data points on either side of it. What is more, the mean curvature of the peaks is of the same order of magnitude as the RMS curvature of the surface; see Fig. 2.15. However, these data once again indicate clearly the considerable influence the sampling interval $\hbar$ has on the absolute values of these quantities; a ten-fold reduction in $\hbar$ can often lead to a change in the derived statistical parameters of as much as a factor of 40.

The dependence of the numerical values of these surface descriptors on the conditions under which they are measured has lead to the introduction of the concept of *functional filtering*.

2.3.4 *Functional filtering*

It is usual practice, indeed in some respects inevitable, to provide some form of filtering on the raw topographic data; this is a necessary condition to obtain finite values of many of the surface parameters commonly used. The numerical values of the height-dependent descriptors depend on the longest wavelengths measured, while those of the texture parameters (slopes, curvatures, and so on) generally depend on the shortest. Information about the surface passes through both high- and low-pass filters and in order to make

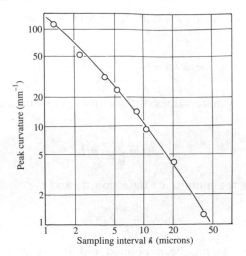

Fig. 2.15   Effect of sampling interval h on the numerical value of the peak surface curvature (from Greenwood 1984).

sensible use of roughness measurements, particularly as predictors of surface behaviour in a real contact problem, we must choose the window of wavelengths between these two filters. This is an example of functional filtering.

Using a skid (see Fig. 2.5), as opposed to an 'absolute' datum, speeds up the setting-up procedure of a stylus instrument considerably; however, it necessarily acts as a high-pass filter removing some elements of the surface waviness. Once removed, this information is, of course, lost and cannot be reconstituted; if these wavelengths are thought to be relevant to the problem being considered then the skid should not be used. In addition to this mechanical filtering, the electrical output of the transducer is also passed through an analogue or, increasingly commonly, a digital, high-pass filter. This has been standardized as having the characteristics of a two-stage C–R circuit: the *cut-off length* is defined by the wavelength at which transmission is just 75 per cent (see Fig. 2.16) and this is usually made equal to the sampled or specimen length L. Provided that this is at least as long as the nominal contact length in the application we have in mind, this is a reasonable procedure, since wavelengths in the surface profile greater than this can only have a minor effect on the contact deformation.

The specification of the low-pass filter is often much more contentious as it amounts to specifying the smallest surface wavelength that it is considered will affect the interaction between the loaded surfaces. The stylus

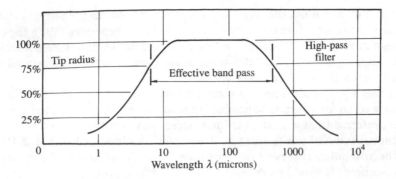

Fig. 2.16 Functional filtering involves the choice of the cut-off lengths at both high and low wavelengths.

of the roughness-measuring instrument itself acts as a form of filter; it has, typically, an effective radius of about 5 microns and thus is unable to resolve fine detail at wavelengths any less than this. Although many conventional styluses are manufactured in the form of truncated pyramids, the same argument applies, and they too act as low-pass filters with effective cut-off lengths of a few microns. This effect places a limit of about 5–10 microns on the sampling interval of most commercial profilometers. The finite size of the stylus also means that even above this scale not all the features of the specimen surface can be faithfully reproduced. In particular peak radii will be overestimated and valleys appear more 'cusp-like' than they are in reality. In practice these errors are only likely to be of significance if the surface contains valleys with trough radii less than about 10 microns or peaks with slopes of more than 45° and experience indicates that few engineering surfaces have slopes of more than about 10° or valleys with radii of curvature much less than 100 microns.

2.3.5 Surface mapping and relocation profilometry

The major disadvantage of stylus instruments is that they provide detail of height variations along a line, which may not be representative of the sample as a whole—this may be especially true if the surface has a pronounced lay. To obtain information about an area some form of scanning technique is required and this can be brought about by mounting the specimen on an indexing x–y table, perhaps controlled by the same microcomputer used to process the data obtained from the transducer. Commercial instruments of this sort are just becoming available. The main problems in the development of this technique have been, firstly, the establishment of an arbitrary flat datum, and secondly, the relatively small size of the scanned area which is limited by the very large quantity of numerical data generated. With the continuing drop in the real cost of computers this is unlikely to remain a serious difficulty. Figure 2.17 illustrates some of the types of three-dimensional plot which this technique can produce.

In moving from the linear examination of a representative portion of the surface to an area scan it must also be realized that a maximum in a stylus trace, referred to as a 'peak', may not necessarily correspond to a true maximum or 'summit' in the surface. This effect is illustrated in Fig. 2.18.

The grid points represent the locations at which the height of the surface is monitored. It would seem reasonable to define a point P (in Fig. 2.18(b)) as a summit if it has a z-coordinate greater than its four nearest neighbours Q_1 to Q_4: thus in the 'contour map' of Fig. 2.18(a) the point A is a true summit. However, consider the points B and C; both satisfy the condition that their height is greater than their four neighbours but nevertheless neither is a *true* summit. The point D would also be interpreted on this criterion as a summit and yet this is clearly not the case: D is the highest point on the ridge or 'col' between the high ground at A and that at B and C: on a traverse from top left to bottom right D would appear as a peak though it is not a summit. Notwithstanding these difficulties of definition, it has been found, over a range of sampling intervals, that in the case of an isotropically random surface if n_p is the number of peaks per unit length and n_s the number of summits per unit area then,

$$n_s \approx 1.8 n_p^2. \tag{2.17}$$

In many cases the mean summit curvature is of the same order as the RMS curvature of the surface as a whole, and furthermore the distribution of summit heights is also close to random with a standard deviation very close to that of the height distribution along a linear track: this is illustrated by the data plotted as circles and crosses in Fig. 2.9.

There are many occasions, particularly in research investigations into the development of worn surfaces, on which it would be advantageous to make repeat traversals of a particular region of the surface, for example, before and after it has been subject to service conditions. In this way the manner in which particular topographical features become modified could be followed in detail. For this technique to succeed it is clearly essential that exactly the same section is traversed each time. This very precise relocation of the specimen is not easy to achieve: even the action of returning the stylus to the start of its travel may lead to a lateral displacement of 8–10 microns which could easily be enough to invalidate the observations. Several research groups have developed arrangements which allow a specimen stage to be relocated to the required degree of accuracy—usually this is achieved by pneumatic loading against a kinematic support. It is also necessary to either

(a)

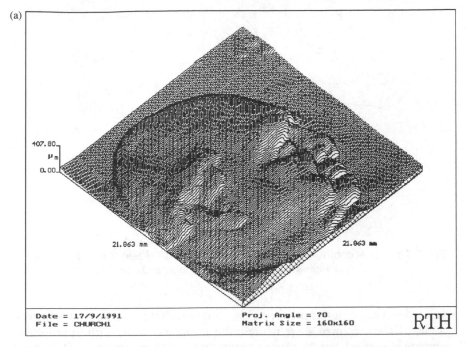

Date = 17/9/1991   Proj. Angle = 70
File = CHURCH1   Matrix Size = 160x160   RTH

(b)

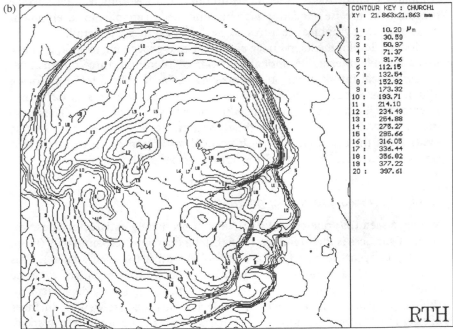

CONTOUR KEY : CHURCH1
XY : 21.863x21.863 mm

1 :	10.20 μm
2 :	30.59
3 :	50.97
4 :	71.37
5 :	91.76
6 :	112.15
7 :	132.54
8 :	152.92
9 :	173.32
10 :	193.71
11 :	214.10
12 :	234.49
13 :	254.88
14 :	275.27
15 :	295.66
16 :	316.05
17 :	336.44
18 :	356.82
19 :	377.22
20 :	397.61

RTH

Fig. 2.17 Typical output of three-dimensional surface profilometry: (a) an orthographic scan of the surface of a Churchill medallion; and (b) contour plot from the same data. (Courtesy of Rank Taylor Hobson Ltd.)

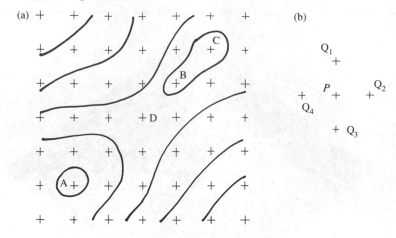

Fig. 2.18 (a) A contour map of a rough surface. (b) Point P is a summit only if it is higher than its four neighbours Q_1 to Q_4.

raise the stylus, or lower the specimen, during the return stroke of the instrument. If the specimen stage is mounted on an x–y table with appropriate computer control then both the techniques of three-dimensional surface mapping and precise relocation may be combined to enable the development of surface topographical features to be followed in great detail.

Modern developments in image capturing and processing techniques have made feasible alternative optical methods of obtaining three-dimensional topographical information on engineering surfaces. These rely on storing digitally the image of the surface formed in a conventional interference microscope illuminated with either white or monochromatic light. The imaging software is capable of providing contour maps, profiles, and oblique views from the captured data with vertical resolution down to a few tens on nanometres and horizontal resolution of better than a micron.

2.3.6 *Surfaces and fractal geometry*

We have seen that one crucial difficulty in producing satisfactory descriptions of surfaces is their tendency to exhibit a degree of self-similarity (more strictly, it appears that surfaces are *self-affine* rather than self-similar). This means that such attributes as mean slopes, or curvatures, or indeed even the true length of a profile (or true area of a surface) are highly dependent on the sampling interval. Whatever the scale at which they are viewed, surfaces are not formed from 'smooth' curves; perhaps it is not even sensible to speak of a mean slope of a curve that is not capable of being differentiated. In this respect real surfaces are similar to mathematical *fractals*, and convincing surface topographies can be constructed on the basis of fractal geometry where the fractal dimension D lies in the range $2.1 < D < 2.5$. The fractal dimension can be thought of in the following way: imagine attempting to measure the true length of a line (perhaps the profile of a surface) with a ruler of length $\hbar$, see Fig. 2.19. As $\hbar$ is made smaller so the apparent or measured length of the profile gets longer as more detail is included: if the profile is 'rough' down to an atomic scale there is virtually no limit to this process. A fractal line, as opposed to a perfect Euclidean line, has the property that its apparent length $L(\hbar)$, depends on the measuring interval $\hbar$ according to a formula of the form

$$L(\hbar) = \text{constant} \times \hbar^{1-D} \tag{2.18}$$

where D is *not* equal to unity (which would be the case for a perfect Euclidean line). As $\hbar$ is reduced so the estimate of length $L(\hbar)$ gets greater and greater. The implications of this way of thinking about surface profiles have still to be fully explored.

2.4 Tribological properties of surfaces

2.4.1 *The physical nature of metallic surfaces*

When metal surfaces are exposed to air then they at once become covered with adsorbed molecules of oxygen and water vapour. With noble metals these surface films may be no more than one or two atoms thick, but with reactive metals a chemical reaction follows which invariably results in the formation of a film of metal oxide. Much work has been carried out by surface physicists and materials scientists on the kinetics of the growth, structure, and topography of such films. While the rate of growth can be influenced by the crystallographic orientation of the underlying material the most important factor in its rate of formation is temperature. Oxides which grow slowly at low temperatures are often relatively smooth, but at higher temperatures, and so increased growth rates, the oxide film may grow from surface nuclei as crystals and 'whiskers' giving a very much rougher topography. In the case of an iron or low-alloy steel surface the oxide layer can consist of a mixture of the various possible oxides of iron, namely Fe_2O_3, Fe_3O_4, and FeO. With some metals the layer consists of two or more distinct films; for example, in the case of copper that part of the film closest to the metal is the lower, cuprous oxide Cu_2O, while the outer most layer is the higher, cupric oxide CuO. With alloys, the surface layer may consist of a mixture of different oxides, that on stainless steels, for example, is a mixture of iron and chromium oxide—it is the latter, Cr_2O_3, which

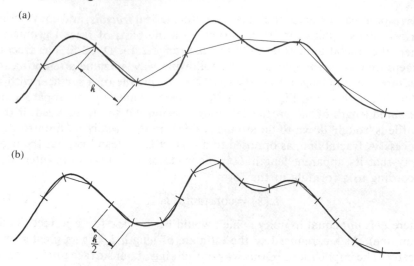

Fig. 2.19   Measuring the length of a surface profile; when the ruler h is reduced by a factor of 2, from (a) to (b), so the estimate of the length of the profile increases, in this case by about 7%.

provides the corrosion protection. With some metals the oxides are extremely tenacious and once a very thin film has formed the surface becomes effectively passivated with no further chemical activity taking place: examples of this are aluminium and titanium. In other cases the oxide has a much more open texture and allows continued diffusion of oxygen to the reactive metal beneath: this means that corrosion can continue even though there is a layer of oxide on the surface: the oxides of iron are typical of this behaviour.

Experimental work in tribology is invariably carried out on prepared surfaces that have either been machined or mechanically abraded and polished. Immediately beneath the inevitable oxide film is a layer of the parent metal which has been heavily deformed or mechanically worked as a result of the method of surface preparation. This region, sometimes known as the *Beilby layer*, may consist of a smeared fudge of metal, metal oxide, and as polishing powder, with an often disrupted crystalline structure which is very different from that of the bulk metal (see Fig. 2.20). Beneath this layer lies material, which although an integral part of the component or specimen, has still been severely sheared or strained by the surface machining or abrasion. It too can have very different mechanical properties from the bulk as a result of this imposed deformation: its grain structure may have become modified and it will almost always have a much greater dislocation density than material deeper into the component. Consequently, when extrapolating from

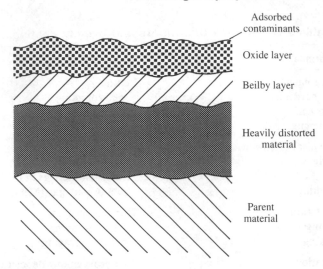

Adsorbed contaminants

Oxide layer

Beilby layer

Heavily distorted material

Parent material

Fig. 2.20 Diagrammatic representation of the structure of a typical polished metal surface. The oxide is 0.01–0.1 micron thick, the Beilby layer also about 0.1 micron in thickness consists of disrupted mixture of parent metal, metalllic oxides and perhaps polishing powder, while the heavily worked layer may extend several microns into the bulk of the material.

experimental observations on surfaces to make predictions about likely behaviour in service it is important to consider carefully many aspects of the conditions under which the surfaces operate, and in which the topography is generated.

2.4.2   Surface engineering

The recognition that a very large number of engineering components can either deteriorate progressively or fail catastrophically through surface-related phenomena, such as wear, fatigue, and corrosion, has led to the establishment of the interdisciplinary subject of *surface engineering*. This can be defined as the application of both traditional and innovative surface technologies to produce a composite material with properties unattainable in either the base or surface materials individually. Broadly speaking these techniques can be divided into those involving the modification (whether by mechanical, thermal, or chemical means) of the existing surface of the component, and those that involve the deposition of additional or overlay material, albeit often in the form of a very thin layer or coating over the bulk substrate. These techniques are summarized in Table 2.5.

A number of these treatments have important applications in tribology.

Table 2.5 Surface treatments and overlay coatings.

Surface treatments	Overlay coatings

Thermal treatments
  induction hardening
  flame hardening
  laser hardening
	Plating
	  electroplating
	  mechanical plating

Thermo-chemical diffusion
  carburizing
  nitriding
  carbo-nitriding
	Weld cladding
	  oxyacetylene
	  tungsten inert gas (TIG)
	  metal inert gas (MIG)

Mechanical treatments
  shot-peening
  cold working
	Thermal spraying
	  flame spraying
	  plasma spraying

Ion implantation	Chemical vapour deposition (CVD)

Laser glazing	Physical vapour deposition (PVD)
	  sputtering
	  evaporation

Improvements in surface mechanical properties can often be made by paying extra attention to the metallurgy and topography of the 'existing' surface. Cold-working processes employed during component manufacture or applied subsequently, like shot-peening, can introduce compressive residual stresses at the surface which tend to resist fatigue crack growth and thus improved wear resistance in situations involving the repeated application of load. The introduction of such surface cold work has a minimal effect on the behaviour of materials at high temperatures, when creep, recovery, and eventually recrystallization destroy the work-hardened surface layer. Thermal treatments, such as induction, flame, or laser hardening rely on heating a thin surface layer of a carbon or alloy steel component into the austenitic range. This surface layer is then rapidly quenched, either by external means or perhaps, as in the case of laser hardening, merely by the presence of the large cool thermal mass of the substrate material, so that it undergoes the martensitic phase transformation; typical hardening depths are in the range 0.5–2 mm as indicated in Fig. 2.21 which also indicates the maximum surface temperatures involved in some other of these processes. Carburizing and carbo-nitriding are thermo-chemical processes relying on the diffusion into the surface of small atoms (of carbon and both carbon and nitrogen respectively) leading, after quenching, to the production of a hard martensitic outer layer or case. In nitrocarburizing and boronizing, surface diffusion leads to the formation of a distinct layer of new compound at the surface with superior tribological properties to those of the bulk.

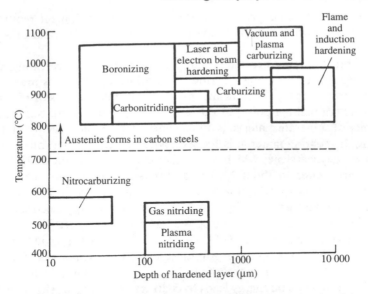

Fig. 2.21 Comparison of the process temperature and depth of hardened material produced by various methods of the surface modification of steels (from Hutchings 1992).

One of the earliest processes for depositing a coating of one metal on another was *electroplating*, and this is still widely used for both decorative and engineering purposes. Chromium electrodeposits used for tribological improvements are hard (850–1250 HV) and typically from 10–500 microns in thickness (much thicker than the chrome plating used for decoration). Nickel films having hardnesses up to 400 HV can also be electrodeposited from acidic solution. Layers up to several millemetres in thickness can be deposited, for example to salvage worn components. There are two classes of coating process in which the coating material is applied in the molten state. These are known as *hard-facing* and *thermal spraying*. In the first of these, the coating material is melted while it is in contact with the substrate by techniques very similar to those used to join metals by welding; the surface of the substrate must be heated to the melting point of the coating material. In the second, the coating material is melted some distance from the substrate and projected towards it in fine droplets; the substrate remains relatively cool and the coating is formed by the solidification of the fine spray as it strikes the target. Hard-facing or welding methods are best suited to the application of thick coatings in the range 1–50 mm and are often used to rebuild surfaces which suffer very high wear or attrition rates, for example in mining and minerals processing machinery. The equipment required is comparatively portable so that repairs can be made under field or service conditions.

In thermal spraying processes, substrate temperatures are rarely above 200°C, even when very-high-melting-point alloys or materials are used to form the droplet spray. This means that not only can coatings of refractory (and thus high-melting-point) metals and ceramics be applied, but that also the choice of substrate material is much wider than for hard-facing. In flame spraying the coating material is fed as a wire or fine powder into a flame, usually of oxyacetylene, which has a temperature of about 3000°C. The droplets are heated to about 2000°C and strike the target surface at about $100\,\mathrm{m\,s}^{-1}$. Plasma-spraying methods are also widely used. The plasma is formed by an inert gas, usually argon, using a high-energy electric are into which the coating material is introduced as a fine powder. The very high temperatures of the arc (15,000°C or more) enable a very wide range of coating materials to be sprayed. Rapid expansion of the hot gas accelerates the molten droplets to more than $250\,\mathrm{m\,s}^{-1}$ and the combination of temperature and kinetic energy leads to coatings of lower porosity than those produced by flame spraying.

In order to achieve enhanced tribological performance, the coating must, of course, remain firmly attached to the substrate, and for this reason correct surface preparation prior to coating is essential. The bonding mechanism of hard-faced coatings is truly metallurgical, whereas for flame- and plasma-sprayed layers it depends more on mechanical interlocking. Even a strong interfacial bond will fail if sufficient stress is applied to it, and an important source of such stress can be the differential thermal contraction between coating and substrate as they cool. Residual stresses from this source can limit the usable thickness of many ceramic coatings to less than 0.5 mm. In order to improve the interfacial strength of some spayed coatings, particularly ceramics, intermediate *bond-coats* are sometimes used.

Methods in which the coating is formed from the vapour phase are divided into two groups. *Chemical vapour deposited*, or CVD, films rely on thermally induced chemical reaction at the surface of a heated substrate with the reagents which form the coating supplied in gaseous form. The reactions may involve the substrate itself, but often do not. Temperatures are in the range 600–1100°C and using mixtures of metal halides, hydrogen, oxygen, nitrogen, hydrocarbons, and boron compounds, coatings of a wide range of metals and their oxides, nitrides, carbides, and borides are possible. In those processes involving *physical vapour deposition*, or PVD, the coating material is transported to the surface in atomic, molecular, or ionic form and derived from physical, rather than chemical, means from a solid, liquid, or gaseous source. Chemical reactions may occur but often they are not involved at all. The substrate is usually very much cooler than in CVD processes, typically 50–500°C above ambient, and this is an attractive feature of the process since, at these levels, it is unlikely to influence adversely the microstructure or properties of the substrate.

The simplest PVD process is *evaporation* which has been used for many years to produce coatings on glass lenses and other optical components. In *sputtering* the energy to transport the material from the source to the substrate is supplied by energetic heavy gas ions which are produced by forming a glow discharge in a low-pressure chamber. Sputtering in an environment incorporating a reactive gas is a versatile process that can be used to deposit thin films of oxides, carbides, and nitrides. In *ion-plating*, a third PVD process, atoms or molecules of the coating material are evaporated from a hot source into a glow discharge formed in argon within which some of the vapour atoms become positively charged. These are then accelerated towards the target which is held at a negative potential of 2–5 kV. If ions with even higher energies strike a surface then they can penetrate comparatively macroscopic distances and change the properties of the material to a depth of perhaps 0.5 micron; this is the principle of the process of *ion implantation*. Ions commonly used in this form of surface engineering include those of nitrogen, carbon, boron, titanium, and aluminium. Their energies lie in the range 50–200 keV giving penetration depths of rather less than a micron. The dosage rates need to be at least of the order of 10^{15} ions mm^{-2} to give practically useful changes in surface properties (this is several orders of magnitude higher than those used to modify the electronic properties of semiconductor materials). Because of this shallow depth, the process causes negligible changes to surface topography or dimensions; it can be applied to ceramics and cermets as well as metals. Typical thicknesses of the surface layers formed by these processes are indicated in Fig. 2.22, and more

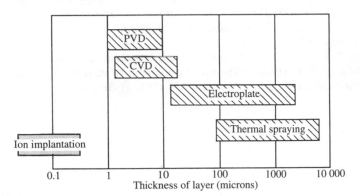

Fig. 2.22 Typical thicknesses of engineering surface layers produced by various plating and deposition techniques.

details of all these aspects of surface engineering can be found in Hutchings (1992), Child (1980), Budinski (1988), and HMSO (1986).

2.5 Further reading

Arnell, R. D., Davies, P. B., Halling, J., and Whomes, T. (1991) *Tribology*, Macmillan, London.

Bowen, D. K. and Hall, C. R. (1975) *Microscopy of materials*. Macmillan, London.

Budinski, K. J. (1988) *Surface engineering for wear resistance*. Prentice Hall, New York.

Child, H. C. (1980) *Surface hardening of steel*. Engineering Design Guide 37, Oxford University Press.

Dagnall, H. (1986) *Exploring surface texture*. Rank Taylor Hobson, Leicester.

Greenwood, J. A. (1984) A unified theory of surface roughness. *Proceedings of the Royal Society*, A**393**, 133-57.

HMSO, (1986) *Wear resistant surfaces in engineering*. UK Department of Trade and Industry, London.

Hutchings, I. M. (1992) *Tribology*. Edward Arnold, London.

Institution of Mechanical Engineers (1967-8) Properties and metrology of surfaces. *Proceedings of the Institution of Mechanical Engineers*, 182 Pt 3K. BS 6741 (1987).

BS6741 Parts I and II. *Surface roughness terminology*. British Standards Institution, London.

Mandelbrott, B. B. (1983) *The fractal geometry of nature*, W. H. Freeman, New York.

Quinn, T. J. F. (1991) *Physical analysis for tribology*. Cambridge University Press.

Sayles, R. S. and Thomas, T. R. (1979) Measurements of the statistical microgeometry of engineering surfaces. *Transactions of the American Society of Mechanical Engineers, Series F, Journal of Lubrication Technology*, **101**, 409-18.

Sherrington, I. and Smith, E. H. (1988) Modern measurement techniques in surface metrology. Part I: stylus instruments, electron microscopy and non-optical comparators. *Wear*, **125**, 271-88; Part II: optical instruments. *Wear*, **125**, 289-308.

Stout, K. J., Davis, E. J., and Sullivan, P. J. (1990) *Atlas of machined surfaces*. Chapman and Hall, London.

Thomas, T. R. (ed.) (1982). *Rough surfaces*. Longman, London.

Williamson, J. B. P. (1967-8) The microtopography of surfaces. *Proceedings of the Institution of Mechanical Engineers*, **182** (3K), 21-30.

3
Contact between surfaces

3.1 Introduction

When two engineering surfaces are loaded together there will always be some distortion of each of them. These deformations may be purely elastic or may involve some additional plastic, and so permanent, changes in shape. Such deflections and modifications in the surface profiles of the components can be viewed at two different scales. For example, consider the contact between a heavily loaded roller and the inner and outer races in a rolling-element bearing. In examining the degree of flattening of the roller we could choose to express the deflections as a proportion of their radii, that is, to view the distortions on a relatively *macroscopic* scale. On the other hand, as we have seen in Chapter 2, at the *microscale* no real surface, such as those of the roller or the race, can be truly smooth, and so it follows that when these two solid bodies are pushed into contact they will touch initially at a discrete number of points or asperities. The sum of the areas of all these contact spots, the 'true' area of contact, will be a relatively small proportion of the 'nominal' or geometric contact area — perhaps as little as only a few per cent of it. Some deformation of the material occurs on a very small scale at, or very close to, these areas of true contact. It is within these regions that the stresses are generated whose total effect is just to balance the applied load. The nominal or apparent contact pressure, that is the normal load divided by the nominal contact area, is very much less than the values of the true, local contact pressures.

In attempting to predict the likely damage to components, or their life under a given set of operating conditions, a knowledge, or at least a realistic estimate, of the true stresses experienced by the material is crucially important. In this chapter we shall develop an analysis of surface contact that can be applied both to macroscopic geometries and to microscopic, or asperity, contacts. We start by investigating the conditions at single tribological contacts between two solids with somewhat idealized, but not unrealistic, surface profiles; in particular, the cases of two cylinders with their axes parallel, and the contact between two spheres. At the macroscopic level these might represent the contact between the rollers, or balls, and the races in a rolling-element bearing while at the microscopic level we can think of modelling the contact between individual surface asperities on two opposing surfaces. We need to establish how the stresses at, and just below, the surfaces depend on the applied load, the shape of the surface profiles, and the material properties of the solids, and we need to do this for both the case when the contact is carrying a simple normal load and when a friction or *traction* load is also present. For more detailed treatments of this topic than are included here the reader is referred to the texts by Johnson (1985) and Hills *et al.* (1993).

3.2 Geometry of non-conforming surfaces in contact

We start by considering the case of a long deformable cylinder pressed into contact with a flat surface which is considered to be rigid. The analysis can subsequently be extended to those situations in which both surfaces are curved and deformable but are still long in the third dimension. Finally, we consider the three-dimensional case of two spheres in contact.

When unloaded, Fig. 3.1(a), the gap z between the flat rigid surface and the profile of the cylinder is obtained by applying Pythagoras theorem, so that

$$z = R - \{R^2 - x^2\}^{1/2} = R - R \left\{ 1 - \frac{x^2}{R^2} \right\}^{1/2}. \tag{3.1}$$

If $x \ll R$, which is generally the case, then we can expand the square root by the binomial theorem to give

$$z = R - R \left\{ 1 - \frac{x^2}{2R^2} + \cdots + \text{higher-order terms in } (x/R) \right\},$$

that is,

$$z \approx \frac{x^2}{2R}. \tag{3.2}$$

In other words we are choosing to model the surface profile as parabolic over the small region at and near the contact.

Now suppose that when the normal load per unit length W/L is applied, the cylinder deforms such that its centre moves a vertical distance δ towards the plane. The true contact area will thus be a long thin rectangular patch, symmetrically disposed about the z-axis. Suppose that this patch is of width $2a$ extending from $x = -a$ to $x = +a$. This is shown in Fig. 3.1(b). Our aim is to relate the size of the patch, described by the magnitude of a, to the normal load intensity W/L and the geometry and material properties of the

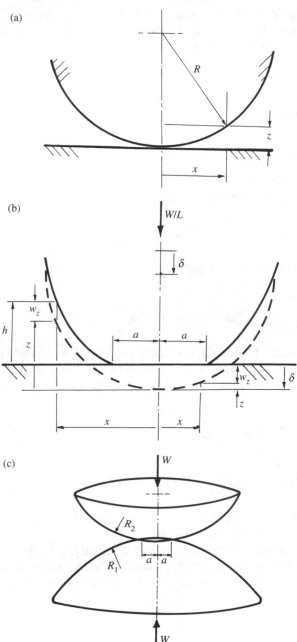

Fig. 3.1 Contact geometry between a long deformable cylinder and a rigid plane surface: (a) unloaded; (b) carrying a load per unit length of *W/L*. (c) Contact between two deformable spheres.

cylinder. In Fig. 3.1(b) the original profile of the cylinder is shown dotted and it is clear that if we consider conditions *within* the contact zone, that is, where $-a < x < +a$, then

$$z + w_z = \delta,$$

where w_z represents the vertical displacement of the *surface* of the cylinder. In this region, where $|x| < a$, $w_z = \delta - z$, that is,

$$w_z = \delta - \frac{x^2}{2R}. \tag{3.3a}$$

Outside the contact zone, where $|x| > a$, there is still free space between the cylinder and the plane, so that

$$w_z > \delta - \frac{x^2}{2R}. \tag{3.3b}$$

and the magnitude of the gap h between the two surfaces is given by the expression

$$h = z + w_z - \delta = \frac{x^2}{2R} + w_z - \delta. \tag{3.4}$$

If now the second surface, previously considered to be rigid and plane, is also capable of deformation, and also has a cylindrical profile (with its axis parallel to the first), then we can write equations corresponding to (3.3a) and (3.3b) as:

within the contact zone
$$w_{z1} + w_{z2} = \delta_1 + \delta_2 - \frac{x^2}{2R_1} - \frac{x^2}{2R_2}, \tag{3.5a}$$

while outside it
$$w_{z1} + w_{z2} > \delta_1 + \delta_2 - \frac{x^2}{2R_1} - \frac{x^2}{2R_2}, \tag{3.6}$$

where suffices 1 and 2 refer to the two surfaces, and δ_2 is the movement of the second cylinder towards the first.

These equations can be written more succinctly by defining $\Delta = \delta_1 + \delta_2$, that is total relative distance of approach of the centres of the two cylinders, and R as the *relative* or *reduced*, radius of curvature where

$$\frac{1}{R} = \frac{1}{R_1} + \frac{1}{R_2}. \tag{3.7}$$

Note that R_1 and R_2 are each *positive* for convex surfaces but would be *negative* for those which are concave. Then, within the contact zone, that is, within the region $|x| \leqslant a$,

$$w_{z1} + w_{z2} = \Delta - \frac{x^2}{2R},$$ (3.8)

while outside it $|x| > a$ and

$$w_{z1} + w_{z2} > \Delta - \frac{x^2}{2R}.$$ (3.9)

The rather more complex case of the contact geometry between two cylinders with *non-parallel* axes is outlined in Appendix 1.

The analogous contact situation in three dimensions involves the contact of two spheres. If we now allow the surfaces to be *spherical*, rather than cylindrical, then the contact patch will be circular in shape, see Fig. 3.1(c), in which it is of radius a. Equations corresponding to (3.8) and (3.9) will now be

$$w_{z1} + w_{z2} = \Delta - \frac{x^2}{2R} - \frac{y^2}{2R} \qquad \text{for} \quad x^2 + y^2 \leqslant a^2$$

or replacing $x^2 + y^2$ by r^2 we can write that within the contact patch, that is, $|r| < a$

$$w_{z1} + w_{z2} = \Delta - \frac{r^2}{2R}$$ (3.10)

while for $|r| > a$,

$$w_{z1} + w_{z2} > \Delta - \frac{r^2}{2R}.$$ (3.11)

The problem we originally posed therefore resolves itself into finding a distribution of pressure across the contact area which will produce a set of surface displacements compatible with the conditions specified above. In the case of two cylinders in contact along a common generator the displacements must satisfy (3.8) and (3.9), while in the case of spheres the relevant conditions are (3.10) and (3.11).

3.3 Surface and subsurface stresses

3.3.1 *Concentrated line loading*

When two solid bodies are brought into loaded contact the stresses developed within each of them may be either entirely elastic or may be sufficiently large for the yield criterion for plastic deformation to be exceeded within one, or possibly both, of them. If the deformation is such that elastic stress fields are predominant then in the case of metals (for which the elastic moduli are high) the strains must be correspondingly small and so the theory of linear elasticity can be applied. The true areas of contact will have dimensions which are small compared with the typical dimensions that characterize the macroscopic shape of the surfaces. In our example this means that the width (or radius) of the contact zone a is small compared with the radii of the curved surfaces so that $a/R \ll 1$. Under these circumstances the distribution of stresses within the contact areas can not be strongly influenced by the conditions distant from them. For the purposes of analysis it is perfectly reasonable to consider that both the solid bodies extend an infinite distance away from the point of contact; in the terminology of continuum mechanics, we can consider each of them to be a *semi-infinite* half-space.

We start with the two-dimensional case by examining the *elastic* stresses and deformations in such a semi-infinite solid loaded over a narrow strip. Our aim is to establish the form of the pressure distribution, and the associated near-surface stresses, which will give rise to the changes in surface shape described by eqns (3.8) and (3.9). A full elastic solution would provide a knowledge of the components of stress and strain at every point within the solid: the stresses must be in equilibrium everywhere within the bulk of the material as well as balancing the applied load on its boundary. Using linear elasticity we can then derive the components of strain from the expressions for stress and to make the overall system acceptable these strains must satisfy the conditions of geometric compatibility. We shall assume, in the two-dimensional picture, that the strip lies parallel to the y-axis and is symmetrical about it, and that the material is in a state of *plane strain*, that is, that $\varepsilon_y = 0$. For this assumption to be justified the thickness of the solid needs to be large compared with the width of the contact zone and this is very often the case in practical contact problems.

In conditions of plane strain, the equations of equilibrium and compatibility can be summarized in Cartesian coordinates as follows (for the derivation of eqns (3.12)–(3.15) see one of the standard texts on strength of materials, for example, Timoshenko and Goodier (1951)). To satisfy equilibrium:

$$\frac{\partial \sigma_x}{\partial x} + \frac{\partial \tau_{xz}}{\partial z} = 0 \quad \text{and} \quad \frac{\partial \sigma_z}{\partial z} + \frac{\partial \tau_{xz}}{\partial x} = 0, \tag{3.12}$$

where σ_x, σ_z are direct stresses and τ_{xz} is the shear stress acting on an element of material at the point (x, z).

For compatibility, the corresponding strains ε_x, ε_z, and γ_{xz} must satisfy the equation

$$\frac{\partial^2 \varepsilon_x}{\partial z^2} + \frac{\partial^2 \varepsilon_z}{\partial x^2} = \frac{\partial^2 \gamma_{xz}}{\partial x \partial z}, \tag{3.13}$$

where the strains are related to the displacements w_x and w_z of a typical particle at coordinates (x, z) by the relations

$$\varepsilon_x = \frac{\partial w_x}{\partial x}; \qquad \varepsilon_z = \frac{\partial w_z}{\partial z} \qquad \text{and} \qquad \gamma_{xz} = \frac{\partial w_x}{\partial z} + \frac{\partial w_z}{\partial x}. \tag{3.14}$$

Linear elasticity links stress and strain via Hooke's law, which for plane strain can be written as

$$\varepsilon_x = \frac{1}{E} \{(1 - v^2)\sigma_x - v(1 + v)\sigma_z\}; \qquad \varepsilon_z = \frac{1}{E} \{(1 - v^2)\sigma_z - v(1 + v)\sigma_x\};$$

$$\gamma_{xz} = \frac{\tau_{xz}}{G} = \frac{2(1 + v)}{E} \tau_{xz}, \tag{3.15}$$

in which E is Young's modulus, G is the elastic shear modulus, and v is Poisson's ratio. Equations (3.13), (3.14), and (3.15) are automatically satisfied if the stresses are derived from an *Airy stress function* $\phi(x, z)$ according to the relations

$$\sigma_x = \frac{\partial^2 \phi}{\partial z^2}; \qquad \sigma_z = \frac{\partial^2 \phi}{\partial x^2} \qquad \tau_{xz} = -\frac{\partial^2 \phi}{\partial x \partial z}, \tag{3.16}$$

provided that the function $\phi(x, z)$ satisfies the biharmonic equation:

$$\left\{ \frac{\partial^2}{\partial x^2} + \frac{\partial^2}{\partial z^2} \right\} \left\{ \frac{\partial^2 \phi}{\partial x^2} + \frac{\partial^2 \phi}{\partial z^2} \right\} = 0. \tag{3.17}$$

If the problem were to be set up in cylindrical polar (r, θ) coordinates the stress function $\phi(r, \theta)$ must satisfy the polar form of the biharmonic equation:

$$\left\{ \frac{\partial^2}{\partial r^2} + \frac{1\partial}{r\partial r} + \frac{1}{r^2} \frac{\partial^2}{\partial \theta^2} \right\} \left\{ \frac{\partial^2 \phi}{\partial r^2} + \frac{1}{r} \frac{\partial \phi}{\partial r} + \frac{1}{r^2} \frac{\partial^2 \phi}{\partial \theta^2} \right\} = 0 \tag{3.18}$$

and the corresponding stresses are derived from the equations

$$\sigma_r = \frac{1}{r} \frac{\partial \phi}{\partial r} + \frac{1}{r^2} \frac{\partial^2 \phi}{\partial \theta^2}; \qquad \sigma_\theta = \frac{\partial^2 \phi}{\partial r^2}; \qquad \tau_{r\theta} = -\frac{\partial}{\partial r} \left\{ \frac{1}{r} \frac{\partial \phi}{\partial r} \right\}. \tag{3.19}$$

The strains ε_r, ε_θ, and $\gamma_{r\theta}$ are in this case related to the displacements w_r and w_θ of a typical particle at (r, θ) by

$$\varepsilon_r = \frac{\partial w_r}{\partial r}; \qquad \varepsilon_\theta = \frac{w_r}{r} + \frac{1}{r} \frac{\partial w_\theta}{\partial z}; \qquad \gamma_{r\theta} = \frac{1}{r} \frac{\partial w_r}{\partial \theta} + \frac{\partial w_\theta}{\partial r} - \frac{w_\theta}{r} \tag{3.20}$$

and Hooke's law becomes

$$\varepsilon_r = \frac{1}{E}\{(1-v^2)\sigma_r - v(1+v)\sigma_\theta\} \; ; \qquad \varepsilon_\theta = \frac{1}{E}\{(1-v^2)\sigma_\theta - v(1+v)\sigma_r\} \; ;$$

$$\gamma_{r\theta} = \frac{\tau_{r\theta}}{G} = \frac{2(1+v)}{E}\tau_{r\theta}. \tag{3.21}$$

The analysis can be started by investigating the stresses and deformations caused by a line load of intensity W/L per unit length applied along the y-axis on the surface of our semi-infinite solid, as illustrated in Fig. 3.2.

The elastic stress field in the (r, θ) plane for this loading can be readily obtained from the Airy stress function (for example, see again Timoshenko and Goodier (1951))

$$\phi(r, \theta) = \frac{-W}{L\pi}r\theta \sin\theta. \tag{3.22}$$

This loading gives rise to a simple radial compressive distribution of stress directed towards O, the point of application of the load; the stresses within the half-space are described by the equations

$$\sigma_r = -\frac{2W}{L\pi r}\cos\theta \qquad \text{and} \qquad \sigma_\theta = \tau_{r\theta} = 0. \tag{3.23}$$

The surface of the half-space is free of stress ($\sigma_r = \tau_{r\theta} = 0$) except at the point of application of the load, where $r = 0$. The theoretically infinite stress

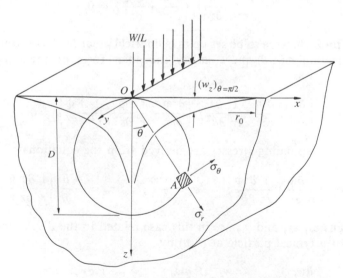

Fig. 3.2 Normal uniform line loading of intensity W/L per unit length on the surface of a semi-infinite solid.

here, suggested by eqn (3.23) which has r in the denominator, is a consequence of assuming that the load is concentrated along a line; in practice the contact area must always have a finite width even if this is exceedingly small. We can also note two other features of eqn (3.23); firstly, that when r is very large the stresses tend to zero and, secondly, that σ_r has a constant magnitude of $-2W/\pi D$ on the circle of diameter D which passes through O (this follows immediately from the geometric relationship that on such a circle $D\cos\theta = r$). As $\tau_{r\theta} = 0$ it follows that σ_r and σ_θ must be *principal stresses*, so that the principal shear stress τ_1 at the point (r, θ) has the value $\sigma_r/2$. Thus, contours of constant maximum shear stress are also circles passing through O. Such trajectories of constant shear stress can be detected experimentally using techniques of photoelasticity, and in such a test the predicted pattern of circles is clearly seen; see Fig. 3.3(a).

The radial compressive stress distribution of eqn (3.23) can, of course, be

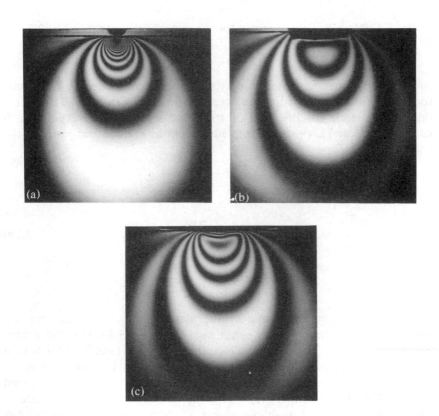

Fig. 3.3 Two-dimensional photoelastic fringe patterns (contours of principal shear stress): (a) concentrated line load; (b) uniform pressure over a finite area; and (c) nominal line contact between cylinders (courtesy of K. L. Johnson).

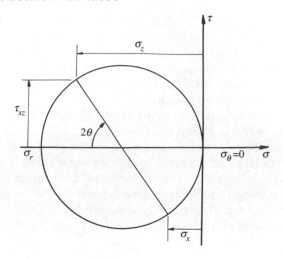

Fig. 3.4 Mohr's circle of stress for the material at point A in Fig. 3.2.

equally well expressed in terms of its Cartesian components. The transformation from one coordinate system to the other can be made by plotting Mohr's circle for the state of stress for a representative point such as A in Fig. 3.2. This procedure is illustrated in Fig. 3.4 and leads to the equations

$$\sigma_x = \sigma_r \sin^2\theta = -\frac{2W}{L\pi}\frac{x^2 z}{(x^2 + z^2)^2}$$

$$\sigma_z = \sigma_r \cos^2\theta = -\frac{2W}{L\pi}\frac{z^3}{(x^2 + z^2)^2} \tag{3.24}$$

$$\tau_{xz} = \sigma_r \sin\theta \cos\theta = -\frac{2W}{L\pi}\frac{xz^2}{(x^2 + z^2)^2}.$$

To find the change in shape of the material under this load we can substitute the stresses given by either eqns (3.23) or (3.24) into the appropriate statement of Hooke's law. Once the strains are established it is then possible to work back to the associated displacements again using either polar or Cartesian components (for example, see again Timoshenko and Goodier (1951)). Of primary interest is the shape of the deformed surface, that is, in polar coordinates the values of the radial and tangential displacements, w_r and w_θ respectively, when $\theta = \pm \pi/2$. Expressions for these quantities can be shown to be

$$w_r(\text{at } \theta = \pm \pi/2) = -\frac{(1-2v)(1+v)W}{2LE} \tag{3.25}$$

and

$$\omega_\theta(\text{at } \theta = +\pi/2) = -\omega_\theta(\text{at } \theta = -\pi/2) = -\frac{(1-v^2)2W}{L\pi E}\ln(r_0/r) \tag{3.26}$$

Equation (3.25) indicates that at all points on the boundary of the solid there is a constant displacement directed towards the origin O. In eqn (3.26) the constant r_0 reflects the choice of the datum from which vertical displacements are measured (since when r is equal to r_0 the vertical surface displacement ω_θ is zero). The essential arbitrariness associated with this choice is an unavoidable feature of problems involving the two-dimensional deformation of an elastic half-space. The shape of the deformed surface is shown in Fig. 3.2 on which the chosen value of r_0 is indicated.

3.3.2   Distributed normal loads

We know that in reality the load W must be distributed over a finite contact area as a pressure p. The local value of this interfacial pressure may vary with position along the x-axis so that p is a function of x, although we shall assume that its distribution is symmetrical about the z-axis; such an arbitrary distribution is shown in Fig. 3.5. We now wish to establish the resultant stress components at any point such as $A(x, z)$ within the bulk of the solid and the vertical displacement of some typical point $C(x, 0)$ on its surface. This can be achieved by considering the total load, which is represented by the area under the distributed pressure, to be made up of an infinite number of elemental line loadings each of which can be analysed by the methods of the preceding section. Here we follow the method of superposing simpler solutions, the 'Green's function method', described more fully in Johnson (1985); an alternative treatment, based on complex potentials and integral transforms, can be found in Gladwell (1980). A typical elemental load of this sort of intensity $p\,ds$ is shown acting at the point $B(s, 0)$ in Fig. 3.5.

The stresses at A due to this effective line loading can thus be written down immediately from eqn (3.24) by replacing x by $x - s$ and W/L by $p\,ds$. Thus, integrating up the effects of all such elemental loads we can write

$$\sigma_x = -\frac{2z}{\pi}\int_{-\infty}^{+\infty} \frac{p(s)(x-s)^2\,ds}{\{(x-s)^2 + z^2\}^2}$$

$$\sigma_z = -\frac{2z^3}{\pi}\int_{-\infty}^{+\infty} \frac{p(s)\,ds}{\{(x-s)^2 + z^2\}^2} \tag{3.27}$$

$$\tau_{xz} = -\frac{2z^2}{\pi} \int_{-\infty}^{+\infty} \frac{p(s)(x-s)\,ds}{\{(x-s)^2+z^2\}^2}.$$

This means that if the shape of the distribution of pressure $p(s)$ is known then, in principle at least, the state of stress at any point in the solid can be evaluated from these equations. In practice, the evaluation of the integrals is straightforward in only a few rather special cases.

The elastic displacements of the surface of the solid can be established in much the same way, that is, by summing the displacements due to all the incremental loads each of magnitude $p\,ds$. Denoting the tangential and vertical displacements of the point C by w_x and w_z respectively, since $w_x = = w_r$ at $\theta = \pm \pi/2$ and $w_z = w_\theta$, we can write using, eqns (3.25) and (3.26), that

$$w_x = -\frac{(1-2v)(1+v)}{2E} \left\{ \int_{-\infty}^{x} p(s)\,ds - \int_{x}^{\infty} p(s)\,ds \right\} \qquad (3.28)$$

$$w_z = \frac{2(1-v^2)}{\pi E} \int_{-\infty}^{+\infty} p(s)\ln\frac{r_0}{|x-s|}\,ds. \qquad (3.29)$$

The step change in displacement at the origin which is implicit in eqn (3.25) necessitates the split in the range of integration of eqn (3.28). Note that once again the equation for the vertical surface displacements contains a constant r_0 representing the datum surface level from which these changes are measured. These equations take on a form which is in some ways more

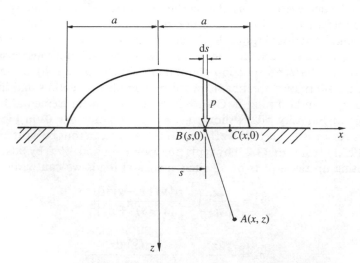

Fig. 3.5 Two-dimensional symmetrically distributed load extending across a strip of width $2a$.

useful for calculation if expressed in terms of displacement gradients, $\partial w_x/\partial x$ and $\partial w_z/\partial x$, that is,

$$\frac{\partial w_x}{\partial x} = -\frac{(1-2v)(1+v)}{E}p(x) \tag{3.30}$$

and

$$\frac{\partial w_z}{\partial x} = -\frac{2(1-v^2)}{\pi E} \int\limits_{-\infty}^{+\infty} \frac{p(s)}{|x-s|}\, ds. \tag{3.31}$$

An important observation can be made from an examination of these equations. Since by definition, on the surface, the strain ε_x is equal to $\partial w_x/\partial x$ then we can equate the two expressions for ε_x from eqns (3.30) and (3.15). On the surface, σ_z, the direct normal stress is equal in magnitude (though opposite in sign) to $p(x)$, the interfacial pressure, so that it follows that

$$(1-v^2)\sigma(x) + v(1+v)p(x) = -(1-2v)(1+v)p(x),$$

that is,

$$\sigma_x = -p(x), \quad \text{which is numerically equal to } \sigma_z. \tag{3.32}$$

Thus we can say that, under *any* system of normal surface pressure, the local in-plane stress, σ_x, at any point under the load is compressive and equal in magnitude to the normal pressure acting at that same point. These locally generated compressive surface stresses are particularly important because they effectively delay the onset of plastic yielding in the uppermost layer of the surface, so adding to the resistance of the surface to plastic deformation.

3.3.3 *Uniform normal pressure*

The simplest example of a distributed normal load is a uniform pressure extending over the range $-a < x < +a$. Suppose this pressure is of magnitude p, as indicated in Fig. 3.6. When eqns (3.27) are applied to this situation the factor p can be taken outside the integral sign since it is a constant. Carrying out the integrations and using the notation of Fig. 3.6, the resultant stresses are

$$\sigma_x = -\frac{p}{2\pi}\{2(\theta_1 - \theta_2) + (\sin 2\theta_1 - \sin 2\theta_2)\},$$

$$\sigma_z = -\frac{p}{2\pi}\{2(\theta_1 - \theta_2) - (\sin 2\theta_1 - \sin 2\theta_2)\} \tag{3.33}$$

$$\tau_{xz} = \frac{p}{2\pi}(\cos 2\theta_1 - \cos 2\theta_2),$$

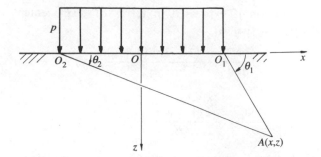

Fig. 3.6   Geometry of a uniform pressure loading of magnitude p.

where θ_1 and θ_2 are the two angles in the figure defined by $\tan\theta_{1,2} = z/(x \mp a)$.

The maximum and minimum direct or principal stresses $\sigma_{1,2}$ in the material can be related to σ_x and σ_z that

$$\sigma_{1,2} = \frac{\sigma_x + \sigma_z}{2} \pm \tfrac{1}{2}\{(\sigma_x - \sigma_z)^2 + \tau_{xz}^2\}^{1/2}.$$

Denoting the angle $\theta_1 - \theta_2$ by α, we can write these conditions as

$$\sigma_{1,2} = -\frac{p}{\pi}(\alpha \mp \sin\alpha). \tag{3.34}$$

The maximum, or principal, shear stress τ_1 has the magnitude $|\sigma_1 - \sigma_2|/2$, that is,

$$\tau_1 = \frac{p}{\pi}\sin\alpha. \tag{3.35}$$

In this case, therefore, contours of constant shear stress will follow the polar curve $\sin\alpha = $ constant, and so should be circles passing through the two extreme positions of the load, O_1 and O_2 in Figs 3.6 and 3.7. Once again photoelastic experiments bear out this pattern of stresses as can be seen from Fig. 3.3(b).

Equation (3.35) indicates that the principal shear stress reaches a *maximum* value along the semicircle on which $\alpha = \pi/2$. We can also note that as a, the width of the loaded area, decreases so the system of subsurface stresses reverts to that under a concentrated load described in Section 3.3.1.

To obtain the displacements of the surface we can use eqns (3.28) and (3.29). Within the contact area, that is, in the region for which $-a < x < +a$,

$$w_x = -\frac{(1 - 2v)(1 + v)}{E}px, \tag{3.36}$$

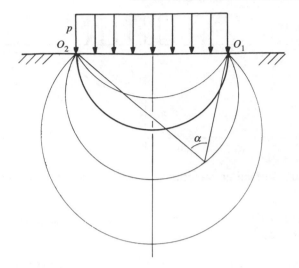

Fig. 3.7 Trajectories of constant shear stress below a uniformly distributed normal pressure. These can be compared to the experimental fringes shown in Fig. 3.3(b).

while outside the contact region, $|x| > a$,

$$w_x = \pm \frac{(1 - 2v)(1 + v)}{E} pa \qquad \text{according as } x < -a \text{ or } x > a. \quad (3.37)$$

The shape of the surface profile itself, that is, the expression for w_z, can be most easily established from the expression developed for the surface gradient, eqn (3.31); thus

$$\frac{\partial w_z}{\partial x} = -\frac{2(1 - v^2)}{\pi E} p \int_{-a}^{+a} \frac{ds}{x - s} \qquad \text{since } p = 0 \text{ for } |x| > a. \quad (3.38)$$

Within the contact zone the evaluation of this integral is not straightforward because of the singularity at $s = x$. This difficulty can be overcome by considering two regions, one extending from $s = -a$ to $s = x - \varepsilon$, and the other from $s = x + \varepsilon$ to $s = +a$, and then allowing the quantity ε to become vanishingly small. This procedure yields the result

$$w_z = -\frac{(1 - v^2)}{\pi E} p \left\{ (a + x)\ln\left(\frac{a + x}{a}\right)^2 + (a - x)\ln\left(\frac{a - x}{a}\right)^2 \right\} + \mathcal{C}$$

$$(3.39)$$

where $\mathcal{C}$ is a constant of integration. Outside the contact zone, that is, at

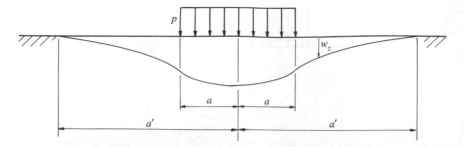

Fig. 3.8 Surface deformations under a uniform normal pressure.

$|x| > a$, there is no difficulty about carrying out the integration leading directly to the result

$$w_z = -\frac{(1-v^2)}{\pi E} p \left\{ (x+a)\ln \left(\frac{a+x}{a}\right)^2 - (x-a)\ln \left(\frac{x-a}{a}\right)^2 \right\} + \mathcal{C}.$$

(3.40)

The constant $\mathcal{C}$ is fixed, once again, by the choice of the datum from which vertical displacements are measured. Figure 3.8 shows the deformed surface profile which has been drawn such that $w_z = 0$ when $x = \pm a'$.

Note that the form of the changes to the surface profile described by eqns (3.39) and (3.40), and illustrated in Fig. 3.8, are *not* of the same form as those described by eqns (3.8) and (3.9) which arose from the contact of two cylinders. The implication of this observation is that the pressure across the contact zone between two initially cylindrical surfaces *cannot* be one of a simple uniform distribution.

3.3.4 *Semi-elliptical pressure distribution: Hertzian contacts*

Line contact

Using the techniques outlined in the previous sections it is possible to evaluate the internal stresses and surface displacements for a variety of pressure distributions in both two- and three-dimensional geometries. One form of distribution of particular and practical importance is that described in two dimensions by the equation

$$p(x) = p_0 \sqrt{1 - x^2/a^2} .$$

(3.41)

The contact pressure rises from zero at the edges of the contact (where $x = \pm a$) to a maximum of p_0 at the centre; the pressure profile is half an ellipse. The total load per unit length of the contact W/L is given by

$$\frac{W}{L} = \int_{-a}^{+a} p(x)\, dx, \qquad (3.42)$$

hence, since $\int_{-a}^{+a} (1 - x^2/a^2)\, dx = \pi/2$, the peak pressure is:

$$p_0 = \frac{2W}{L\pi a}. \qquad (3.43)$$

The *mean* pressure p_m over the contact strip is equal to $W/2aL$ and thus

$$p_m = \frac{\pi}{4} p_0. \qquad (3.44)$$

The shape of the elastically deformed surface can be found by substituting this semi-elliptical pressure distribution into eqns (3.28) and (3.29). Writing $\xi = |x/a|$, so that within the contact zone $\xi < 1$, the vertical displacement w_z of the surface is given by the expression

$$w_z = \frac{2W(1 - \nu^2)}{L\pi E} (\xi^2 + C), \qquad (3.45)$$

while outside the contact area, that is, for $|\xi| > 1$

$$w_z = \frac{2W(1 - \nu^2)}{L\pi E} \left\{ \ln(\xi + \sqrt{\xi^2 - 1}) + \frac{1}{2(\xi + \sqrt{\xi^2 - 1})^2} + C + \frac{1}{2} \right\}$$
$$(3.46)$$

where C is a constant. The shape of the profile of a plane half-space subject to a load of this form is sketched in Fig. 3.9(c). The constant C has been chosen so that $w_z = 0$ at $\xi = 3$.

This distribution of normal pressure is of special interest because the surface deflections of the half-spaces arising from it can be made consistent with the conditions relating to the contact of two cylinders, that is, to eqns (3.8) and (3.9), provided that a and W are related by the equation

$$a^2 = \frac{4WR}{L\pi} \left\{ \frac{1 - \nu_1^2}{E_1} + \frac{1 - \nu_2^2}{E_2} \right\}.$$

This condition may be written more concisely by defining E^*, the *contact modulus*, by the relation

$$\frac{1}{E^*} = \frac{1 - \nu_1^2}{E_1} + \frac{1 - \nu_2^2}{E_2} \qquad (3.47)$$

so that

$$a^2 = \frac{4WR}{L\pi E^*} \quad \text{and thus} \quad p_0 = \sqrt{\frac{E^* W}{RL\pi}}. \qquad (3.48)$$

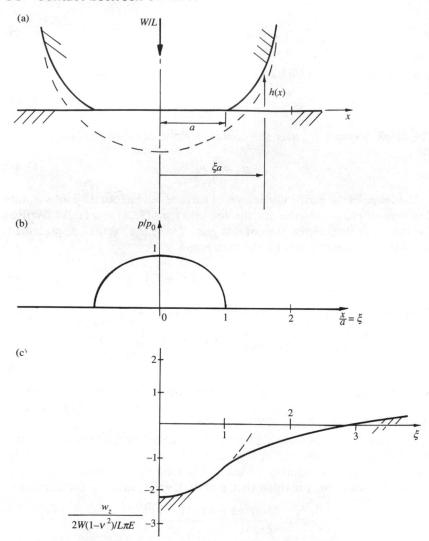

Fig. 3.9 Surface deformation arising from semi-elliptical contact stresses: (a) contact between an equivalent cylinder and a plane extends over the region $-a < x < +a$; (b) semi-elliptical pressure distribution with peak value p_0 on the centre line; and (c) the resulting surface deformation of the plane half-space.

Some texts work in terms of the *reduced* elastic modulus E' defined as being equal to $2E^*$.

The stresses within the material under this surface loading can be found by substituting this semi-elliptical pressure distribution into eqn (3.27). Integration along the z-axis leads to expressions for the principal stresses σ_x and σ_z, namely

$$\sigma_x = -p_0 \left\{ \left(1 + \frac{2z^2}{a^2}\right) \left(1 + \frac{z^2}{a^2}\right)^{-1/2} - 2\frac{z}{a} \right\}$$

$$\sigma_z = -p_0 \left\{ 1 + \frac{z^2}{a^2} \right\}^{-1/2}. \tag{3.49}$$

The magnitude of the principal shear stress τ_1 is then given by

$$\tau_1 = p_0 \left\{ \frac{z}{a} - \frac{z^2}{a^2} \left(1 + \frac{z^2}{a^2}\right)^{-1/2} \right\}.$$

This expression has a maximum value of $0.30\,p_0$ below the surface at the point where $z = 0.78a$. The variations in σ_x, σ_z, and τ_1 (as fractions of p_0) with depth z are plotted in Fig. 3.10(a), and contours of constant values of τ_1 are shown in Fig. 3.10(b). These may be compared with the experimental photoelastic fringe pattern for contact of this nature shown in Fig. 3.3(c) when once again there is good general agreement.

When considering two initially cylindrical components pressed together we may well be interested in the *magnitude* of the gap between them, that is, dimension h in Fig. 3.9(a). Obviously $h = 0$ over the region $-a < x < + a$, that is, $-1 < \xi < + 1$; outside the contact strip, when $|\xi| \geqslant 1$, the gap h is given by the equation

$$\frac{h}{a^2/2R} = \xi(\xi^2 - 1)^{1/2} - \ln\{v + (\xi^2 - 1)^{1/2}\}. \tag{3.50}$$

Often we are principally concerned with the shape of the profile relatively close to the contact zone, that is, at values of ξ only just a little greater than unity. Under these conditions this expression can be satisfactorily approximated by the relation

$$\frac{h}{a^2/2R} = \frac{4\sqrt{2}}{3} (\xi - 1)^{3/2}, \tag{3.51}$$

where $\xi - 1$ is a measure of the distance of the point of interest from the edge of the contact. The correspondence between these two formulae can be judged from Fig. 3.11.

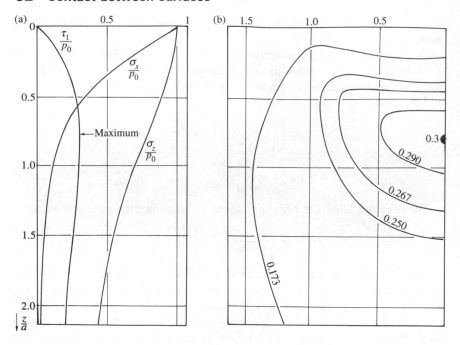

Fig. 3.10 Elastic contact of cylinders: (a) subsurface stresses along the axis of symmetry Oz; and (b) contours of constant principal shear stress. These can be compared with those of Fig. 3.3(c).

Point contact

The semi-elliptical form of pressure distribution can be extended into the third dimension, and so the analysis made applicable to the contact of two spheres, by writing eqn (3.41) as

$$p(r) = p_0\sqrt{1 - r^2/a^2}, \qquad \text{where} \qquad r^2 = x^2 + y^2 \qquad (3.52)$$

and where a now becomes the radius of the contact circle. If W is the total load on the contact spot, then

$$W = \int_0^a 2\pi r p(r)\, \mathrm{d}r = \frac{2}{3} p_0 \pi a^2. \qquad (3.53)$$

The mean pressure is $p_\mathrm{m} = W/\pi a^2$, and hence in this case

$$p_\mathrm{m} = \frac{2}{3} p_0. \qquad (3.54)$$

The deflections of the surfaces, both inside and outside the contact, due

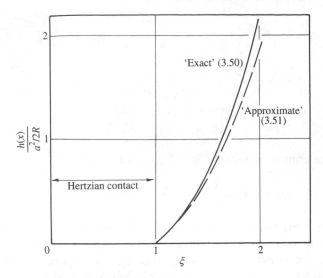

Fig. 3.11 Comparison of eqns (3.50) and (3.51) close to the edge of the contact region. The discrepancy is comparatively small in the region $1 < \xi < 2$.

to this pressure loading are consistent with eqns (3.10) and (3.11) provided that

$$a^3 = \frac{3WR}{4E^*},\tag{3.55}$$

where R is the reduced radius of contact as defined in (3.7) and E^* is the contact modulus as defined by eqn (3.47). The normal surface displacements within the contact patch, that is, for the region $r < a$, are then given for each body by

$$w_z = \frac{1 - v^2}{E}\frac{\pi p_0}{4a}\left(2a^2 - r^2\right)$$

and from this it follows that Δ, the mutual approach of two relatively distant points (for example, the centres of the two spheres) is given by

$$\Delta = \left\{\frac{9W^2}{16RE^{*2}}\right\}^{1/3} = \frac{a^2}{R} = \frac{a\pi p_0}{2E^*}.\tag{3.56}$$

The surface and subsurface stresses below each surface can be evaluated in much the same way as for cylinders in contact. On the surfaces, within the contact circle

$$\sigma_r = \frac{1-2v}{3}\frac{a^2 p_0}{r^2}\left\{1-\left(1-\frac{r^2}{a^2}\right)^{3/2}\right\} - p_0\left\{1-\frac{r^2}{a^2}\right\}^{1/2}$$

$$\sigma_\theta = -\frac{1-2v}{3}\frac{a^2 p_0}{r^2}\left\{1-\left(1-\frac{r^2}{a^2}\right)^{3/2}\right\} - 2vp_0\left\{1-\frac{r^2}{a^2}\right\}^{1/2} \qquad (3.57)$$

$$\sigma_z = -p_0\left\{1-\frac{r^2}{a^2}\right\}^{1/2}.$$

Outside the contact area, that is, when $r > a$,

$$\sigma_r = -\sigma_\theta = \frac{(1-2v)a^2}{3r^2} \qquad \text{and} \qquad \sigma_z = 0. \qquad (3.58)$$

These distributions are plotted in Fig. 3.12(a) for a material with Poisson's ratio v equal to 0.3, a value typical of many metals. The circumferential stress σ_θ is everywhere compressive, while the radial stress is compressive over very nearly the whole of the contact patch, and tensile everywhere outside it. The greatest value of this tensile stress is $(1-2v)p_0/3$ and occurs at the very edge of the contact area. This tensile stress component can be practically very important in those circumstances in which one of the surfaces is brittle and thus susceptible to failure by tensile cracking.

Within the bulk of the material the stresses along the z-axis are given by

$$\sigma_r = \sigma_\theta = -(1+v)p_0\{1-\frac{z}{a}\arctan\frac{a}{z}\} + \tfrac{1}{2}p_0\{1+\frac{z^2}{a^2}\}^{-1}$$

$$\sigma_z = -p_0\{1+\frac{z^2}{a^2}\}^{-1}. \qquad (3.59)$$

Along this axis, σ_r, σ_θ, and σ_ζ are the principal stresses and so the maximum shear stress τ_1 is of magnitude $\tfrac{1}{2}|\sigma_z - \sigma_\theta|$. The distribution of τ_1 below the surface is shown in Fig. 3.12(b) again for the case $v = 0.3$: the maximum value is 0.31 p_0 (or 0.47 p_m) and occurs at a depth of $0.48a$.

An analysis of stresses generated around the contact point of two loaded curved surfaces was published by Heinrich Hertz in 1882 and represents one of the most important milestones in the history of tribology. At the time he was studying the optical fringes in the gap between two glass lenses and was concerned about the possible influence of the elastic deformation of the surfaces arising from the contact pressure between them. His analysis covered the general case of contact between doubly curved surfaces when the contact patch will be elliptical in shape. The two examples we have considered, cylinders and spheres, can be thought of as two special cases of this more general treatment. Theory of this sort finds extensive practical application in many branches of tribology: in a wide variety of elastic contact

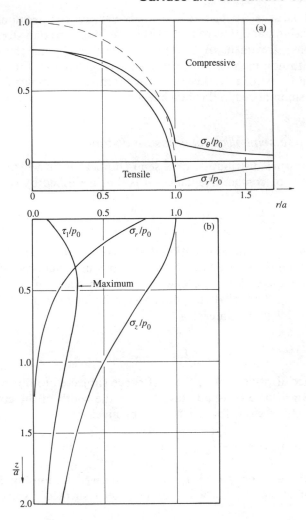

Fig. 3.12 Elastic contact of spheres: (a) surface stresses—Hertz pressure shown dotted; and variation of the (b) subsurface stresses along the axis of symmetry with depth.

problems it is usual to picture the loads varying in a semi-elliptical or 'Hertzian' fashion and thus to talk of the resulting Hertzian stresses. For a more complete treatment of Hertzian analysis, including the contact of cylinders with non-parallel axes, the reader is referred to Johnson (1985), Chapter 4. For ease of reference the most often used formulae for Hertzian contact are summarized in the data sheet for this chapter (see Appendix 5).

3.3.5 *Inverse semi-elliptical pressure distribution*

A second pressure distribution of a semi-elliptical form which is of value in the analysis of contact phenomenon is that described by the equation

$$p(x) = \frac{p_0}{\sqrt{1 - x^2/a^2}}. \tag{3.60}$$

This is the *inverse* of the distribution described by eqn (3.41). This pressure profile has a *minimum* value of p_0 on the axis of symmetry, when $x = 0$, and rises to a theoretically infinite value at the edges of the contact, as illustrated in Fig. 3.13. If W/L is once again the load per unit length of the contact then, by simple integration,

$$p_0 = \frac{W}{L\pi a}. \tag{3.61}$$

The surface displacement of the half-space subjected to this distribution can be shown to be a constant, say Δ, over the width of the contact strip and outside this region, for $|x| > a$, to be given by

$$w_z = \Delta - \frac{2(1 - \nu^2)}{\pi E} \ln \left\{ \frac{|x|}{a} + \left(\frac{x^2}{a^2} - 1 \right)^{1/2} \right\}. \tag{3.62}$$

The shape of the deformed surface has also been plotted in Fig. 3.13 where the value of Δ has been chosen to give zero surface displacements at $x = \pm 3a$.

It is clear from an examination of Fig. 3.13 that there is a practical application for which this distribution can be considered immediately relevant, namely the indentation of an elastic half-space by a perfectly rigid, frictionless punch which is infinitely long in the y-direction and of width $2a$ in the x-direction. Since the punch is rigid, the surface of the elastic solid must remain flat beneath it, recovering to its undeformed height in the surrounding elastic hinterland. Of course, this is a simplification of the real situation — no real punch can be truly rigid (though it might have a modulus very much greater than that of the substrate) and no real interface can be truly frictionless. In addition, even if these conditions are approached, no substrate could sustain the infinitely large stresses at the ends of the contact

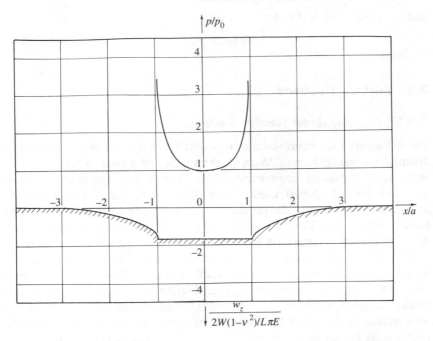

Fig. 3.13   The inverse semi-elliptical pressure distribution and the resultant surface displacements.

which are implied by eqn (3.60): within a small, local region the yield
criterion must be exceeded leading to localized plastic flow. Nevertheless,
this analysis does suggest the likely form of the pressure distribution and
does, as we shall see, turn out to be of considerable importance in the
analysis of contacts carrying both normal and tangential loads.

When extended into the three dimensional case the inverse semi-elliptical
pressure distribution described by the equation

$$p(r) = \frac{p_0}{\sqrt{1 - r^2/a^2}} \tag{3.63}$$

will likewise generate uniform normal displacements across the contact circle
of radius a. The distance of mutual approach $\varDelta$ of two relatively distant
points is then given by

$$\varDelta = \frac{a\pi p_0}{E^*}, \tag{3.64}$$

and the total load W by the equation

$$W = 2p_0 \pi a^2. \tag{3.65}$$

3.4   Surface tractions

3.4.1   *Stresses under tractive loadings*

The situation so far examined has been that set of stresses and deformations arising from purely normal loads on the surface of a semi-infinite body. As soon as a tangential component of loading is applied, usually known as a *traction*, additional stresses and their associated displacements are generated. We now consider the state of stress arising from a purely tangential or tractive load. The combination of normal and tractive forces can then be examined by superposing the solutions of the two simpler problems.

We start by considering a concentrated line load of intensity F/L per unit length acting along the y-axis tangentially to the surface. This is shown in Fig. 3.14. This load will produce a radial stress field similar to that arising from a force normal to the surface though rotated through 90°. As long as we continue to measure θ from the line of action of the force then the polar expressions for stress will be of the same form as eqns (3.23), namely

$$\sigma_r = -(2F/L\pi r)\cos\theta \qquad \sigma_\theta = \tau_{r\theta} = 0. \tag{3.66}$$

Contours of constant shear stress are now semicircles through O as indicated. Ahead of the applied force, as we might expect, σ_r is compressive while behind the line of action of F/L it is tensile.

In Cartesian coordinates these equations become

$$\sigma_x = -\frac{2F}{L\pi}\frac{x^3}{(x^2+z^2)^2}; \qquad \sigma_z = -\frac{2F}{L\pi}\frac{xz^2}{(x^2+z^2)^2};$$

$$\tau_{xz} = -\frac{2F}{L\pi}\frac{zx^2}{(x^2+z^2)^2}. \tag{3.67}$$

Taking account of the changes in the definition of θ, the equations for the surface displacements derived earlier still stand. If there is no rigid body rotation of the half-space, and no vertical displacement of points on the z-axis, then the surface displacements are

$$w_r\,(\text{at }\theta=0) = -w_r\,(\text{at }\theta=\pi) = \frac{-(1-v^2)}{L\pi E}2F\ln\frac{r}{r_0}$$

$$w_\theta\,(\text{at }\theta=0) = w_\theta\,(\text{at }\theta=\pi) = \frac{(1-2v)\,(1+v)}{2LE}F. \tag{3.68}$$

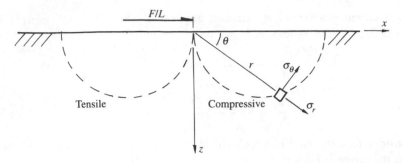

Fig. 3.14   Concentrated tangential line load on a semi-elliptical half-space: contours of constant shear stress are semicircles.

These equations show that the whole surface ahead of the force is *depressed* by an amount which is proportional to F while the surface which is behind *rises* by an equal amount. Once again the tangential surface displacements vary logarithmically, the datum value chosen fixing the value of r_0.

The effects of a distributed traction across the contact area can be established by methods entirely analogous to those used for distributed pressures in Section 3.3.2. Suppose that the traction stress at position $x = s$ is equal to $\tau(s)$. Then

$$w_x = \frac{2(1-\nu^2)}{\pi E} \int_{-a}^{+a} \tau(s) \ln \frac{r_0}{(x-s)}\, ds$$

$$w_z = \frac{(1-2\nu)(1+\nu)}{2E} \left\{ \int_{-a}^{x} \tau(s)\, ds - \int_{x}^{a} \tau(s)\, ds \right\}. \qquad (3.69)$$

If this pair of equations, which relate surface displacements to surface traction distributions, is compared to eqns (3.28) and (3.29) an immediate similarity is apparent: in particular, if the distributions of the normal pressure $p(s)$ and tangential traction $\tau(s)$ are of the same shape then the surface displacements in the x-direction due to $\tau(s)$ will be the same as those in the z-direction due to $p(s)$, that is,

$$(w_x)_\tau = (w_z)_p \qquad \text{and} \qquad (w_z)_\tau = -(w_x)_p \qquad (3.70)$$

where the suffices p and τ refer respectively to the cases of normal pressure and tangential loading. We can now apply these ideas to distributions of traction analogous to the cases of distributed normal loads considered in Sections 3.3.3 and 3.3.4.

Once again it can in some cases be useful to work in terms of the displacement gradients $\partial w_x / \partial x$ and $\partial w_z / \partial x$. By analogy with eqns (3.30) and (3.31) we can write

$$\frac{\partial w_x}{\partial x} = -\frac{2(1-v^2)}{\pi E} \int_{-\infty}^{+\infty} \frac{\tau(s)}{x-s} \, ds$$

$$\frac{\partial w_z}{\partial z} = \frac{(1-2v)(1+v)}{E} \tau(x) \tag{3.71}$$

Consider first the case in which the shear stress has a constant value of τ_0 over the range $-a < x < +a$. The stresses within the half-space will be given by

$$\sigma_x = \frac{-\tau_0}{2\pi} \{ 4 \ln (r_1/r_2) - (\cos 2\theta_1 - \cos 2\theta_2) \}$$

$$\sigma_z = \frac{\tau_0}{2\pi} (\cos 2\theta_1 - \cos 2\theta_2) \tag{3.72}$$

$$\tau_{xz} = -\frac{\tau_0}{2\pi} \{ 2(\theta_1 - \theta_2) + \sin 2\theta_1 - \sin 2\theta_2 \}$$

where r_1 and r_2 are distances from the edges of the loaded area to the point (x, z). Fig. 3.15 shows the distribution of the stress component σ_x as a multiple of the stress τ_0 at the surface. The discontinuity in shear stress at the edge of the loaded area has a striking effect on the logarithmic term in the first of these equations and leads to an infinite elastic stress. In practice local plastic flow will be induced so vitiating this, but in the case of *oscillating* tractive loads these high levels of reversing stress can contribute significantly to the failure of surfaces through the mechanism of fretting (see Section 5.2.6).

3.4.2   *Sliding of elastic contacts*

In tribological problems we are usually concerned with the transmission of tractive as well as normal loads. As an example, consider the case of two *identical* cylinders pressed together along a common generator by a normal force W/L per unit length and then subjected to a tangential force F/L. In the analysis we shall assume that the pressure distribution and contact width at the surface can be derived from the Hertzian analysis, and are thus given by eqns (3.41) and (3.48). Furthermore, we shall suppose that these remain unchanged by the subsequent application of the tractive force. The general situation is illustrated in Fig. 3.16. Eventually F/L will reach such a value that gross sliding occurs and we will have reached the situation of *limiting friction*. The contact can then be characterized by μ, the ratio of the tangential force to the normal force. The numerical value of this ratio, the

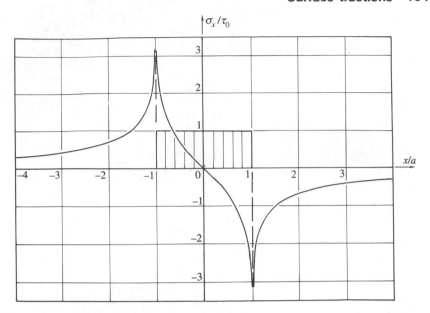

Fig. 3.15 Distribution of the surface stress component σ_x under a uniform tractive load of τ_0 applied over the region $-a > x > a$. There are stress singularities at the ends of the contact.

coefficient of friction, depends on many aspects of the nature and condition of the surfaces and is discussed more fully in Chapter 4. Even if F/L is less than its critical value, so that there is no gross sliding, the application of atraction load will still lead to shear deformations *within* each cylinder as indicated by the distortion of the centre line in Fig. 3.16. We can note that since the cylinders are identical, symmetry of the deformation within each is preserved. Points within the contact zone will undergo tangential displacements w_{x1} and w_{x2} relative to points distant from the surface (such as the centres of the cylinders) which move tangentially through effectively rigid body displacements δ_1 and δ_2.

Let points A_1 and A_2 denote two points at the interface which were in contact prior to the application of F/L. If the absolute displacements, that is, relative to the origin O, are s_1 and s_2 respectively, then

$$s_1 = w_{x1} - \delta_1 \text{ and } s_2 = w_{x2} - \delta_2. \tag{3.73}$$

Thus s, the component of 'slip' between A_1 and A_2, is of magnitude $|s_1 - s_2|$ where

$$s = |s_1 - s_2| = |(w_{x1} - w_{x2}) - (\delta_1 - \delta_2)|. \tag{3.74}$$

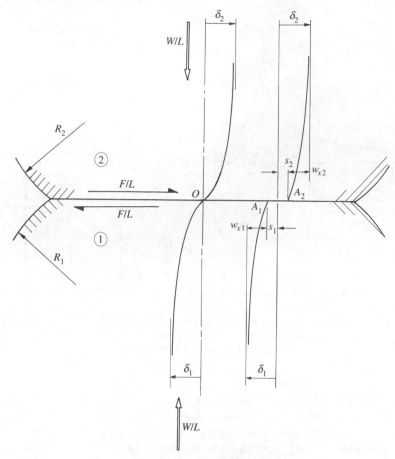

Fig. 3.16 Contact zone between two identical cylinders, surfaces 1 and 2, carrying both a normal load W/L and a traction F/L per unit length of the contact.

In the case of no slip $s = 0$ so that

$$w_{x1} - w_{x2} = \delta_1 - \delta_2 = \delta_x. \qquad (3.75)$$

The right-hand side of eqn (3.75), that is, δ_x, denotes the relative tangential displacement of the two bodies as a whole and is independent of the position of points A_1 and A_2 within the 'stick' region. Thus we can generally say that *within a sticking region* all points undergo the same tangential displacement. In our particular case, since the elastic modulii of bodies 1 and 2 are identical we can in addition say that their relative displacement δ_x is equally divided between them, that is, that

$$w_{x1} = -w_{x2}. \qquad (3.76)$$

Note that this would not be the case if their elastic properties differed.

We thus now are faced with the problem of finding a distribution of tangential traction on each half-space consistent with a constant tangential displacement across the contact zone. To solve this problem we can make use of the analogy between pressures and normal displacements and tractions and tangential displacements which was encapsulated in eqn (3.70). We have shown, in section 3.3.3, that an *inverse semi-elliptical* pressure distribution gives rise to a *constant* normal surface displacement and so we can immediately say that the distribution of surface traction required to give a constant *tangential* displacement is also inverse semi-elliptical, that is, that

$$\tau(x) = \frac{F}{aL\pi(1 - x^2/a^2)^{1/2}}.$$ (3.77)

This traction distribution is shown in Fig. 3.17, as curve A, and rises to theoretically infinite values at $x = \pm a$. Of course, these values of traction cannot be sustained in reality since the maximum value that τ can have at any point is μp where p is the local pressure. Thus towards the edges of the contact we might expect to have some *slip*, whereas towards the centre, where the traction is low but the normal pressure high, we anticipate a region of *no-slip* or *stick*.

The points of transition between the sticking and slipping regions can be

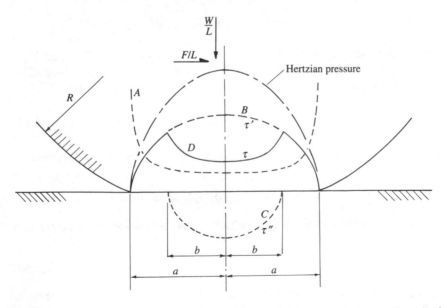

Fig. 3.17 Contact between cylinders with parallel axes: surface tractions and tangential displacements due to the application of the tangential traction F/L.

established for a particular value of F/L in the following manner. Consider first the case in which F/L is increased to reach the value $\mu W/L$, so that the surfaces are just on the point of gross sliding. The traction distribution $\tau'(x)$ will be

$$\tau'(x) = \mu p_0 (1 - x^2/a^2)^{1/2} \tag{3.78}$$

where $p_0 = 2W/L\pi a$. This distribution is shown as curve B in Fig. 3.17. From eqn (3.70) the surface tangential displacements will be distributed in the same way as the normal displacements under a Hertzian pressure, that is, parabolically. If no slip occurs at the mid-point, where $x = 0$, then by combining eqns (3.73) and (3.5) we may write that in surface 1

$$w'_{x1} = \delta_1 - \frac{(1 - v^2)}{Ea} \mu p_0 x^2 \tag{3.79}$$

where the dashed quantities refer to the traction $\tau'(x)$, the value of $v = v_1 = v_2$, and $E = E_1 = E_2$. A similar expression of opposite sign applies to surface 2. These distributions of tangential displacement satisfy eqn (3.76) only at the origin and so we can conclude that elsewhere in the contact the surfaces *must* slip.

Now consider the superposition of a second distribution of traction described by

$$\tau''(x) = -\frac{b}{a} \mu p_0 \left\{ 1 - \frac{x^2}{b^2} \right\}^{1/2} \tag{3.80}$$

acting over a strip of width $2b$ (where $b < a$) shown as curve C in Fig. 3.17. The tangential displacements within the region $-b < x < +b$ produced by this distribution will, by analogy with the previous example, be

$$w''_{x1} = \delta''_1 + \frac{b}{a} \frac{(1 - v^2)}{Eb} \mu p_0 x^2. \tag{3.81}$$

Thus within the central strip of width $2b$ the net surface displacement is given by

$$w_{x1} = w'_{x1} + w''_{x1} = -\delta'_1 + \delta''_1 = \text{constant} \tag{3.82}$$

and for the opposing surface, for which an equal traction of opposite sign is applied,

$$w_{x2} = w'_{x2} + w''_{x2} = -\delta'_2 + \delta''_2 = \text{constant}. \tag{3.83}$$

Thus the conditions for no slip are satisfied across the region $-b < x < +b$. Furthermore within this region the resultant total traction is given by

$$\tau'(x) + \tau''(x) = \mu p_0 \left\{ \left(1 - \frac{x^2}{a^2} \right)^{1/2} - \frac{b}{a} \left(1 - \frac{x^2}{b^2} \right)^{1/2} \right\} \qquad (3.84)$$

which is everywhere less than μp_0. The extent or width of the sticking zone, that is, the ratio b/a, is dependent on the magnitude of the applied tangential load F, since

$$\frac{F}{L} = \int_{-a}^{+a} \tau(x) \, dx = \int_{-a}^{+a} \tau'(x) \, dx + \int_{-b}^{+b} \tau''(x) \, dx \qquad (3.85)$$

that is, $F/L = \mu W/L - \mu(b^2/a^2)W/L$, or

$$\frac{b}{a} = \left\{ 1 - \frac{F}{\mu W} \right\}^{1/2}. \qquad (3.86)$$

The physical behaviour of the contact is summarized by this equation. If the normal load W/L is kept constant while the tangential force on the junction F/L is varied then, as soon as F/L increases above zero, microslip begins at the edges of the contact. As F/L grows so this slipping region spreads inwards according to eqn (3.86). When F/L approaches the limiting value of $\mu W/L$ the sticking region has degenerated to a line at $x = 0$. When F/L reaches magnitude $\mu W/L$ gross sliding at the interface is initiated. The change in the relative size of the sticking region with the increase in the applied tangential load is illustrated in Fig. 3.18.

3.5   Loading beyond the elastic limit

3.5.1   Plastic deformation

Returning to the problem of two cylinders carrying only a normal load, we have shown that the maximum shear stress under a Hertzian contact of width $2a$ occurs at a depth of $0.78a$, and is of magnitude $0.30p_0$, which is approximately equal to $0.20W/aL$ where W/L is the total load per unit length on the contact and p_0 is the peak Hertzian pressure. There must be some value of W for which the material of the weaker cylinder starts to deform *plastically* rather than elastically; to estimate this critical value, W_Y, we need to apply an appropriate yield criterion to the material of the surface. The yield of ductile metals is usually taken to be governed by either the Tresca maximum shear stress criterion or the von Mises strain energy criterion. Suppose a specimen of the metal in question yields in simple tension at a stress Y. The principal stresses are $\sigma_1 = Y$ and $\sigma_2 = \sigma_3 = 0$. The maximum shear stress is equal to one-half of the greatest difference between the principal stresses and so is equal to $Y/2$. The Tresca criterion thus

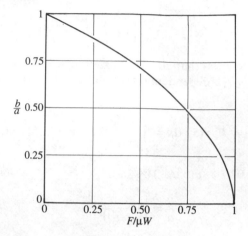

Fig. 3.18 Variation in the relative extent of the sticking region b/a with applied tangential load.

suggests that in pure shear the material will yield at a shear yield stress whose magnitude k is given by

$$k = Y/2. \tag{3.87}$$

On the other hand, the von Mises strain energy criterion depends on the value of the expression

$$(\sigma_1 - \sigma_2)^2 + (\sigma_2 - \sigma_3)^2 + (\sigma_3 - \sigma_1)^2.$$

In pure shear the principal stresses are $(k, -k, 0)$ so that the von Mises condition predicts k and Y are related by $6k^2 = 2Y^2$: it follows that

$$k = Y/\sqrt{3}. \tag{3.88}$$

Although careful experiments on metallic specimens tend to support the von Mises criterion, the difference in the predictions of the two is not large ($<15\%$) and in view of the spread of material properties likely in practice it is often considered quite acceptable to use whichever criterion leads to greater algebraic simplicity.

In the case of the two-dimensional contact of cylinders the condition of plane strain deformation ensures that the stress component σ_y is the intermediate principal stress. Applying the Tresea criterion thus involves equating the maximum principal shear stress (which from Section 3.3.4 has been shown to be equal to $0.30p_0$) to k or $Y/2$. Hence, the critical value of the peak pressure p_0^Y is given by

$$p_0^Y = 3.3k = 1.67\, Y. \tag{3.89}$$

The corresponding value of the mean pressure p_m^Y is thus, from eqn (3.44), given by

$$p_m^Y = \frac{\pi}{4} p_0^Y \approx 2.6k = 1.3Y.$$

The load per unit length of the contact W^Y/L required to bring material to the point of yield can be found by substituting the value of p_0^Y into the expression for W/L in eqn (3.48), hence

$$\frac{W^Y}{L} = \frac{\pi R}{E^*} p_0^{Y2} = 8.76 \frac{RY^2}{E^*}. \qquad (3.90)$$

Even when some yielding has taken place the scale of the changes of shape must be small. This is because initial yield has occurred *beneath* the surface, so that the plastic zone is still totally surrounded by a region in which the stresses and strains are still elastic. This limits the extent of plastic deformation, since the plastic strains must be of the same order as the adjacent elastic strains. The material displaced by the flattening of the contact is accommodated by an *elastic* expansion of the surrounding hinterland. This is an example of *contained* plasticity.

If the normal load on the contact is increased still further, beyond W^Y/L, then the plastic zone grows until eventually it breaks out at the free surface. Once this occurs the constraint on the scale of the plastic deformation offered by the elastic hinterland is much reduced; the plastic strains can become very much larger with significant changes in the surface profile. In an analysis of this situation, in contrast to the elastic–plastic case, it is reasonable to consider the elastic strain components to be negligibly small in comparison to the plastic. Provided it does not strain harden to any appreciable extent, the material may now be idealized as a *rigid perfectly plastic* solid which flows under a constant shear flow stress k.

The case can now be analysed by one of the standard methods for dealing with large-strain plastic-forming processes; for example the *slip line field* technique in two dimensions, or the approximate, but more readily applied, *upper bound* method in both two- and three-dimensional geometries; see for example Johnson and Mellor (1964) or Hosford and Caddell (1983). The upper bound method depends on postulating some mechanism of plastic flow which is compatible with both the external constraints on the body and with volume constancy. Equating the rate at which the external load does work to the rate at which energy is absorbed within the material provides an upper bound on the true maximum load the contact can sustain.

In the case of a single, loaded line contact a possible collapse mechanism consisting of a system of rigid triangular blocks of material sliding over one another is illustrated in Fig. 3.19(a). Also shown, in Fig. 3.19(b), is the

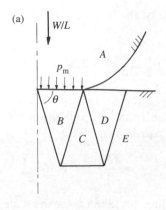

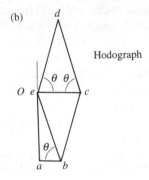

Fig. 3.19 Possible collapse mechanism for a cylinder loaded against a flat. In (a) B, C, and D are rigid blocks sliding over one another so that the interfacial sliding velocities are given in (b) by the vectors bc, cd, and so on.

relationship between the velocities of individual regions of material and hence the magnitude of the velocity discontinuities at the interfaces.

The upper bound theorem provides the inequality

$$p_{\mathrm{m}} \times u_0 \leqslant k \sum s_{\mathrm{i}} \times \Delta u_{\mathrm{i}} \qquad (3.91)$$

where p_{m} is the mean pressure required to maintain plastic flow as the upper surface indents the lower at a speed u_0. The material flow stress in shear is k; the area of a typical internal interface over which there is a tangential velocity discontinuity of magnitude Δu_{i} is s_{i}, and the summation is made over all the n such internal interfaces. The load per unit length of the contact is then $W/L = 2a \times p_{\mathrm{m}}$. Values of s_{i} can be established from the geometry of the figure and values of Δu_{i} read off from the velocity diagram or *hodograph*. Expressing both as a function of θ we have, in this case, the relation that

$$p_m = 2(2\operatorname{cosec}2\theta + \cot\theta). \tag{3.92}$$

The value of θ which provides the *minimum* value of p_m and so gives the lowest upper bound solution for this pattern of deformation, occurs when

$$\tan\theta = \sqrt{2},$$

and hence

$$p_m = 5.66k = 2.83Y. \tag{3.93}$$

When the upper bound technique is applied to other plastic indentation geometries it is found that although the numerical value of the ratio p_m/Y depends to some extent both on details of the geometry and the frictional conditions on the surface it is always close to the value 3. Thus we can see that after subsurface yielding is initiated, at an interfacial mean pressure of about Y, there is a transitional range, involving values of p_m between Y and $3Y$, when the plastic flow is constrained by the surrounding elastic material and within which the elastic strains cannot be neglected in comparison with the plastic before the plastic zone breaks free at the outer surface. These three ranges of loading—elastic, elastic–plastic, and fully plastic—are characteristic of nearly all engineering structures subjected to loadings of increasing severity. The pressure p_m can be thought of as a measure of the indentation hardness $\mathcal{H}$ of the material, that is, the smallest value of normal pressure required to bring about significant plastic deformation under a rigid indenter; see Section 1.2.1. It immediately follows from equation (3.93) that there is a simple but useful numerical relation between indentation hardness and flow stress, namely that

$$\mathcal{H} \approx 5.66k. \tag{3.94}$$

3.5.2   Elastic–plastic contact

While problems in the intermediate elastic–plastic region are, in principle, capable of solution, these are difficult to obtain in practice in all but the simplest of geometries. Difficulties arise in finding solutions which satisfy both the equations of equilibrium and compatibility on both sides of the boundary between elastically and plastically stressed material. This is because the size and shape of the elastic–plastic boundary is not known a priori. In the case of the contact between a flat surface and a comparatively blunt indenter, that is, one with an included angle of more than about 140° (whether it be a two-dimensional wedge, or a three-dimensional sphere, cone, or pyramid), the experimental observation that the subsurface displacements are approximately radial from the point of first contact, with roughly cylindrical or spherical contours of equal strain, has led to a simplified approximate model.

We can think of the contact surface of the harder indenter as being encased in a semicylindrical 'core' of radius a; see Fig. 3.20. Within this core the material is in a state of pure hydrostatic compression, say at pressure $\bar{p}$ while outside it stresses and displacements have radial symmetry and are taken as being equal to those in an infinite elastic perfectly plastic solid containing a cylindrical or spherical cavity under pressure. The expressions for the stresses and displacements arising from such a cavity are known from the theory of elasticity (see, for example Johnson (1985)). The elastic–plastic boundary lies at a radius $c(>a)$ while at the interface $r = a$, between the core and the elastic–plastic zone, two conditions must be satisfied. Firstly, the hydrostatic stress in the core must equal the radial stress component in the plastic region, and secondly, when the indenter moves down through a further increment of penetration the volume it displaces must be accommodated by the radial displacements at $r = a$, in other words the core is treated as being incompressible.

The value of $\bar{p}$ which is consistent with both these conditions for cylindrical line contacts is given by

$$\frac{\bar{p}}{Y} \approx \frac{1}{\sqrt{3}}\left\{1 + \ln\left(\frac{4E\tan\vartheta}{3\pi Y}\right)\right\} + \mathfrak{D}, \tag{3.95}$$

where ϑ is the value of the angle between the local tangent and the surface

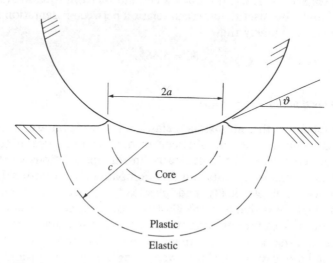

Fig. 3.20 Elastic–plastic contact between a rigid indenter and a deforming material.

at the edge of the contact and $\mathcal{D}$ is a (small) numerical term which accounts for the variation in ϑ during the course of the indentation. If the geometry of the contact is that of a wedge on a flat then geometric similarity is maintained and the term $\mathcal{D}$ has the value zero.

Very similar arguments can be used in the case of a point, rather than a line, contact. Here, once again for full plasticity the mean contact pressure is close to three times the yield strength of the material. Within the elastic–plastic regime

$$\frac{\bar{p}}{Y} \approx \frac{2}{3} \left\{ 1 + \ln \left(\frac{E\tan\vartheta}{3\pi Y} \right) \right\}. \tag{3.96}$$

Thus in both cases the pressure beneath the indenter is a function of the single non-dimensional group $E\tan\vartheta/Y$. This group can be interpreted as the ratio of the strain *imposed* by the indenter (which is related to its 'sharpness' or the value of $\tan\vartheta$) to the strain *capacity* of the material as measured by the elastic strain at yield (i.e. Y/E); it is thus a measure of the severity of the loading. Of course, the stress in the material immediately below the indenter is not under purely hydrostatic stress—in the case of an axisymmetric indentation the value of the mean pressure at the surface is approximately $\bar{p} + \frac{2}{3}Y$, so that this is the best estimate of p_m. It follows that the three regions of behaviour, elastic, elastic–plastic, and fully plastic, can be displayed on a graph of p_m/Y versus $E\tan\vartheta/Y$. The elasticity of the indenter can be taken into account by replacing E by E^* (defined as in eqn (3.47)). Such a 'map' of material behaviour is shown in Fig. 3.21. First yield for a spherical indenter occurs at about $p_m/Y = 1.1$ and for a cone at about $p_m/Y = 0.5$. The fully plastic state is reached when the group $E^*\tan\vartheta/Y$ has a numerical value in the range 30–40.

The same gradual transition from purely elastic contact to fully plastic behaviour will occur when the contact is subject to a tangential load applied simultaneously with the normal pressure. The presence of a surface traction moves the point of initial yield from the Oz-axis of symmetry and brings it closer to the surface as shown in Fig. 3.22(a) in which computed shear stress contours have been drawn for a traction force whose magnitude is one-third of the total normal load, that is, $F/W = \frac{1}{3}$. When this distribution is compared to Fig. 3.10 it can be seen that the application of the tangential load has both increased the magnitude of the value of the maximum subsurface shear stress (in this case by about 20%) and moved its position (point C) to bring it both closer to the surface and away from the axis of symmetry. Once the magnitude of the maximum principal shear stress has been found its value can be set equal to the flow stress k to give a prediction of first yield. Figure 3.22(b) shows the photoelastic fringes for an experimental contact carrying both normal and tractive loads. There is good agreement between the predicted pattern and that observed.

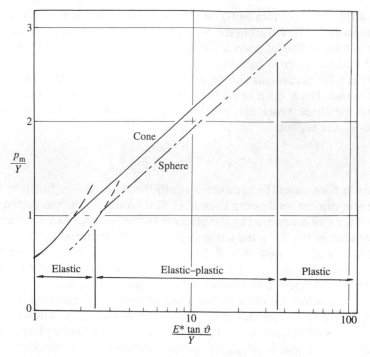

Fig. 3.21 Indentation of a half-space by spheres and cones. p_m/Y is the ration of the mean contact pressure to the material yield stress and $E^*\tan\vartheta/Y$ is a measure of the severity of loading (from Johnson 1985).

As the frictional traction at the surface gets larger some stage is reached when the position of maximum shear occurs actually *at* the surface rather than beneath it. This effect is illustrated in Fig. 3.23 curve A in which the ratio of the maximum Hertz pressure to the shear flow stress, that is, to p_0/k which is just sufficient to cause initial plastic flow is plotted against the traction or friction or traction coefficient F/W. Figure 3.23 is thus effectively a second map which can be used to determine both the likelihood of plastic deformation and whether this is at or below the surface. It applies to a single contact at which the severity of the loading, described by both the ratio of the peak Hertzian pressure to the uniaxial yield strength and the traction coefficient F/W, is known. In the case of a line contact there is a slight complicating factor in that it is possible for deformation to occur by spread of material in the out-of-plane, or Oy-direction; however, this mode of lateral flow is favoured only over a fairly narrow range of traction coefficients. Also seen in Fig. 3.23 as curve B is the corresponding curve for the

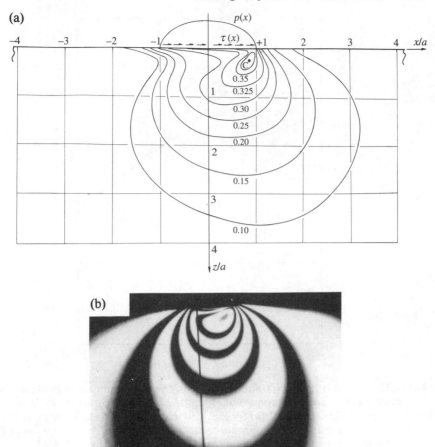

Fig. 3.22 Contours of principal shear stress τ_{max}/k below a sliding Hertzian contact; traction is from left to right. (a) Computed for $F/W = \frac{1}{3}$ (from Suh 1989; and (b) observed photoelastic fringes (from Cameron 1966).

von Mises yield criterion and as curve C that for point contact using von Mises.

Figure 3.23 illustrates an important general point: if the friction or traction coefficient is less than about 0.3 then the plastic zone beneath a sliding contact is contained within the bulk and does not break through to the free surface. The plastic strains must therefore be comparatively small and of the order of the elastic strains. But if the traction coefficient exceeds 0.3 then

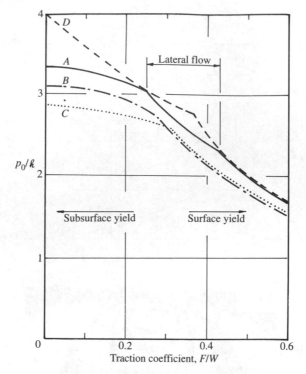

Fig. 3.23 Effect of surface tractions F/W on the maximum Hertz pressure p_0 for either yielding and shakedown of material of shear flow stress k. Curve A corresponds to line contact, first yield, Tresca; curve B to line contact, first yield, von Mises: curve C to point contact, first yield, von Mises; and curve D to line contact, after shakedown, Tresca (from Johnson 1985).

the flow field does extend to the surface for both line and point contacts. Having lost the constraining effect of the elastic boundary it follows that the plastic strains can now become much larger.

3.6 Repeated contacts; the role of residual stresses and shakedown

In many practical contacts the loaded surfaces have to withstand many repeated passes of the load. If, in the first pass, the elastic limit is exceeded (so that the operating point of the contact is above either curves A or B in Fig. 3.23) an element of plastic deformation will take place, and thus when the load is removed some system of residual stresses will remain in the material. In the second pass of the load the material is subjected to the combined action of the applied contact stresses and the system of residual stress left behind from the first pass. Generally speaking, such residual stresses are protective, in the sense that they will make yielding less likely in the second pass than was the case in the first. It is possible that after a few passes the residual stress system builds up to such an extent that the applied load can be carried entirely elastically. This process is known as *elastic shakedown*. We can investigate whether or not shakedown is likely by appealing to an important theorem in stress analysis; this states that if *any* distribution of residual stress can be found which, together with the elastic stresses due to the load, constitute a system of stress which is everywhere within the elastic limit of the material then the system *will* shakedown. The strength of the theorem lies in the fact that the system of residual stress that we use for the calculation does not have to be the real one – as long as it is a possible one. The important practical implication is that, providing a suitable set of residual stresses is possible, then the material will be successful in finding them.

When this argument is applied to the case of a repeated Hertzian contact between cylinders, the condition for elastic shakedown to occur turns out to be that the peak Hertzian pressure p_0 must be no greater than $4.00k$. Since we know, from eqn (3.89), that using the Tresca criterion the critical value of p_0 for yield on the first application of the load is $3.3k$ this means that if the repeated load is such that $3.3k < p_0 < 4.0k$ then, although there will be some yield in the first cycle of operation, shakedown *will* occur. As the total load W/L on the contact is proportional to $(p_0)^2$ it follows that the ratio of the shakedown limit W^S to the elastic limit W^Y is given by

$$\frac{W^S}{W^Y} = \left\{\frac{4.0}{3.3}\right\}^2 = 1.47. \tag{3.97}$$

In other words the load on such a contact can be nearly 50% greater than that needed to bring about yield on the first cycle before conditions are so severe that continuous plastic deformation will be produced on every subsequent cycle.

A similar analysis can be made of a Hertzian line contact which also carries tractive load F/L. As the traction on the surface is increased, so the limiting value of the maximum Hertz pressure for which shakedown is possible reduces. This effect is illustrated in Fig. 3.23 curve D, from which it can be seen that as the traction grows, so the allowable interval between the load for first yield and that for shakedown becomes narrower.

The effect of strain hardening can also lead to an apparent shakedown phenomenon irrespective of the contact geometry. With repeated deformation the effective value of k, the material shear strength, may rise in the surface layers, so that a load which initially exceeds the shakedown limit subsequently lies below it. Further refinements of the shakedown argument are possible for materials which strain harden in particular ways: for example, in a material which exhibits some form of non-linear strain hardening (which is typical of many engineering alloys) the situation of incremental plastic collapse or *plastic ratchetting* can arise. Here gross deformation of the surface and close-to-surface material may accumulate after thousands or even millions of repeated loading cycles each of which contributes an element of plastic strain. Shakedown can also play a part in rolling contacts, such as a ball on a flat, in which with repeated traversals an originally plane surface becomes grooved so increasing the contact area and reducing the contact stresses: see Section 11.4.

3.7   Contact of rough surfaces

3.7.1   *Nominally flat surfaces*

Having established the modes of behaviour of a single contact spot we are now in a position to proceed to consider what happens when two real surfaces, that is, surfaces carrying a large number of asperities with a range of summit heights, are brought into loaded contact. The first point to note is that the elastic stresses depend on the *relative* profiles of the surfaces, described by the reduced radius of curvature R and the contact modulus E^*. This means that, for the purposes of analysis, we can imagine all the deformable surface imperfections to be concentrated on one surface, while the second surface is considered to be both rigid and plane. The situation is illustrated in Fig. 3.24.

Using the notation introduced in Chapter 2, we suppose the rough surface to consist of N hills or asperities whose heights z above the mean level vary in some statistical fashion. This distribution can be described by the probability density function $\phi(z)$ which must be such that

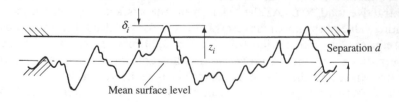

Fig. 3.24   Contact between rough surfaces modelled by a deformable rough surface and a rigid plane.

$$\int_{-\infty}^{+\infty} \phi(z)\,dz = 1, \tag{2.4}$$

so that all the summits are included. Following, for simplicity, the treatment of Greenwood and Williamson (1966) we assume that the tips or summits of the asperities are spherical, and furthermore, that they all have the same characteristic radius of curvature R_s. When a normal load is applied the rigid surface moves toward the mean level of the rough surface so that, when the separation is d, it will have made contact with all the asperities for which $z > d$. The number n of such contacts will be given by

$$n = N \int_{d}^{\infty} \phi(z)\,dz. \tag{3.98}$$

A typical summit, originally of height z_i, must thus have been compressed by an amount δ_i where

$$\delta_i = z_i - d, \tag{3.99}$$

and the contact area between its deformed shape and the flat surface will be a small circular patch, say of area A_i and radius a_i so that

$$A_i = \pi a_i^2. \tag{3.100}$$

So far we have not specified whether the response of the material is elastic or plastic. In the case of a rough surface which exhibits perfect plasticity, deformation of each contact spot will occur at the same normal pressure p_m (see Section 3.5.1) and so it follows immediately that as long as p_m remains constant the total real area of contact between the two solids will be in direct proportion to the applied load W. The actual number of individual contact points will, however, depend very much on the nature of the function $\phi(z)$.

The rather more taxing question is, What is the situation when the contact is *purely* elastic? This can occur either because the normal load is insufficiently large to cause significant plastic flow, or perhaps more realistically, because after repeated loading, the shakedown mechanism has given rise to elastic conditions. We can now, in such circumstances, treat each asperity contact spot as a separate Hertzian contact between a sphere and a flat. The relationship between the radius of the contact area a_i and the compression δ_i is given by eqn (3.56) with δ_i replacing Δ, so that

$$a_i^2 = \delta \times R_s = (z_i - d)R_s \tag{3.101}$$

hence

$$A_i = \pi \times a_i^2 = \pi R_s(z_i - d). \tag{3.102}$$

The total real area of contact A is the sum of all the n contacts, so that

$$A = \Sigma A_i = N\pi R_s \int_d^\infty \phi(z)(z-d)\,\mathrm{d}z. \qquad (3.103)$$

Now the load W_i on this typical contact is related to the contact area by equation (3.55) so that

$$W_i = \frac{4E^*}{3R_s} a_i^3 = \frac{4}{3}E^* R_s^{1/2}\delta_i^{3/2}, \qquad (3.104)$$

and the total load W is thus given by summing up all n such contacts, so that

$$W = \frac{4}{3}NE^* R_s^{1/2} \int_d^\infty \phi(z)(z-d)^{3/2}\,\mathrm{d}z. \qquad (3.105)$$

The numerical values of n, A, and W given by eqns (3.98), (3.103), and (3.105) clearly depend on the form of the function of $\phi(z)$. In their original work Greenwood and Williamson considered a random or Gaussian distribution of summit heights and evaluated these integrals numerically; however, they also suggested that an analytical treatment is possible if we make use of the observation that whatever the precise form of the bell-shaped distribution of $\phi(z)$ it is only the most prominent asperities, that is, those of large z, which will take part in the surface interactions. Within this region we can reasonably model $\phi(z)$ by an exponential expression of the form

$$\phi(z) = \lambda\exp(-\lambda z), \qquad (3.106)$$

where λ is a suitable constant. A comparison of this function with the Gaussian distribution is shown in Fig. 3.25 with the value of λ set at $2/\sigma_s$, where σ_s is the standard deviation of the summit heights; this gives agreement at $z/\sigma_s = 1.12$.

Adopting this model the total number of contact points becomes

$$n = N\lambda \int_d^\infty \exp(-\lambda z)\,\mathrm{d}z; \qquad (3.107)$$

the real area of contact A is given by

$$A = N\pi R_s \lambda \int_d^\infty (z-d) \times \exp(-\lambda z)\,\mathrm{d}z, \qquad (3.108)$$

and the total load is

$$W = \frac{4}{3}NE^* R_s^{1/2}\lambda \int_d^\infty (z-d)^{3/2} \times \exp(-\lambda z)\,\mathrm{d}z. \qquad (3.109)$$

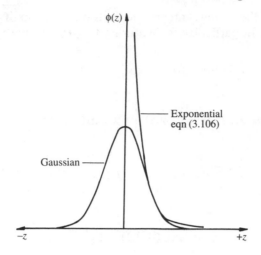

Fig. 3.25   The exponential approximation to the upper region of the Gaussian distribution of surface roughness.

Each of these integrals involves some function of the term $z - d$, that is, they are all of the form

$$\int_d^\infty f(z - d) \times \exp(-\lambda z)\,dz,$$

where $f(z - d)$ represents the appropriate function in eqns (3.107)–(3.109), namely $(z - d)^0$, $(z - d)^1$, and $(z - d)^{3/2}$. If we replace $z - d$ by the variable t (and thus dz by dt), and note that as z goes from d to infinity, so the variable t goes from zero to infinity, then all the expressions involve the integral

$$\int_0^\infty f(t) \exp\{-\lambda(t +)d)\}\,dt, \tag{3.110}$$

that is,

$$\int_0^\infty f(t) \exp\{-\lambda d\} \exp\{-\lambda t\}\,dt.$$

The first exponential term can be taken outside the integration sign, so that the required integral is equal to

$$\exp(-\lambda d) \int_0^\infty f(t) \exp(-\lambda t)\,dt.$$

The value of the definite integral depends on the form of $f(t)$ but *not* on the value of d. In particular, substituting for $f(t)$, we have

$$n = N\lambda\exp(-\lambda d) \int_0^\infty \exp(-\lambda t)\,\mathrm{d}t,$$

$$A = N\pi R_s\lambda\exp(-\lambda d) \int_0^\infty t \times \exp(-\lambda t)\,\mathrm{d}t, \qquad (3.111)$$

$$W = \frac{4}{3}NE^*R_s^{1/2}\lambda\exp(-\lambda d) \int_0^\infty t^{3/2} \times \exp(-\lambda t)\,\mathrm{d}t,$$

so that

$$n = N\lambda\exp(-\lambda d) \times C_0 \qquad (3.112a)$$

$$A = N\pi R_s\lambda\exp(-\lambda d) \times C_1 \qquad (3.112b)$$

$$W = \frac{4}{3}NE^*R_s^{1/2}\lambda\exp(-\lambda d) \times C_{3/2} \qquad (3.112c)$$

where C_0, C_1 and $C_{3/2}$ are constants. The key interpretation of eqns (3.112) is that the number of points of contact n, and the real contact area A, are both *proportional* to the applied load W. As the separation of the surfaces decreases, although the area of any particular contact spot increases, more asperities make contact so that the average size of the patches of real contact, that is, A/n, remains constant. Provided the surface contains a *range* of surface roughnesses (which we have approximated by the exponential function) the direct proportionality between applied load and area of contact extends throughout the elastic range of material response as well as the plastic regime.

Two other observations can be made on the basis of these results. Firstly, if we now apply a tangential traction to the contact we know that sliding will be opposed by shear forces generated at the points of contact. The mechanisms underlying this frictional resistance are discussed more fully elsewhere but it would not seem unreasonable to suppose that the maximum available frictional force F depends upon the product of the area of true contact A and some characteristic interfacial shear strength τ, so that $F = A \times \tau$. But since we have shown that A is proportional to W it follows that F also depends directly on W, so that the ratio F/W, that is, the coefficient of friction μ, is a constant and will be characteristic of the surfaces and independent of the size of the load they carry. This is therefore in accord with one of the empirical 'laws of friction'. In fact, this is still predicted even if τ is not the same at all the contact points but is dependent on the local value of pressure – provided our use of the exponential approximation for the height distribution is justifiable (see Problem 3.9(c)).

A second consequence of the area of real contact being proportional to the load is that the actual mean contact pressure is constant. Its value p_m is equal to the load W divided by the area of contact A, and, from eqn (3.112), is thus given by

$$p_m = \frac{W}{A} = \frac{4}{3\pi} \frac{E^*}{R_s^{1/2}} \frac{C_{3/2}}{C_1},$$ (3.113)

and is therefore independent of the separation of the surfaces. When the numerical values of the groups C_1 and $C_{3/2}$ are substituted, equation (3.113) becomes

$$p_m \approx 0.39 E^* \sqrt{\frac{\sigma_s}{R_s}}.$$ (3.114)

Since we are modelling each asperity contact as being equivalent to that between a sphere and a plane we can use the results of the Hertz analysis (Section 3.3.4) to predict those conditions which will just lead to the initiation of plastic deformation. In such a Hertzian contact the maximum shear stress occurs beneath the surface and is of magnitude $0.47 p_m$. Setting this shear stress equal to the flow stress of the material k, we can see that the critical value of p_m to cause yielding is equal to $2.15k$. Since the flow stress is simply related to the hardness of the material $\mathcal{H}$ (by eqn (3.94)) we can say that there is likely to be some plastic yielding if

$$p_m \geq \frac{2.15 \times \mathcal{H}}{5.66}.$$

Combining this with the expression for p_m from (3.114) the condition for the onset of plasticity becomes that

$$\frac{E^*}{\mathcal{H}} \sqrt{\frac{\sigma_s}{R_s}} \geq C$$ (3.115)

where C is a constant which, although depending to some extent on the form of the height distribution, is of the order of unity. The non-dimensional group on the left hand side of (3.115) is sometimes known as the *plasticity index* Ψ, and neatly describes the likely deformation characteristics of a rough surface. If its value is very much *less* than unity then asperity deformation is likely to be entirely elastic; on the other hand, if Ψ is significantly greater than 1 the response is predominantly plastic. The factor $\sqrt{\sigma_\sigma/R_s}$ is associated with the mean slope ϑ of the surface and can be obtained, approximately at least, from the profilometry data. Of course, a profilometer measures the heights of many points on the surface, not just those of summits, but it can be shown that for an isotropic surface having a height distribution which is close to Gaussian, with a standard deviation σ, the standard deviation of the heights of the summits σ_s, is numerically close to σ. There is a clear similarity between the plasticity index Ψ and the group $E^*\tan\vartheta/Y$ introduced in Section 3.5.2.

3.7.2 *Elastic contact of rough curved surfaces*

In our treatment of the contact of rough surfaces we have so far assumed that we are dealing with two nominally flat surfaces in contact. Most tribological contacts are not of this form but are nearly always between non-conforming or curved surfaces. The question then naturally arises, under what circumstances can the surface roughnesses on such loaded curved surfaces be safely neglected in calculating the contacts stresses (using for example the Hertzian analysis) and under what circumstances can they not? For the situation to be amenable to quantitative treatment it is necessary for the two ranges of scales involved to be very different: these are typified on the one hand by the radius of curvature of the mean level of the surfaces, say R, and on the other by height and spatial distribution of the asperities. The nominal contact area will contain many deforming asperities and these then act as a compliant layer lying over the surface of the body. Contact extends over a larger area than would be the case if the solids were perfectly smooth. The geometry is illustrated in Fig. 3.26(a); here, for the sake of clarity, the surface roughness (characterized by the standard deviation or RMS roughness σ) has been transferred to the plane surface which is in contact with a smooth surface of radius R.

The details of the quantitative solution are beyond the scope of this discussion (they can be found in Johnson (1985)) but the conclusions can be conveniently expressed if we define a non-dimensional parameter χ to describe the influence of the surface roughness by the relation

$$\chi = \left\{ \frac{\sigma R}{a^2} \right\} = \sigma \times \left\{ \frac{16RE^{*2}}{9W^2} \right\}^{1/3}, \qquad (3.116)$$

where a is the contact radius for smooth surfaces according to the Hertz analysis. The smoother the surfaces then the smaller is the value of χ (since it depends on σ). The influence of χ on the effective pressure distribution for point contacts can be seen in Fig. 3.26(b) which illustrates the extent of the contact zone for two values of χ which correspond to smooth ($\chi = 0.05$) and rough ($\chi = 2$) engineering surfaces. In this plot the pressure is normalized by the maximum Hertz pressure for smooth surfaces, and the radial distance by the Hertzian contact radius a; the dotted curve represents the Hertzian semi-ellipse. With a rough surface the peak pressure is much below the normal Hertzian value and falls more or less asymptotically to zero over

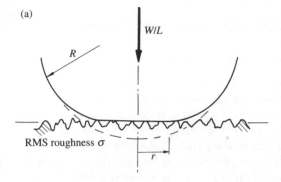

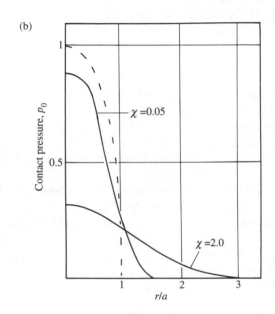

Fig. 3.26 Contact of a smooth elastic sphere with a rough surface: (a) basic geometry; and (b) effect of the contact parameter χ (from Greenwood and Tripp 1967).

an area very much greater than the Hertzian area. However, if χ is less than about 0.1 both the pressure curve and the contact radius are really very close to Hertzian values. We can thus draw the important conclusion that the Hertz theory for perfectly smooth surfaces can be used to calculate the contact stresses on real engineering surfaces without introducing numerical errors of more than a few per cent, provided the non-dimensional group $(\sigma R/a^2)$ is less than about 0.1.

3.8   Thermal conditions in sliding contacts

3.8.1   *Stationary heat source*

In order to maintain a pair of loaded surfaces in relative sliding motion some form of external agency must do mechanical work to overcome the forces of interfacial friction. This input of mechanical work ultimately appears as an increase in the internal energy of the system and thus, nearly always, as an increase in temperature (the alternative, a change of phase with no associated increase in temperature may be important in some circumstances — for example, if one of the surfaces is a frozen solid such as ice). The temperature gradients set up between areas of the surface and the bulk can lead, through the mechanism of differential thermal expansion, to the introduction of additional stresses and deformations which can themselves modify the contact geometry and stress fields. Steady state conditions, if they exist at all, can involve a complex interaction between these mechanical and thermal effects. In what follows we shall consider both how best to recognize those situations in which thermal effects and the generation of surface temperatures are likely to be significant, and how to make first-order estimates of the resultant temperature increases. More detailed analysis of the various problems in heat conductivity that are necessarily involved can be found in Johnson (1985), Barber (1991) and especially the classic work on thermal analysis, Carslaw and Jaeger (1959).

We start by considering the two-dimensional problem, and imagine a line source which spontaneously generates a quantity of thermal energy, of magnitude H per unit length, at time $t = 0$. The line source lies along the Oy-axis in the surface of a semi-infinite solid initially at uniform temperature Θ_0. The half-space has density ρ, specific heat c, and thermal conductivity K; these material properties can often be conveniently grouped together into the *thermal diffusivity* κ, defined by the relation

$$\kappa = K/\rho c. \tag{3.117}$$

Heat is conducted radially through the solid (we neglect any heat lost from the free surface) so that the distribution of temperature Θ is symmetrical about the y-axis. The equation relating Θ to position r and time t, given in a non-dimensional form, is

$$\frac{\Theta - \Theta_0}{\Theta_0} = \frac{1}{2\pi(t/T)} \exp\left\{\frac{(r/b)^2}{4(t/T)}\right\} \tag{3.118}$$

where $T = H/LK\Theta_0$ and $b^2 = H/\rho c\Theta_0$ so that $\kappa = b^2/T$. This relationship is shown in Fig. 3.27 in which the non-dimensional temperature rise is plotted against position r/b for various values of the parameter t/T. When t is equal

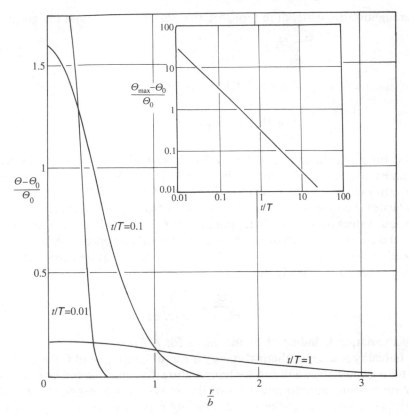

Fig. 3.27 Temperature rise under an instantaneous line heat source of intensity H versus distance r for various times. The inset shows the decay of the peak temperature at $r = 0$.

to zero, that is, at the instant at which the source is applied, the temperature on the Oy-axis reaches a theoretically infinite value but this falls rapidly with time: the area under each curve represents the energy input per unit length and thus remains constant with time. The variation of Θ_{max}, the temperature at $r = 0$, with t/T is shown in the inset. Figure 3.27 can be used to estimate the thermal conditions in a tribological contact in which there is a very rapid input of thermal energy, as might occur as a result of a very sudden change in the interfacial frictional conditions.

Perhaps a more realistic set of circumstances is modelled by a continuous source of thermal energy acting at the boundary of the half-space. Supposing initially that this is a point source of intensity $\dot{H}$, constant with time and applied at $t = 0$, acting at the origin, the expression for the temperature throughout the material in terms of the position r and time t is given by

$$\frac{\Theta - \Theta_0}{\Theta_0} = \frac{1}{2\pi (r/b)} \left\{ 1 - \text{erf}\left[\frac{(r/b)^2}{4(t/T)} \right] \right\} \qquad (3.119)$$

where $b = \dot{H}/K\Theta_0$; $T = \rho c \dot{H}^2/K^3\Theta_0^2$ (so that again $\kappa = b^2/T$) and

$$\text{erf}(\zeta) = \frac{2}{\sqrt{\pi}} \int_0^\zeta \exp(-\xi^2) \, \text{d}\xi.$$

This function, the *error function*, arises in several branches of engineering mathematics and is sketched in the inset to Fig. 3.28; when $\zeta > 2$ the value of erf(ζ) can be taken as unity; when $\zeta < 0.5$ then erf(ζ) $\approx \zeta$. Figure 3.28 indicates the form of the variation of $(\Theta - \Theta_0)/\Theta_0$ with distance (r/b) for various values of the non-dimensional time t/T. The infinite temperature at $r = 0$ arises from the assumption of a point source of energy. After a sufficiently large time, that is, when $t/T \gg 1$, the temperature throughout the solid tends to the steady distribution

$$\frac{\Theta - \Theta_0}{\Theta_0} = \frac{1}{2\pi (r/b)}. \qquad (3.120)$$

This variation is indicated by the curve for $t/T = \infty$.

In reality sources of thermal energy must, of course, be of finite area: the resulting subsurface and surface temperature distributions can be estimated by superposing appropriate point or line solutions. For example, for a circular heat source of radius a and intensity $\dot{h}$ per unit area, the *mean* surface temperature $\bar{\Theta}$ within the contact is given by

$$\bar{\Theta} - \Theta_0 = \frac{8}{3\pi} \frac{\dot{h}a}{K}, \qquad (3.121)$$

While for a square contact patch $2a \times 2a$, the corresponding equation is

$$\bar{\Theta} - \Theta_0 = 0.946 \frac{\dot{h}a}{K}. \qquad (3.122)$$

3.8.2   Moving heat sources

Tribological contacts usually involve relative sliding at the interface; the source of thermal energy, the result of frictional mechanisms within the contact patch, must be in relative motion to at least one of the two loaded half-spaces. In order to investigate the temperature distribution we therefore need to consider the thermal conditions below a *moving* source of heat. If we are dealing with steady state conditions it is rather more convenient to imagine the source of heat to be at rest and the half-space beneath it to be moving

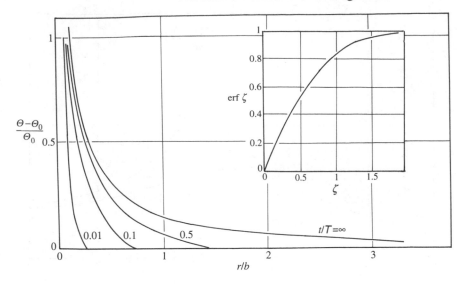

Fig. 3.28 Temperature rise in a half-space due to a continous point source of heat of intensity $\dot{H}$. The quantities b and T are defined in the text. The inset shows the form of the error function.

at speed, say U. In the two-dimensional case of an infinitely long source of energy, of intensity $\dot{h}$ per unit area, of width $2a$, lying symmetrically about the Oy-axis as shown in Fig. 3.29, the surface temperature can be shown to be of the form

$$\Theta - \Theta_0 = \frac{\dot{h}a}{K\mathrm{Pe}^{1/2}}f(\mathrm{Pe}, X) \tag{3.123}$$

where $\mathrm{Pe} = Ua/2\kappa$ and $X = Ux/2\kappa$, so that the parameter X traverses the numerical range zero to the value of the group Pe through the contact. The parameter Pe is known as the *Peclet number* and can be interpreted as the ratio of the speed of the surface to the rate of thermal diffusion into the solid. At large values of Pe (say in excess of 5), during the time taken for the surface to traverse the heated zone, relatively little thermal energy can diffuse more than a very short distance into the solid — heat flow is thus essentially to the surface. Surface temperature distributions for various values of Pe are plotted in Fig. 3.29. The maximum temperatures occur towards the rear of the contact where the material has had a greater exposure to the heat source.

When Pe is very large, eqn (3.123) takes the form

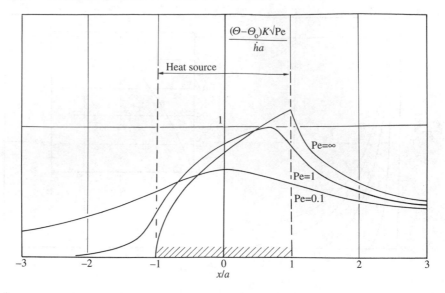

Fig. 3.29 Steady state surface temperature profiles in a half-space moving under a heated band of width 2*a* for various values of the Peclet number Pe.

$$\Theta - \Theta_0 = \frac{2}{\sqrt{\pi}}\frac{\dot{h}}{K}\sqrt{\frac{\kappa(x+a)}{U}}, \qquad (3.124)$$

so that at the exit of the contact, where $x = a$, the maximum temperature Θ_{max} is given by the equation

$$\Theta_{max} - \Theta_0 = \frac{2}{\sqrt{\pi}}\frac{\dot{h}}{K}\sqrt{\frac{2\kappa a}{U}} = \frac{2}{\sqrt{\pi}}\frac{\dot{h}a}{K}Pe^{-1/2}. \qquad (3.125)$$

Θ_{max} is sometimes known as the *flash* temperature and can be important in the initiation of some mechanisms of surface failure. The *mean* or average surface temperature $\bar{\Theta}$ is given by

$$\bar{\Theta} - \Theta_0 = \frac{4}{3\sqrt{\pi}}\frac{\dot{h}a}{K}Pe^{-1/2}. \qquad (3.126)$$

When Pe is small, that is, $Pe \ll 1$, the situation can be adequately described by the equations for a static energy source. In the transition range, Fig. 3.30 illustrates the effect of the variation of the Peclet number on the mean surface temperatures under a square source of dimensions 2a × 2a.

In practice the heat input $\dot{h}$ is often the result of frictional sliding so that

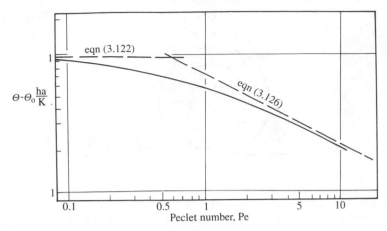

Fig. 3.30 Effect of Peclet number Pe on mean surface temperatures under a square heat source.

its value depends on the normal pressure p, the coefficient of friction μ, and the relative sliding velocity U according to

$$\dot{h} = \mu p U. \tag{3.127}$$

Despite the fact that p varies across the contact it is often sufficiently accurate in making estimates of peak and mean temperature rises to take p, and thus $\dot{h}$, as constant across the contact area. However, it is important to note that in eqns (3.123)–(3.125) the velocity U is that of the material *relative* to the contact patch. If the contact patch is itself moving, say with velocity u_c, then in evaluating the Peclet numbers for the two solids we must consider their velocities *relative* to the contact patch rather than the absolute values. This can be done by subtracting the velocity u_c from the absolute velocities. For example, consider the situation of Fig. 3.31 which shows a disc 2 of radius R and thickness L rotating counter-clockwise at a rate ω. It is pressed by a force W into contact with a plane stationary surface 1 and simultaneously moved, so that its centre is translating at velocity u. The contact patch which we suppose is of width $2a$ must always be vertically below the centre of the disc, on the line of action of W, and so is also moving from left to right with speed u; in this case therefore $u_c = u$. The appropriate relative velocity for body 1 is thus $0 - u$, that is, of magnitude u, while for body 2 the appropriate velocity is equal to $(u + R\omega) - u = R\omega$.

The Peclet numbers for bodies 1 and 2 are thus respectively

$$\mathrm{Pe}_1 = \frac{ua}{2\kappa_1} \quad \text{and} \quad \mathrm{Pe}_2 = \frac{R\omega a}{2\kappa_2}. \tag{3.128}$$

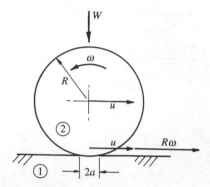

Fig. 3.31 The half-space 1 is at rest, disc 2 rotates with angular speed ω, and translates at velocity u so that there is slip at the contact patch which has width $2a$.

The choice between the high-speed solution, eqn (3.124) and a steady state form, such as eqns (3.121) or (3.122) will depend on the magnitudes of Pe_1 and Pe_2. The energy input per unit are $\dot{h}$ is given by the expression

$$\dot{h} = \mu W(u + R\omega)/2aL. \tag{3.129}$$

Whether viewed from the point of view of either body 1 or 2 the relative surface sliding speed is still $(u + R\omega)$.

One problem remains to be solved, namely the partition of the thermal energy between the two solids. Suppose η to be the proportion of $\dot{h}$ conducted into body 1, so that $1 - \eta$ is the proportion conducted into body 2. To determine the numerical value of η we can equate the *maximum* temperatures generated on each; these must be the same since the surfaces are in intimate contact. Supposing, in the example above, that Pe_1 and Pe_2 are both greater than 5. Then from (3.125) $\Theta_{1,max} = \Theta_{2,max}$ that is,

$$\frac{\eta}{K_1\sqrt{Pe_1}} = \frac{1 - \eta}{K_2\sqrt{Pe_2}},$$

whence

$$\eta = \frac{1}{1 + \{K_2\sqrt{Pe_2}/K_1\sqrt{Pe_1}\}}. \tag{3.130}$$

When the surface layers of a component are heated above the temperature of the bulk, differential thermal expansion will lead to a reduction in the radius of curvature of the surface so that initially flat surfaces will become convex. This, in turn, will tend to concentrate the applied load into smaller contact patches, so increasing the contact pressure within them, leading to an increase in the thermal energy input per unit area $\dot{h}$. This process cannot go on indefinitely but is generally limited by the phenomenon of wear at the points of intense contact; this will be much accelerated by the concentrations in load and temperature. In steady running conditions there will be some form of dynamic equilibrium between these two effects but the details of this are difficult to predict. However, it is worth noting that measurements of the profile of a worn component made after disassembly and at uniform ambient temperature must be interpreted very carefully if an effort is to be made to deduce the state of affairs under dynamic running conditions. Areas of the surface which are depressions under uniform temperatures may, in fact, be high spots of the profile under operating conditions. Discussion of the interactions between thermal and mechanical loadings can be found in Ling (1973), Burton (1980), Johnson (1985), and Barber (1991).

3.9 Further reading

Barber, J. R. (1991) *Elasticity*. Kluwer Academic Publishers, Dordrecht.

Burton, R. A. (1980) *Thermo-elastic effects in sliding contact*, North Western University, Evanston, Il.

Cameron, A. (1966) *Principles of Lubrication*, Longman, London.

Carslaw, H. S. and Jaeger, J. C. (1959) *Conduction of heat in solids*. 2nd Edition, Oxford University Press.

Gladwell, G. M. L. (1980) *Contact problems in the classical theory of elasticity*. Sijhoff and Noordhoff, Amsterdam.

Greenwood, J. A. and Williamson, J. B. P. (1966) Contact of nominally flat surfaces. *Proceedings of the Royal Society*, **A295**, 300–30.

Greenwood, J. A. and Tripp, J. H. (1967) *The elastic contact of rough surfaces. Transactions of the American Society of Mechanical Engineers, Journal of Applied Mechanics*, **34**, 153–67.

Hills, D. A., Nowell, D., and Sackfield, A. (1993) *Mechanics of elastic contacts*. Butterworth-Heinemann, Oxford.

Hosford, W. F. and Caddell, R. M. (1983) *Metal forming*. Prentice-Hall, New Jersey.

Johnson, K. L. (1985) *Contact mechanics*. Cambridge University Press.

Johnson, W. and Mellor, P. B. (1964) *Engineering plasticity*. Van Nostrand, New York.

Ling, F. F. (1973) *Surface mechanics*. John Wiley, New York.

Suh, N. P. (1986) *Tribophysics*, McGraw-Hill, Englewood Clifts, NJ.

Timoshenko, S. P. and Goodier, J. N. (1951) *Theory of elasticity*. McGraw-Hill, New York.

4
The friction of solids

4.1 The genesis of solid friction

4.1.1 *The laws of friction*

Friction is the resistance encountered when one body moves tangentially over another with which it is in contact. The work expended against friction is often *redundant*, that is, it makes no useful contribution to the overall operation of the device of which the bodies are part, and ultimately must be dissipated as waste heat. Consequently, in most tribological designs our aim is to keep these frictional forces as small as possible. Of course there are exceptions to this general rule, occasions when sufficient friction is essential to continued progress and there are many practical devices which rely on the frictional transmission of power: automobile tyres on a roadway, vehicle brakes and clutches, as well as several of the variable-speed transmission systems now finding wider application. When two objects are to be held together, the only alternative to methods which rely on friction is the formation of some sort of chemical or metallurgical bond between them. The development of this sort of technique — adhesives and 'superglues', and even welding and brazing — are relatively recent; 'traditional' forms of fixing rely almost exclusively on friction. A nail hammered into a piece of wood is held in place by frictional effects along its length; if the frictional interaction were substantially reduced, the nail would be squeezed out. Similarly, the grip between a nut and a bolt depends on adequate friction between them. Any woven or knitted fabric depends on the friction between adjacent threads — there are no glues or adhesives; it would be impossible to knot together two pieces of very-low-friction string.

As children we soon learn some of the qualitative laws of friction. That, for example, it is much easier to slide a light object across a horizontal floor than a very much heavier one. But it is soon apparent that as well as the mass (or more correctly the weight) of the object, the material of which it is made is of significance. A block of ice skids over the floor much more readily than a block of wood of the same weight. Observation also tells us that in general it is much easier to move a heavy weight if it is supported by rollers or wheels than if it slides on non-rotating skids — the forces of rolling friction are generally much less than those associated with sliding friction.

In a modern internal combustion engine perhaps 15–20 per cent of the power produced goes to overcoming the friction between the various internal moving parts; in an aero-engine the percentage losses may be much smaller but, nevertheless, still represent a significant component of the fuel consumption. However, in many situations this primary effect may be less important in the long run than the damage to machine components and assemblies by the wear that is the usual companion of friction. Higher friction invariably means the generation of higher surface temperatures and this almost always exacerbates the rates of surface damage and wear. Not only do engineers wish to avoid the cost of providing replacements for prematurely worn components, but also that of shutting down the machine for maintenance and repair, which, especially if unexpected or unplanned, may be of even greater significance.

Our experience of the way in which sliding objects behave can be distilled into the two empirical *laws of friction*. It must be remembered that the precision of these laws is often not high and we must not be surprised to find cases in which they do not apply; situations in which these laws are apparently 'broken' do not necessarily imply any contravention of a much more fundamental 'law' of nature, such as that of the conservation of momentum or of universal gravitation. Although appreciated qualitatively for many centuries, the laws of friction were first stated quantitatively by the French engineer Guillaume Amontons in 1699, and have since become known by his name.

The first of Amontons' laws states that the friction force F between a pair of loaded sliding surfaces is proportional to the normal load W that they carry: the tangential force required to slide a metal block along another surface is proportional to the weight of the block. If the mass of the block is doubled then so is the force required either to initiate, or to maintain, the sliding motion; this is illustrated in Fig. 4.1(a) and (b). The constant of proportionality between F and W that this observation implies is universally known as the *coefficient of friction*, μ. Note that the value of μ depends on the nature of both the solid surfaces; it is meaningless to talk about 'the coefficient of friction of steel' we must specify the counterface and the conditions under which the surfaces are operating.

The second of Amontons' laws states that the friction force F between the two solid bodies is independent of the *apparent* area of contact between them: returning to our sliding block, we find experimentally that the value of the frictional force does not depend on the area of the contact face of the block — the same tangential force is required to maintain sliding whether the block sits on one of its larger faces or one of the smaller ones; see Fig. 4.1(c).

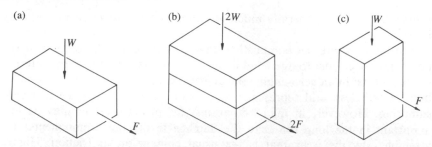

Fig. 4.1   Amontons' laws of friction: (a) a tangential force F is required to slide the block which has weight W; (b) doubling the weight of the block doubles the friction force; and (c) the value of F remains the same even when the apparent or nominal area of contact is changed.

When Amontons presented these laws to the French Royal Academy of Sciences the first law was accepted without question, but the second was greeted with great doubt and scepticism, and the members of the Academy demanded that the experimental work be repeated. These further tests confirmed Amontons' conclusion that friction was independent of apparent area, and this law has been widely used by engineers and designers ever since.

4.1.2   Physical basis of the laws of friction

The problem of explaining why moving a mass over an apparently flat, horizontal surface should require a significant expenditure of energy had exercised the minds of enquiring scientists and philosophers long before Amontons formulated his laws. He recognized that the surfaces with which he was dealing were not completely smooth, and postulated that the origin of frictional effects was intimately involved in some way with either the work that had to be done in lifting the normal load over the surface asperities or in bending them down or breaking them off.

Some years after this, an English investigator, John Desaguliers, criticized this surface roughness explanation on the observational grounds that when surfaces were made smoother, far from decreasing friction, the resistance to motion was often made greater. He suggested that this apparently paradoxical behaviour was attributable to areas of *adhesion* within the general contact zone. However, he was unable to reconcile this picture with Amontons' experimental laws — in particular, the fact that an adhesional explanation would seem to suggest that doubling the area of contact would lead to double the degree of adhesion and so twice the friction force, while in practice friction is effectively independent of contact area. Some years later, the French engineer and scientist Charles Coulomb also considered the possibility of adhesion but similarly dismissed it as a significant contributor to friction on much the same grounds. He felt that friction was due to the work done in dragging one surface up the roughnesses on the other. His picture of friction depended on the asperities of the two surfaces becoming interlocked, rather like two pieces in a jigsaw puzzle. Under a tangential load the surfaces were imagined to move apart and it was the work done against gravity that this movement involved that accounted for the frictional resistance. Coulomb's work in turn was criticized in 1804 by Professor Leslie of Edinburgh University who pointed out that if friction is due solely to the overcoming of gravity then, if there is no net increase in elevation of the moving body, there can equally well, on average, be no friction. While parts of the body may be sliding up roughnesses there will be parts elsewhere sliding down, so that although the friction force may fluctuate it must do so about zero.

In 1750 the prodigious Swiss mathematician Leonhard Euler also adopted this model of rigid interlocking asperities and showed that if these are assumed to be triangular and of slope α then the coefficient μ is equal to $\tan\alpha$. It was, in fact, Euler who introduced the use of the symbol μ for the ratio of the tangential to the normal force. Some years before this, in 1737, Bélidor, a French professor of mathematics, had also considered the geometry of surface roughness and had modelled a surface as an array of spherical asperities; this concept, albeit in combination with a statistical distribution of heights or sizes is still very much in vogue with tribologists.

With our understanding of the geometry of surfaces we can now re-examine Amontons' laws. We know, for example, that when two surfaces come into contact they actually touch at a discrete number (although perhaps a large number) of contact spots and that the total resistance to tangential motion is the sum of the resistances over all these small areas of true contact. If we could say that the resistance force per unit area over all these spots, that is, the specific resistance to shearing τ, is a constant, then the total force F is given by

$$F = A \times \tau, \tag{4.1}$$

where A is the true area of contact. We need make no assumptions about whether the conditions at a particular spot are either elastic or plastic, only that each area is representative of the whole. To achieve Amontons' laws all we now require is a linear relationship between load W and area A, that is, that

$$A \propto W. \tag{4.2}$$

The influence of the normal load on the area of contact between two solids has been discussed in Section 3.7 where we showed that a linear relation of the form of eqn (4.2) will arise when either all the contacts involve plastic deformation of the softer surface, or the surfaces contain a sufficient spread of asperity heights so that it does not matter whether any individual spot is elastic or plastic. In many tribological pairs the value of the plasticity index Ψ (see Section 3.7) indicates that the contact area contains both elastic and plastic contacts: it is the nature of the statistical distribution of surface geometry that explains the validity of the second of Amontons' laws. If both expressions (4.1) and (4.2) hold then it follows that F/W, that is, μ, is constant. Numerical values of μ are not only associated with a particular pair of materials but also with the physical conditions of sliding, in so far as these affect the value of the shear stress τ in eqn (4.1). The importance of the surroundings within which sliding occurs can easily be demonstrated by two simple examples. The coefficient of friction of a rubber tyre on a *dry* road surface is close to unity, while if the road is wet the value of μ may be reduced to very much less than half that value. A brass block sliding on a steel plate in ordinary atmospheric conditions without any particular care being taken to clean the surfaces will produce a value of μ typically close to 0.5. If the experiment is repeated with very clean surfaces (as would occur, for example, in a high vacuum or space environments) the coefficient of friction is observed to be very much greater, perhaps reaching 10 or even more.

To understand why μ takes on a particular value in any given set of circumstances really means being in a position to predict the value of τ in eqn (4.1). Despite much careful experimentation and analysis of this deceptively simple problem there is still no foolproof method of establishing the values of τ and thus μ between two surfaces from a knowledge of their individual mechanical and physical properties and their local surface topographies. In the last resort, the engineering designer can only achieve peace of mind if tests are carried out on specimens of the actual materials it is planned to use under the very same conditions that are likely to be met in practice. This is clearly not always feasible, either physically or economically, and this fact continues to stimulate much work on the underlying and fundamental mechanisms of both friction and wear.

As surfaces slide over one another mechanical energy is dissipated, and so as well as a physical mechanism for the *generation* of the friction force, we also need a convincing mechanism of energy *loss*. It has become conventional to divide the mechanisms of friction which can account for these two features of the problem into two classes, those effects associated with *adhesion* and those involving *deformation*.

4.1.3   *Adhesion*

We have seen that two loaded surfaces touch at a discrete number of contact points and that the real area of contact is very much smaller than the nominal area. The pressure at these points of true contact is therefore correspondingly higher than the nominal value. This implies that the outermost atoms of one surface can be in very close proximity to those of the counterface and this is especially true if the surfaces are relatively clean and uncontaminated by extraneous surface layers. If we wish to separate the surfaces sufficient force must be applied to pull or slide them apart. This argument applies whatever the nature of the materials but is particularly significant in the case of ductile metals. For example, if we take two pieces of gold, typical of a noble metal on which there is no outer oxide film, and press them together they will effectively form one piece as the surfaces become 'cold welded' together. The junction will be comparable in strength to the parent metal and will be capable of exhibiting very high normal and shear strengths. Of course, gold is hardly a common engineering material and the surfaces of more practical components are usually covered by relatively thick films of oxide (if not also of oil and grease) which, even under the very high pressures generated at asperity contacts, much reduce the degree of adhesion and prevent this very dramatic state of affairs (see Section 2.4). However, adhesion can still play a vital part in surface interactions; the development of an explanation of friction in which the forces of adhesion are important owes much to the work of Professors Bowden and Tabor (Bowden and Tabor 1950; 1964; 1973) and such a model is able to explain a remarkable number of experimental and practical observations.

In its simplest form the adhesional theory of friction supposes that because real contact is made only at the tips of the asperities then, even at relatively low loads, the true contact pressures are sufficiently high to bring about local plastic deformation. The load W is related to the true area of contact A by a material property which represents the yield pressure of the softer surface; it will be numerically very close in value to the indentation hardness $\mathcal{H}$. Thus,

$$W \approx A \times \mathcal{H}. \tag{4.3}$$

As a result of this severe, although localized, plastic deformation, strong adhesive junctions are formed and if there is to be tangential motion of the surfaces these must be sheared. The total friction force F is made up of the product of the area A and the specific shear strength which any junction can show. Setting aside any effects of work or strain hardening, it is impossible for any junction to exhibit a shear strength greater than that of the bulk of the weaker material which, in the notation of Section 3.5, was designated k; it thus follows that

$$F = A \times k \tag{4.4}$$

so that

$$\mu = \frac{F}{W} = \frac{k}{\mathfrak{K}}. \tag{4.5}$$

Equation (4.5) is in accord with Amontons' laws of friction, namely that μ is a constant independent of load, and further, through the physical argument that has led up to it, provides an explanation of why the frictional force F is independent of the nominal or apparent area of contact between the two solids. In addition to this, the coefficient of friction has been shown to be associated with the ratio of two strength properties of the weaker solid, k and $\mathfrak{K}$. Although the *absolute* values of these quantities can vary widely from one material to another, for many their ratio is more or less constant and generally close to about 0.2; consequently, this picture further explains why there is a comparatively narrow spread in the numerical values of μ found in practice between dry solid surfaces; see Table 4.1.

However, there are problems with reconciling some of our other observations with this appealingly straightforward picture. For example, equation (4.5) suggests that the value of μ depends only on two (related) properties of the weaker solid and will be essentially independent of the nature of

Table 4.1 Typical friction coefficients of unlubricated materials at low speeds in normal atmospheres against a mild steel counterface. The values are very dependent on the precise conditions of the test and should be taken as being only broadly representative. (Data from Bowden and Tabor (1964).)

Slider	μ
Metals	
mild steel	0.55–0.8
aluminium	0.5
copper	0.8
cadmium	0.4
chromium (hard plate), hard steel	0.4
indium	2
lead	1.2
phosphor bronze	0.3
α-brass (Cu–30 per cent Zn)	0.5
leaded α/β brass (Cu–40 per cent Zn)	0.2
grey cast iron	0.4
white-metal bearing alloys	0.5–0.8
Non-metals	
brake and friction materials	0.4–0.7
carbon-based materials	0.2

the harder counterface. Things are not quite so simple in practice as we commonly observe that the other member of the sliding pair does influence the value of μ. While it is true that the range of values of coefficients of friction found under sliding in normal atmospheric conditions is small, they are often greater than 0.2, and the situation can be very different if the metal surfaces are thoroughly cleaned or if the sliding takes place in a vacuum. Under these circumstances it is almost impossible to slide one metal surface over another and the value of μ can increase by an order of magnitude; the very high frictional forces then observed indicate that the true area of contact is very much greater than this simple theory would predict. To accommodate these observations we need a mechanism by which the islands of true contact at which the overall shear strength is generated can enlarge, that is, one of *junction growth*.

This behaviour can be understood if we remember that individual asperities deform under the combined action of both normal and shear stresses. If the local shear stress increases then the associated normal pressure necessary to maintain plasticity of the junction reduces; if the load stays the same then a reduction in pressure demands an increase in the true area supporting the load so that the junction must 'grow'.

Junction growth

When a single metallic asperity is loaded normally in contact with a hard counterface the mean pressure p_m will be given by the expression

$$p_m = \frac{W}{A},\tag{4.6}$$

where A is the area of contact and W the load. Plastic behaviour of the contact will be initiated at some particular value of p_m which will be related to the material flow stress in shear k by a relation of the form of eqn (3.93), that is

$$p_m = \sqrt{\alpha_2}k,\tag{4.7}$$

where α_2 is a constant whose value depends on the geometry of the contact and the particular yield criterion that we adopt but is likely to be of the order of 30 or so: eqn (3.93) would suggest a value to 32.

Now suppose that a tangential force F is applied to the contact and that this produces a uniform shear stress τ over the interface. Both theory and experiment suggest that the condition for continued plastic flow will have the form

$$p^2 + \alpha_1 \tau^2 = \alpha_2 k^2,\tag{4.8}$$

where p is the required value of the normal pressure and α_1 is a second constant which may not necessarily be numerically equal to α_2. The implication of eqn (4.8) is that the application of τ causes the critical value of pressure

p to fall to below the value of p_m which was required when the contact carried a normal load only. This is in effect the same as saying that if the normal load W remains constant then the maintenance of plasticity allows the area of the interface to 'grow' from its initial value of A to some larger value: this is the phenomenon known as junction growth.

If the junction is formed between relatively clean surfaces then the value of τ may approach k so that it is clear from equation (4.8) that we might require only a very modest value of p to maintain plastic flow; under these high frictional conditions there is likely to be significant junction growth and the coefficient of friction can increase to very large values. Rearranging equation (4.8) we have

$$\mu = \frac{\tau}{p} = \frac{m}{\sqrt{1 - m^2 \alpha_1 / \alpha_2}}, \tag{4.9}$$

in which the ratio τ/k representing the relative strength of the interface has been replaced by the *friction factor*, m. Figure 4.2 shows the effect on μ of the value of m when α_2 is equal to 32 but for the values of the ratio $\alpha_1/\alpha_2 = 1$ and 0.9. The curves show that if we have a strong degree of

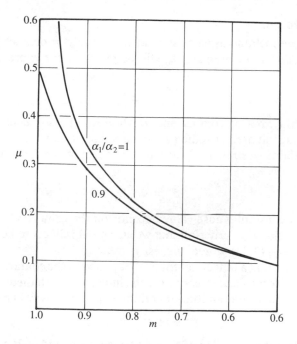

Fig. 4.2 Overall coefficent of friction μ for an asperity junction as a function of the friction factor m. If m is close to unity the value of μ can grow to very large values.

adhesion at the interface (i.e. say $m > 0.95$) then the coefficient of friction can reach very high values; they also illustrate the dramatic reduction in the coefficient of friction associated with relatively small reductions in the interfacial shear strength. For example, if α_1/α_2 is equal to 1, only a 10 per cent reduction in the value of τ/k or m (from unity to 0.9) reduces μ from the very high values associated with clean surfaces to less than 0.4, a value not dissimilar to that of real surfaces.

Figure 4.3 shows some experimental confirmation of this effect: plotted are friction curves obtained in an experiment in which a nickel asperity was slid on a tungsten surface. Under normal atmospheric conditions the coefficient of friction was about 0.4; when the apparatus was evacuated this rose only slightly. However, when the surfaces were degassed (i.e. heated to drive off any loosely held adsorbed contaminants) the resulting clean metal surfaces adhered strongly (i.e. gave high values of the ratio τ/k) and consequently very high values of μ. Admitting only a very minute trace of oxygen would be expected to 'contaminate' the junction by the formation of local patches of oxide and so effectively reduce the value of τ and consequently lower the overall coefficient of friction.

The preceding argument can be used to explain the tribological rule of thumb that it is poor practice to design a contact in which two like materials slide on one another. Once any initially protective low-shear-strength surface film has been displaced there is likely to be extreme junction growth and associated surface damage to both components. When the two surfaces are of different materials it can be argued that the degree of

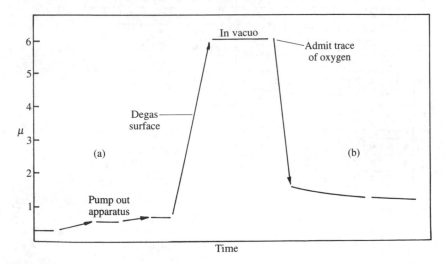

Fig. 4.3 The effect of absorbed films of oxygen and other contaminants on the friction between metal surfaces (from Bowden and Tabor 1950).

potential adhesion in the absence of any contaminating layers is likely to be associated with their mutual solubility in either the solid or liquid phases. The compatibility chart of Table 4.2, derived from the binary phase diagrams of the respective elements, can be used as a first guide to the choice of preferred friction pairs.

4.1.4 Deformation

When we examine a pair of dry sliding surfaces there is often scoring and surface damage on at least one of them, which makes the direction of relative sliding very obvious. The plastic work that has gone into this ploughing deformation must be provided by the agent that is initiating or maintaining tangential motion and so it makes a contribution to the frictional force and so to the coefficient of friction. In the simplest picture of surface interactions this force is simply *added* to any arising from adhesional effects.

A real asperity is likely to have a complex geometry but for the purpose of modelling can be conveniently thought of as having the shape either of a cone (or perhaps a pyramid), Fig. 4.4(a), or a sphere Fig. 4.4(b). In either

Table 4.2 Chart showing the relative mutual solubilities of pairs of pure metals, deduced from their binary phase diagrams (from Rabinowicz 1980).

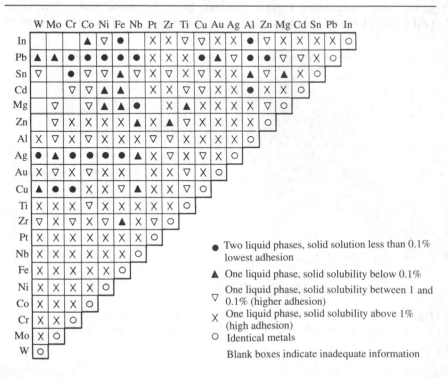

	W	Mo	Cr	Co	Ni	Fe	Nb	Pt	Zr	Ti	Cu	Au	Ag	Al	Zn	Mg	Cd	Sn	Pb	In
In			▲	▽	●			X	X	▽	▽	X	X	●	▽	X	X	X	X	O
Pb	▲	▲	●	●	●	●	●	●	X	X	X	●	▲	▽	●	●	▽	▽	O	
Sn	▽		●	▽	▽	▲	▽	X	▽	X	▽	X	X	▲	▽	▲	X	O		
Cd			▽	▽	▲	▲		X	X	▽	▽	X	X	●	X	X	O			
Mg	▽		▽	▲	▲	●		X	▲	X	X	X	X		▽	O				
Zn	▽		X	X	X	X	▲	X	▲	▽	X	X	X	X	O					
Al	X	▽	X	▽	X	X	X	▽	▽	X	X	X	X	O						
Ag	●	▲	●	●	●	●	▲	X	▽	X	▽	X	O							
Au	X	▽	X	▽	X	X		X	X	▽	X	O								
Cu	▲	●	●	X	X	▽	▲	X	X	▽	O									
Ti	X	X	X	▽	X	X	X	X	X	O										
Zr	▽	X	▽	X	▽	▲	X	▽	O											
Pt	X	X	X	X	X	X	X	O												
Nb	X	X	X	X	X	X	O													
Fe	X	X	X	X	X	O														
Ni	X	X	X	X	O															
Co	X	X	X	O																
Cr	X	X	O																	
Mo	X	O																		
W	O																			

● Two liquid phases, solid solution less than 0.1% lowest adhesion

▲ One liquid phase, solid solubility below 0.1%

▽ One liquid phase, solid solubility between 1 and 0.1% (higher adhesion)

X One liquid phase, solid solubility above 1% (high adhesion)

O Identical metals

Blank boxes indicate inadequate information

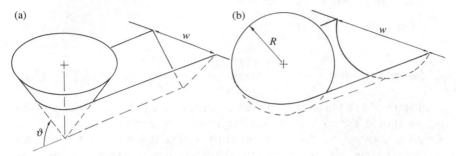

Fig. 4.4 A surface asperity modelled as (a) a cone, and (b) a sphere.

case a groove will be swept out as the asperity is dragged over the surface. To a first approximation we could imagine that the tangential force F required to do this is represented by the product of the cross-sectional area of the groove A' and some pressure p' which is that needed to displace material in the surface. Thus

$$F = A' \times p'. \tag{4.10}$$

In the case of a cone A' is equal to $w^2 \times \tan\vartheta$, where w is the width of the groove and the angle ϑ represents the slope or 'roughness' of the asperity; ϑ is the complement of the semi-angle of the cone. We can thus write, for a cone, that

$$F = \frac{w^2}{\cot\vartheta} \times p', \tag{4.11}$$

while for a spherical asperity the corresponding expression is

$$F = R^2 \left\{ \arcsin\frac{w}{2R} - \frac{w}{2R}\sqrt{1 - \frac{w^2}{4R^2}} \right\} \times p'. \tag{4.12}$$

The *depth* of penetration, and thus the area supporting the normal load W, will be related to the penetration hardness $\mathcal{H}$ of the substrate. Since, once motion starts, only the front half of the indenter actually supports the load, we can write that

$$\text{for a cone} \quad W = \frac{\pi w^2}{2}\cot^2\vartheta \times \mathcal{H} \tag{4.13}$$

$$\text{for a sphere} \quad W = \frac{\pi w^2}{4} \times \mathcal{H}. \tag{4.14}$$

Thus, by dividing equation (4.11) by (4.13) or (4.12) by (4.14) we can obtain expressions for the coefficient of ploughing friction associated with either cones or spheres, namely for cones $\mu = \dfrac{2}{\pi} \tan\vartheta \times \dfrac{p'}{\mathcal{H}}$ (4.15)

for spheres $\mu = \dfrac{2}{\pi} \left\{ \left(\dfrac{2R}{w} \right)^2 \arcsin \dfrac{w}{2R} - \sqrt{\left(\dfrac{2R}{w} \right)^2 - 1} \right\} \times \dfrac{p'}{\mathcal{H}}$. (4.16)

In spite of the uncertainty of the exact numerical value of the pressure p' we should expect it to be similar to the indentation hardness. If for simplicity, we assume that p' is equal to $\mathcal{H}$ then it is clear that in the case of conical asperities the coefficient of friction depends solely on the slope of the asperity ϑ and is not changed by the scale of the indentation; the relationship between μ and ϑ is shown graphically in Fig. 4.5(a). Examination of real engineering surfaces (see Chapter 2) suggests that even for quite rough surfaces the effective value of the angle ϑ is likely to be small, that is, less than 10°, and thus we can see from the figure that the contribution of the ploughing frictional term is going to be similarly comparatively small, introducing an additional coefficient of friction, over that due to adhesion, of less than 0.1.

On the other hand, in the case of a spherical asperity the value of the coefficient of friction depends on the term $w/2R$ and so changes as the sphere digs deeper into the deformed surface. Figure 4.5(b) shows that if $w/2R > 0.2$ then ploughing makes a significant contribution to friction. This ploughing contribution can be especially important when the contact is contaminated by free particles: these might be the debris removed in previous encounters or contamination introduced into the contact area from the outside environment. Such particulate material is likely to be both larger and more angular than the surface roughnesses on the two opposing surfaces and therefore in a position to make a more significant contribution to the frictional resistance to motion. The coefficient of two metals sliding together can sometimes be almost halved by ensuring that the wear debris that is formed is removed as soon as possible and not allowed to become trapped in the interface.

If the dimension R is very much greater than the width of the track w, then eqn (4.16) reduces to

$$\mu \approx \frac{4}{3\pi R} \sqrt{\frac{W}{\pi \mathcal{H}}}$$ (4.17)

so that for an asperity of a given size R carrying a fixed normal load W the contribution to the overall coefficient of friction due to ploughing is proportional to $1/\sqrt{\mathcal{H}}$ and so will be of increasing importance with softer materials.

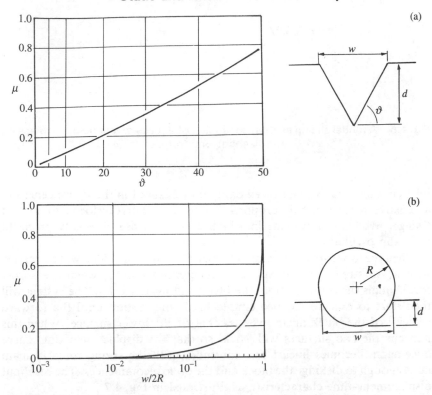

Fig. 4.5 The coefficient of ploughing friction versus (a) ϑ the roughness angle for conical indenters, and (b) the ratio of groove width w to asperity diameter $2R$ for spheres.

4.2 Static and kinetic friction: stick–slip effects

If we carry out in practice the simple demonstrations of the laws of friction illustrated in Fig. 4.1 then another characteristic of virtually all dry sliding frictional contacts becomes apparent: the frictional force required to *initiate* motion is more than that needed to maintain the surfaces in the subsequent relative sliding. In other words the *coefficient of static friction* μ_s is greater than that of *kinetic friction* μ_k; sometimes this is known as the third of Amontons' laws of friction. This feature, together with the inevitable natural elasticity of any mechanical system, can often lead to the troublesome phenomenon of *stick–slip motion*.

Figure 4.6 represents, in its simplest form, the essential mechanical features of the sliding system: the spring of stiffness k allows for the natural elasticity or compliance of the loading mechanism, while λ models its

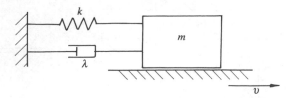

Fig. 4.6   A model sliding system consisting of a mass m, spring of stiffness m, and dashpot with rate λ.

inherent damping. The idealized damper or dashpot in the figure generates a resistive force which is proportional to the relative velocity of its end fixings. Mechanical systems in which λ is small are especially prone to stick–slip oscillation.

When the lower specimen moves from left to right the block will be carried with it, so stretching the 'spring'; this continues until, when the spring force equals the static friction, the block will begin to slip. The system will then start to execute damped simple harmonic motion until the forward velocity of the block again matches that of the lower surface. When this happens the two surfaces will 'stick' so that the displacement time curve once again becomes linear; this continues until the spring force is again large enough to dislodge the block and the cycle recommences. The resultant displacement–time characteristic is illustrated in Fig. 4.7.

These fluctuations can be a serious problem in practical situations and measures are usually taken to eliminate or suppress them. One philosophy is to choose the friction pair so that μ_s is no greater than μ_k; the only really practical way of doing this is by covering one of the component surfaces by a soft boundary lubricant film (see Chapter 9). The disadvantage of this method is that this applied surface film has a finite life and will be gradually worn away so that there is always the danger that stick–slip motion will be once again initiated. Alternatively, we can design the mechanical system so that the amplitude of any such oscillations would be very small; this can be done by reducing the system compliance (i.e. making the spring as stiff as possible) and/or increasing both the component inertias and the system damping. Experience has shown that if the potential stick–slip amplitude is less than about 10^{-2} mm (which is typical of the dimensions of a single contact spot on engineering surfaces) then this form of troublesome instability may be effectively eliminated, albeit at the cost of making the assembly generally more massive, and thus more costly. The full description and analysis of stick–slip or dynamic friction phenomena actually pose severe practical and computational difficulties; see, for example, Oden and Martins (1985).

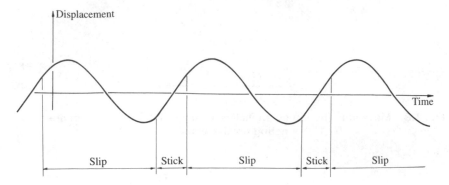

Fig. 4.7 The displacement–time curve associated with stick–slip motion. Displacement increases linearly with time during periods of 'stick'; when 'slipping' the mass performs simple harmonic motion.

4.3 Measurement of friction

The engineering designer will often want some guidance on the probable frictional forces between loaded machine components and turn for practical advice to the tribologist; he or she will often rely on experimentally obtained data. The essential feature of any device designed to measure frictional interaction is a means of applying a known normal load between the two test surfaces which simultaneously carry a measurable tangential force; it must be possible to increase this force until either detectable, or steady, relative motion is achieved. Many ingenious set-ups have been used to investigate particular circumstances, for example, those at extremes of temperature or environmental conditions. The simplest method is the inclined plane test illustrated in Fig. 4.8. Here one of the materials is in the form of a flat plate on which rests a block of the second. If the inclination of the plate to the horizontal at which slip starts is f then

$$\mu_s = \tan f. \qquad (4.18)$$

For evaluating friction during continuous sliding it is usual to measure the frictional force by a suitable dynamometer (i.e. a very stiff calibrated spring) although the normal load is often still most conveniently provided by dead-weight loading. Probably the most common form of friction-measuring rig used in tribology laboratories, which can also be used to measure the associated degree of wear and surface damage, is one based on the pin-on-disc configuration illustrated in Fig. 4.9(a).

In its simplest form the pin is held stationary while the disc rotates beneath it, so that any particular point on the resultant wear track on the disc comes

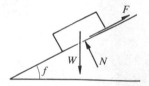

Fig. 4.8 Measuring friction by an inclined plane. The forces shown are those acting on the block.

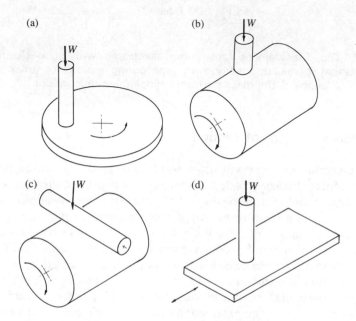

Fig. 4.9 Friction-measuring devices: (a) pin-on-disc; (b) pin-on-cylinder; (c) crossed cylinders; and (d) reciprocating rig.

into contact with the pin over and over again—a *multiple-pass* arrangement. In a rather more sophisticated apparatus the pin may be arranged to move radially during the course of the experiment so that it continuously encounters fresh disc surface. This arrangement provides *single-pass* conditions; however, the speed of the drive to the disc must also be continuously varied if a steady sliding speed is to be maintained. Alternative standard arrangements are of pin-on-cylinder and crossed cylinders as illustrated in Fig. 4.9(b) and (c). If the pin is given a spherically profiled tip then the contact conditions will be *non-conformal* (they are inherently so in a crossed-cylinder machine). On first loading contact is made at a single point so that the corresponding pressure is extremely high; naturally, with time a small wear flat or scar develops which reduces this pressure. Non-conformal contact geometry can be used either to simulate conditions in very heavily loaded contacts (such as gear teeth), or provide accelerated tests on the friction and wear of a number of candidate materials for a particular application. If the profiles of the two rubbing surfaces are carefully matched before the experiment is initiated, so that the contact is *conformal*, then the test can be used to simulate the behaviour of such components as plane or thrust bearings, face seals, or even clutches and brakes.

The importance of environmental effects on friction has already been noted. Some friction pairs are particularly susceptible to the presence or absence of oxygen, water vapour, or other gases and vapours, and so for scientific investigations the friction test may need to be carried out in an enclosed environmental chamber. Space environments can be simulated by running under high-vacuum conditions. However, if pin-on-disc tests are run in the presence of a liquid lubricant, measurements of friction and wear are liable to be very variable and exhibit considerable experimental scatter. The explanation for this lies in the rather poorly controlled hydrodynamic conditions at the interface between the two components. Very small deformations of the apparatus, due to such effects as pressure or thermal loading, may be sufficient to generate varying amounts of hydrodynamic lift and so give large fluctuations in μ. For the same reason care must be exercised when using a reciprocating rig, Fig. 4.9(d), in the presence of a lubricant when the degree of hydrodynamic lift will vary with the velocity of sliding. Many other forms of friction-measuring apparatus have been developed for specialized application involving either different forms of relative motion or environmental conditions; lubricated tests are rather more often carried out either on a disc machine or one of the commercial instruments described in Section 9.5.

4.4 Friction of non-metallic materials

The friction of non-metallic materials can be viewed in the same light as that of metals; again it can be ascribed to two sources: adhesion and deformation. However, because of the fundamental differences between both the structure and the properties of metals and non-metals, we need to examine once again the micromechanics that are likely to be involved in surface interactions.

4.4.1 *Adhesion effects*

At the most fundamental level, the net force at the interface between two solid materials distance h apart depends on the competing effects of the local forces of attraction and repulsion between individual atoms or molecules in the two contacting surfaces. A pair of *perfectly smooth* surfaces brought together would have an equilibrium separation of h_0 at which the forces of attraction just balance those of repulsion. At this very fine atomic or molecular scale, the variation of the force per unit area, that is, the pressure, $p(h)$, as a function of separation h can be expressed by the relation

$$p(h) = - \mathcal{A}h^{-n} + \mathcal{B}h^{-m}, \qquad (4.19)$$

where $\mathcal{A}$, $\mathcal{B}$, m, and n are material constants, and $m > n$.

When h is equal to h_0 then $p(h) = 0$; if $h > h_0$ then the solids attract each other, while they repel if h is less then h_0. Although these forces can be large, their range of action is exceedingly small. It is usual to express their effect, and thus the potential adhesive strength of the junction, as the work required to separate the surfaces, that is, to change their separation from h_0 to $h \rightarrow \infty$, and then to ascribe a surface energy γ to each newly created surface. If the solids are dissimilar then the 'pull-off' work $\Delta \gamma$ required to separate the junction becomes

$$\Delta \gamma = \gamma_1 + \gamma_2 - \gamma_{12}, \qquad (4.20)$$

where γ_1 and γ_2 refer to the intrinsic surface energies of the two solids and γ_{12} refers to that of the interface. Numerical values of surface energies can be found by quite independent experimental techniques, but when these are substituted into an expression such as (4.20) the theoretical pull-off term is found to be very much greater than that usually observed; this remains true even when great care is taken in the experiments to remove any contaminating surface films. The explanation for this general lack of adhesional effects involves the scale of the roughnesses present on real surfaces; in general, asperity heights are likely to be very large compared with the distances over which these forces act. Adhesive junctions formed on loading between the lower asperities are pushed apart on removing the load by the compression effects at the higher asperity pairs. If this 'jacking mechanism' is to some extent suppressed then adhesion effects can be significant; for example, almost atomically flat surfaces can be produced on specimens of carefully cleaved mica and, when these are brought together, real and apparent areas become almost the same. In these circumstances measurable amounts of adhesion are observed. At the other end of the hardness range, very-low-modulus materials, such as both natural and synthetic rubbers, can deform sufficiently to accommodate moderate amounts of roughness on the counterface. Again under pressure real and apparent contact areas become similar and once again significant amounts of adhesion are observed.

We can analyse the problem following the argument of Johnson, Kendall and Roberts (1971), by considering an axisymmetric point contact in much the same way as a classical Hertzian contact but additionally allow for surface adhesion effects. The surface deformations, $w_{z1} + w_{z2}$, must be such as to satisfy the compatibility condition of eqn (3.10), that is,

$$w_{z1} + w_{z2} = \Delta - \frac{r^2}{2R} \qquad \text{for } r^2 \leq a^2, \tag{3.10}$$

where Δ is the approach of distant points in the bodies. We have seen that the Hertz semi-elliptical pressure distribution (eqn (3.52)) is an acceptable solution to this elasticity problem, that is, that

$$p(r) = p_0\{1 - r^2/a^2\}^{1/2}. \tag{3.52 bis}$$

However, in addition, to account for the effects of adhesion pulling the surfaces together we could add to this load some multiple of the *inverse* semi-elliptical distribution (eqn (3.60)) which gives a constant normal surface displacement, that is,

$$p(r) = p_0'\{1 - r^2/a^2\}^{-1/2}. \tag{3.60 bis}$$

The values of the constants p_0 and p_0' can be related to the applied normal load W and the size of the contact spot a by the relation

$$W = \int_0^a 2\pi r\{p(r) + p'(r)\}\mathrm{d}r.$$

Using equations (3.53) and (3.65) we can then write that

$$W = \frac{2}{3} p_0 \pi a^2 + 2p_0' \pi a^2. \tag{4.21}$$

Applying the condition that the total energy associated with the system (that is, the energy associated both with elastic straining and surface energy) must be a minimum under equilibrium conditions, leads to the further conditions that

$$p_0 = \frac{2aE^*}{\pi R} \qquad \text{and} \qquad p_0' = 2\sqrt{\frac{\gamma E^*}{\pi a}}, \tag{4.22}$$

where γ is the surface energy per unit area of each surface. It follows by substitution in eqn (4.21) that we can write

$$W - \frac{4E^*a^3}{3R} = -4\sqrt{\pi E^* \gamma a^3}. \tag{4.23}$$

When the two bodies carry a normal *compressive* load, local adhesive forces pull the surfaces together so that the contact area is greater than would be predicted by simple Hertz analysis; note that eqn (4.23) reduces to the Hertz eqn (3.55) when $\gamma \to 0$. Reducing the load to zero leaves the surfaces adhering together with a residual contact radius. If a *tensile* pull-off load is applied then the contact shrinks; when $\mathrm{d}W/\mathrm{d}a = 0$ the situation becomes unstable and the surfaces separate. If a_B then represents the size of the contact spot at this condition and W_B the tensile *adhesion* or *pull-off load*, then a_B is given by

$$a_B = \left\{ \frac{9\pi\gamma R^2}{4E^*} \right\}^{1/3}, \tag{4.24}$$

and

$$W_B = 3\pi\gamma R. \tag{4.25}$$

Equation (4.23) can then be conveniently rewritten as

$$\frac{W}{W_B} - \left\{ \frac{a}{a_B} \right\}^3 = -2 \left\{ \frac{a}{a_B} \right\}^{3/2} \tag{4.26a}$$

and this relation between contact radius and load is plotted in Fig. 4.10. Point B represents the pull-off conditions. Also shown in the figure for comparison as the chained line is the load curve for Hertzian point contact which, in this nomenclature, has the equation

$$\frac{a}{a_B} = \left\{ \frac{W}{W_B} \right\}^{1/3}. \tag{4.26b}$$

Confirmation of the essential validity of this approach has been provided not only by experimental observation on soft materials but also by recent powerful computer simulations of individual asperity contacts involving large, but finite, numbers of interacting atoms.

4.4.2 Friction of rubbers and elastomers

When a rigid counterface with a smooth surface and a large radius of curvature is traversed over a rubber surface the adhesional effects described above can play an important role in the generation of the frictional resistance to motion. Relative motion at the interface is then often principally due to 'waves of detachment' which flow across the contact patch from the leading edge. These waves, sometimes known after their first observer as *Schallamach waves*, originate from a form of elastic instability associated with the zone of compressive stress adjacent to the leading edge of the slider (Schallamach, 1957 and 1971). Rather than gross sliding at the interface the surface displacements move in folds or buckles across the contact. Before the rubber can buckle it must first be peeled from the hard track and the energy required to bring about this separation at the interface accounts for

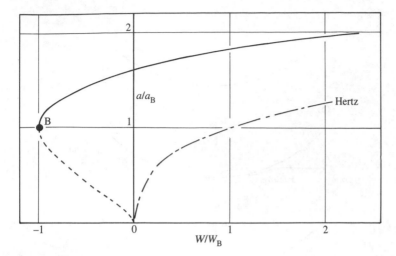

Fig. 4.10   Variation of contact radius *a* with load *W* for a single point contact between surfaces showing significant effects of surface adhesion. a_B is the contact spot size at which conditions just become unstable under a tensile load W_B. The behaviour of a Hertzian contact which exhibits no adhesive effects is shown as the chained line (from Johnson *et al.* 1971).

much of the frictional resistance. This mechanism is illustrated in Fig. 4.11 at the right-hand or leading edge. Simultaneously with the generation of Schallamach waves at the leading edge of the contact, further peeling apart of the interface can occur at the trailing edge; this too is illustrated in Fig. 4.11. This phenomenon is rather akin to interfacial crack propagation, as indicated in Fig. 4.11(b). Local recovery and slip can lead to reattachment of the rubber at the outermost edge of the contact, Fig. 4.11(c), and this is followed by the growth of 'healing cracks' as the surfaces become reattached, Fig. 4.11(d). The cyclic process can then be recommenced when the local stresses become high enough to once again propagate the initial interface crack. This phenomenon leads to significant variations in the value of the frictional force *F* with time (or displacement). The overall coefficient of friction in such dry-sliding situations depends to some extent on the nature of the rubber but can be very high, that is, in excess of 2. Adhesion between the rubber surface and the counterface leads to the generation of tensile stresses in the free surface behind the slider and these may be enough to rupture the rubber into an array of tears at right angles to the direction of sliding.

Both natural and synthetic rubbers consist of long molecular chains principally of carbon and hydrogen atoms. These are cross-linked by the introduction of sulphur atoms during the vulcanizing process in which the

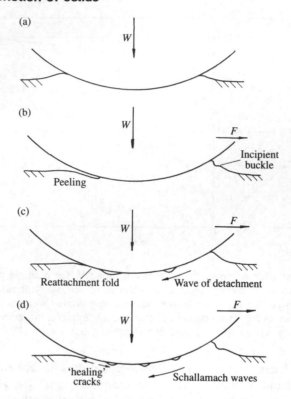

Fig. 4.11 Frictional effects between a rubber surface and a hard smooth counterface: (a) normal loading only; (b) normal and tangential loading, buckling at the leading edge, and 'crack propagation' at the trailing edge when force F is applied to the upper counterface; (c) and (d) Schallamach waves of detachment propagate from the leading edge, waves of reattachment occur at the trailing edge (from Barquins 1984).

viscous latex fluid is converted into a solid material. In a specimen of rubber in its natural state the molecular chains are tangled and coiled together in a random fashion. Under stress the chains become more nearly parallel and aligned, and the material may begin, as a result of this regularity of structure, to exhibit some almost crystalline features of behaviour. When the stress is removed this crystallinity is lost and the chains slip back to their original chaotic arrangement. If the molecules are comparatively 'smooth' then this recovery can be achieved with relatively little loss of energy so that the rubber shows a high degree of resilience. On the other hand, some rubbers have a molecular structure with large side groups attached to the main chain. During stretching and recovery work must be done in dragging these groups over one another so that significant energy is lost over each cycle of loading and unloading. Such a rubber will have a low resilience but a high hysteresis. In practice, most rubbers used in engineering components also contain a good deal of 'filler', usually consisting of fine particles of carbon. The presence of such a filler increases the hardness of the rubber and its resistance to wear and usually its hysteretic losses. When the term 'hardness' is used with respect to an elastomer it implies something rather different from its meaning with respect to plastically deforming metals. An elastomer, suffers no permanent plastic indentation and thus conventional hardness is an inappropriate idea to apply to its behaviour. (The hardness of an elastomer is usually measured by a falling weight instrument, such as the Shore scleroscope, in which the rebound height from the surface of a small falling mass is measured.)

If the radius of curvature of the slider becomes small, so that it becomes almost needle-like, then the surface interaction is rather different. Such an asperity can penetrate comparatively deeply into the rubber and although, as the asperity moves over the surface of the rubber, the stress concentration is greatest on its front face, no cutting occurs here because failure is prevented by adhesional effects. The rubber tears instead where it first loses contact with the asperity, the tears proceeding more or less at right angles to the direction of maximum stress along the dotted lines shown in Fig. 4.12(a). When the asperity has passed, the rubber relaxes and the region around the tears turns back, through almost 90°, and thus gives rise to a torn groove in the direction of sliding. A small lip is often formed above the original specimen surface which may become detached so contributing to material wear.

One standard test for rubbers and elastomers (ASTM D-2228 Rubber Abrasion Resistance) involves a wheel of the test material being pressed into radial contact against a sharp loaded blade, as shown in Fig. 4.12(b). With time, transverse ridges are formed on the worn rubber surface whose spacing is characteristic of the physical properties (principally the hardness) of the elastomer; the worn surface is very different from that which would be formed on a metal or ceramic wheel where the ridges would form parallel to the sliding direction. A very similar pattern of abrasion is found both on elastomer surfaces worn against experimental arrays of asperities as well as in real situations such as heavily loaded vehicle tyres driven over concrete road surfaces (Moore 1972).

4.4.3 *Friction of polymers*

In contrast to metallic materials, which are essentially incompressible in the plastic state, many polymers have relatively low bulk moduli and increase their density by a significant factor under the action of hydrostatic pressure.

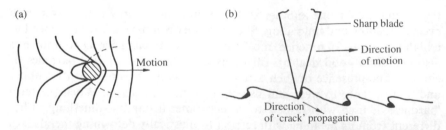

Fig. 4.12   (a) Tearing of a rubber surface by a sharp asperity (shown hatched); as the asperity moves through the surface in the direction indicated the rubber tears along the dotted line. (b) Friction and wear test using a transverse blade in an elastomer leave a series of transverse ridges.

This change in intermolecular spacing can lead to two tribologically significant effects in considering a single heavily loaded asperity contact. The first of these arises from the fact that the material shear stress k becomes a function of the local normal pressure p. In particular, experimental evidence suggests that to a satisfactory degree of modelling k and p are related by an equation of the form

$$k = k_0 + \alpha p \qquad (4.27)$$

where k_0 and α are material constants. The second effect that this change in density can have is to make the contact area A a non-linear function of the load W; again, empirically, it has been found that

$$A \propto W^{2/n} \qquad \text{where} \qquad 2 < n < 3. \qquad (4.28)$$

Since the friction traction F is equal $k \times A$ it follows that

$$\mu = \frac{F}{W} = k_0 W^{(2-n)/n} + \alpha. \qquad (4.29)$$

If $n > 2$ then this relationship leads us to expect a decrease in μ as the load W is made larger. Although in the bulk this effect is usually small, it is thought that it can be of rather more significance when polymers are applied as thin protective or boundary lubricant layers. Figure 4.13 illustrates the effect in the case of contact between nylon and steel specimens. For this material pair, the coefficient of friction first reaches a maximum with increasing load before subsequently falling away.

The critical load, that is, that producing the greatest frictional force, decreases with increasing sliding speed and is associated with the *glass transition* of this particular polymer.

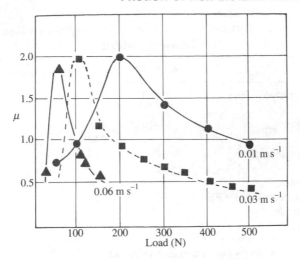

Fig. 4.13 Relationship between the coefficient of friction μ and the contact load W for three sliding speeds for nylon on steel, (from Yamaguchi 1990).

The mechanical behaviour of polymers is critically dependent on temperature: at low temperatures they behave in an essentially brittle manner, showing little elongation before fracture which is often associated with the existence of a degree of crystallinity. As the temperature is raised, polymers do not in general melt, or undergo a distinct phase change as in the case of metals, but rather soften over what can sometimes be a comparatively broad range of temperatures. They lose any crystallinity they previously may have possessed and become amorphous and glassy; the temperature required to bring about this change in structure and behaviour is known as the glass transition temperature. The data of Fig. 4.13 were obtained at ambient temperature; if, for a given polymer, plots of μ versus sliding speed at various temperatures are made then such curves are often seen to be segments of a single 'master curve', which describes the coefficient of friction as a function of the sliding velocity at a selected reference temperature. The master curve is derived from the experimental results by translating the curve obtained at temperature Θ horizontally by an amount, usually written as $\log_{10}\alpha_T$, so that they fit the reference curve at the chosen reference temperature. Over a limited temperature range, from the glass transition temperature Θ_g to approximately $\Theta_g + 100°C$, it has been found that the data for a range of polymers can be fitted by the expression

$$\log_{10}\alpha_T = \frac{-17.4(\Theta - \Theta_g)}{51.6 + (\Theta - \Theta_g)} . \qquad (4.30)$$

Table 4.3 Critical temperatures for polymers (data from Tanner (1985)).

Polymer	Θ_g	Θ_m (melting temperature)	Θ_u
High-density polyethylene	−110	134	404[†]
Polystyrene			
Polypropylene	−10	165	387[†]
Nylon 6–6	50	240	
Polymethyl methacrylate			
PMMA	90–100		
Natural rubber	−73	28	
Polytetrafluoroethylene PTFE	−150	327	400[*]
Polyvinylchloride	87		130
Polycarbonate	150		

[†]Temperature at which polymer loses half its mass in 30 mins.
[*]Degrades to a monomer above this temperature.

Equation (4.30) is known as the Williams–Landel–Ferry shift equation (see Ferry, 1981) and was originally formulated to deal with the creep and stress-relaxation behaviour of visco-elastic polymers exposed to varying temperatures and rates of loading. At first it was thought that the numerical values were *universal* constants applicable to all polymers but more recent work has shown that a better fit is obtained when the constants are specially chosen for each. A selection of these constants for a number of polymers is shown in Table 4.3: Θ_u represents an upper temperature above which the polymer should not be heated.

4.4.4 Friction of ceramics and cermets

Ceramics

The elements of the periodic table are conventionally divided into two categories: metals and non-metals. Metallic elements are those whose atoms easily lose their outer electrons and become positive ions. They are characterized by being good conductors of heat and electricity and are generally relatively chemically active. Atoms of non-metals hold their outer electrons more firmly and will, in fact, accept or share additional ones. Ceramic materials are generally considered to be those compounds composed of a combination of a metal and a non-metal, for example, alumina Al_2O_3 or magnesia MgO. These typical ceramics are both chemically simple and have a regular crystallographic structure. More complicated ceramics include clays (which when fired have important engineering applications as electrical insulators) and the complex oxides known as *spinels*. Amorphous materials, that is, those not possessing a regular or repeated crystallographic structure, such as glasses, are also included in the family of ceramics. Some elements such as silicon, Si, can exhibit both metallic and non-metallic behaviour; in quartz, SiO_2, silicon behaves as a true metal, its behaviour is rather less fundamentally metallic in silicon carbide SiC or silicon nitride Si_3N_4, and is distinctly non-metallic in such compounds as $MoSi_2$: the first two of these compounds would certainly be classified as ceramics.

Ceramics are generally more stable both thermally and chemically than metals and also much harder. Practical polycrystalline *engineering ceramics* are usually much stronger in compression than tension (in contrast to metals which have comparable tensile and compressive strengths) largely because of the great number of inclusions and microcracks which they inevitably contain and which act as local very severe stress raisers. It is principally the presence of these imperfections that limits their structural application as their presence can reduce toughness to such values that ceramic components have an unacceptably high chance of sudden, catastrophic failure under an unforeseen impact or step change in load. The ceramics of potentially greatest tribological interest are probably those based on alumina Al_2O_3, silicon nitride Si_3N_4, silicon carbide SiC, and zirconia ZrO_2; see for example Kato (1990) and Table 4.4. The raw material from which such components are manufactured is invariably a finely graded powder (usually a mixture of several ceramic phases) with a characteristic particle size of less than a micron. The production route normally involves isostatic compaction followed by sintering, or some other heat treatment, at temperatures in the range of 1200–1800°C. For many applications alumina-based ceramics, of which there are a large number of commercial grades, are more than adequate in terms of physical properties. Their deficiencies when compared with the very best ceramics include a relatively poor resistance to thermal shock and relatively limited mechanical strength or thermal stability at very high temperatures. There is currently considerable interest in the toughening processes which can be exploited with zirconia ceramics. While a *fully* stabilized zirconia has few property advantages over alumina (and many disadvantages) partial stabilization followed by a heat treatment (to precipitate a metastable tetragonal phase in the basic cubic matrix) to form *partially stabilized zirconia*, PSZ, can lead to a marked improvement in strength and particularly toughness values.

Silicon nitride is a covalent solid and cannot readily be sintered (it decomposes rather than melts at about 2800°C); however there are several means of producing useful engineering components from this material. Reaction-bonded silicon nitride, RBSN, components are made by first forming a preform of silicon powder which is then heated in a nitrogen-rich atmosphere; subsequent heat treatments can reduce the porosity and improve the mechanical properties. Simple shapes can be formed by hot pressing silicon

Table 4.4 Mechanical properties of some fine-grained engineering ceramics.

Material	Elastic modulus (GPa)	HV Vickers hardness (GPa)	Fracture toughness K_{IC} (MPa m$^{0.5}$)
Alumina Al_2O_3	~390	~22	~5
Silicon nitride Si_3N_4	210	10(RBSN)–18(HPSN)	~5
Silicon carbide SiC	480	~28	~3
Zirconia (PSZ) ZrO_2	~160	~12	up to 10

nitride powder but a flux such as magnesia (MgO) or yttria Y_2O_3 must be incorporated. The resultant hot-pressed silicon nitride HPSN has good mechanical properties but is difficult to machine and the introduced sintering aids cause a fall-off in properties above 1000°C. *Sialons* are based on silicon nitride in which partial substitution by aluminium and oxygen for the silicon and nitrogen has occurred. This produces a family of ceramic alloys which can be tailored to suit particular applications. Silicon carbide, SiC, is available in both reaction-bonded and pure sintered forms. Its high hardness can, in appropriate circumstances, lead to its being a particularly attractive tribological material. It is widely used as one of the rubbing elements in mechanical seals (see Section 10.5.2) and as a sleeve material in the plain hydrodynamic bearings incorporated in hermetically sealed pumps; these use the service fluid as the lubricant which is sometimes contaminated by considerable quantities of abrasive particles.

Fine-grained ceramics also have great potential as tribological materials in rolling bearings, being both less sensitive to both elevated temperatures and particulate contamination. In addition, because of their higher elastic moduli, typically at least 50 per cent greater than steels, rolling bearings utilizing ceramic elements or races possess improved stiffness.

Table 4.5 shows some values of friction and wear coefficients observed in the unlubricated sliding of pins of these ceramics against discs of the same material. The values of the friction coefficients for alumina, silicon nitride, silicon carbide, and zirconia in air are within the range 0.44–0.9; this is very close to the range of oxidized metals in air (Table 4.1). However, there is a substantial difference between these two classes of materials under high-vacuum conditions when values of μ for the ceramics are not greatly increased as opposed to the enormous growth in frictional resistance that is typical of more ductile metals. Virtually no junction growth can occur at the friction conjunction of two ceramics, so that although the interface may be weakened by the presence of surface layers or contaminants, the effect is much less dramatic than with metallic materials. When a ceramic pin is run against a metallic counterface both adhesive transfer and the generation of wear particles make the understanding of the interaction particularly difficult.

The friction of ceramics can be substantially reduced from the values observed in dry air by the formation of surface films generated by some form of tribo-chemical reaction. The presence of water or water vapour may often reduce the coefficient of friction but can have conflicting effects on the wear rate; for example, the wear rate of alumina is reduced but that of zirconia is increased. Because of the brittleness of ceramics, various types of cracks are generated by friction in the vicinity of the friction track. Figure 4.14 shows the form of the cracks that can be left in a ceramic surface after the passage of a loaded indenter; as the indenter passes over the surface these

Table 4.5 Friction coefficient μ and wear coefficient k of some ceramic materials (data from: Kato (1990)).

Ceramic	Vacuum μ	Vacuum k $mm^3 N^{-1} m^{-1}$	Air μ	Air k $mm^3 N^{-1} m^{-1}$	V (velocity), p (pressure), R (radius)
Alumina	0.98	1.17×10^{-4}	0.7	4×10^{-7}	$V = 300$ mm s^{-1}, $R = 6.4$ mm $W = 9$ N
			0.5	8×10^{-6}	$V = 400$ mm s^{-1}, $R = 4$ mm $W = 10$ N
Silicon carbide	0.84	4.9×10^{-5}			$V = 300$ mm s^{-1}, $R = 6.4$ mm $W = 9$ N
			0.8	1.5×10^{-6}	$V = 20$ mm s^{-1}, flat $W = 40$ N, $p = 0.1$ MPa
PSZ			0.9	3×10^{-4}	$V = 20$ mm s^{-1}, flat $W = 40$ N, $p = 0.1$ MPa
			0.6	7×10^{-4}	$V = 400$ mm s^{-1}, $R = 4$ mm $W = 10$ N
Silicon nitride	0.85	15×10^{-4}	0.75	1.3×10^{-3}	$V = 1$ mm s^{-1}, $R = 3$ mm $W = 10$ N, $p = 0.6$ GPa
	0.68	6.5×10^{-5}	0.44	3×10^{-6}	$V = 150$ mm s^{-1}, flat $W = 10$ N, $p = 3.1$ MPa
			0.8	1.5×10^{-6}	$V = 20$ mm s^{-1}, flat $W = 40$ N,

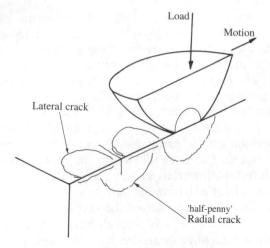

Fig. 4.14   The pattern of cracks induced by friction in brittle materials (from Blau and Lawn 1984).

can grow both during the loading and unloading phases experienced by the surface and near-surface material elements. Radial cracks are oriented normal to the specimen surface, on so-called *median* planes, and have a characteristic *half-penny* shape with their centres on the track of the asperity. A second set of cracks, called *lateral* cracks, often form subsurface saucer-shaped defects, which meet the surface some way from the friction track itself. The production of these cracks is important since it is their nucleation and growth that gives rise to the formation of wear fragments (see also Fig. 5.11).

Cermets

A cermet is a heterogeneous body composed of two or more intimately mixed but separate phases, at least one of which is a ceramic, and the other a metal which is generally known as the *binder*. These materials are also somewhat misleadingly known as *hard metals*. The aim of such a mixture is to combine the strength and toughness of the metal binder or matrix with the hardness and thermal resistance of the ceramic. Cermets are formed by mixing, pressing, and sintering the starting phases which are in the form of fine powders. The member of this family of materials with by far the most widespread application is that consisting of tungsten carbide incorporated in a cobalt matrix. Other, less common, cermets can be formed between the carbides of titanium, chromium, tantalum, and niobium (either individually or mixed) in a suitable binder (usually cobalt, although chromium, nickel, or nickel-molybdenum have been used). The properties of cermets are affected by the composition and grain size of the hard constituent as well as by the composition and percentage of the binder: the higher the proportion of binder the less brittle or tougher is the material. On the other hand, increasing the proportion of binder will generally decrease the hardness, density, modulus, and wear resistance of the finished product so that there is a trade-off between hardness (i.e. wear resistance) and toughness or resistance to impact damage. Tungsten-based carbides can be used in oxidizing conditions up to about 550°C and in inert atmosphere up to c. 850°C. However, care must be taken over the choice of the counterface material as with ferrous materials at high temperatures (or sliding speeds) there can be considerable interfacial diffusion leading to rapid loss of material and surface disintegration of the cermet. Either titanium-based carbides, or a coated grade (i.e. a WC-cobalt-based material carrying a thin inert coating, often of alumina or another ceramic) may give much superior performance. An extremely important application for these materials is as cutting tool bits; for machining metals, the binder usually represents less than 10 per cent of the mixture and ceramic-coated tools are now the norm. The higher-binder-content (9–12 per cent) materials are used for heavier shock applications, for example those involving intermittent rather than continuous cutting, while uncoated materials with 16–20 per cent cobalt are used for rock cutting drills, crusher rolls, ball mill liners, and so on.

4.5 Further reading

Armstrong-Hébourney, B. (1991) *Control of machines with friction.* Kluwer Academic Press, Boston.

Barquins, M. (1984) On a new mechanism in rubber friction. *Wear,* **97**, 111.

Blau, P. J. and Lawn, B. R. (eds.) (1984) *Micro–indentation techniques in materials science and engineering.* ASTM STP889 Philadelphia.

Bowden, F. P. and Tabor, D. (1950) *The friction and lubrication of solids: Part I.* Oxford University Press, Oxford.

Bowden, F. P. and Tabor, D. (1964) *The friction and lubrication of solids: Part II.* Oxford University Press, Oxford.

Bowden, F. P. and Tabor, D. (1973) *Friction; an introduction to tribology.* Heinemann, London.

Buckley, D. H. (1981) *Surface effects in adhesion, friction, wear an lubrication.* Elsevier, Amsterdam.

Ferry, J. D. (1981) *Viscoelastic properties of polymers.* Wiley, New York.

Halling, J. (1976) *Introduction to tribology,* Wykeham Publications, London.

Johnson, K. L., Kendall, K., and Roberts, A. D. (1971) Surface energy and the contact of elastic solids. *Proceedings of the Royal Society,* **A324**, 301–13.

Kato, K. (1990) Tribology of ceramics, *Wear,* **136**, 117–33.

Kragelskii, I. V. (1965) *Friction and wear*. Butterworth, London.

Kragelskii, I. V., Dobychim, M. N., and Kornbalov, V. S. (1982) *Friction and wear: calculation methods*. Pergamon Press, Oxford.

Moore, D. F. (1972) *The friction and lubrication of elastomers*. Pergamon Press, Oxford.

Neale, M. J. (ed.) (1973) *Tribology Handbook*, Butterworth, London.

Oden, J. T. and Martins, J. A. C. (1985) Models and computational methods for dynamic friction phenomena. *Computer methods in applied mechanics and engineering*, **52**, 527–634.

Rabinowicz, E. (1965) *Friction and wear of materials*, John Wiley, New York and London.

Rabinowicz, E. (1980) in *Wear Control Handbook* (ed. M. B. Petersen and W. O. Winer) p. 475, ASME, New York.

Schallamach, A. (1957) Friction and abrasion of rubber. *Wear*, **1**, 384–417.

Schallamach, A. (1971) How does rubber slide? *Wear*, **17**, 301–12.

Suh, N. P. (1986) *Tribophysics*. Prentice-Hall, Englewood Cliffs, New Jersey.

Society of Tribologists and Lubrication Engineers (1987) *Tribology of ceramics* Park Ridge, Il.

Tanner, R. I. (1985) *Engineering rheology*, Oxford University Press.

Yamaguchi, Y. (1990) *Tribology of plastic materials*, Elsevier, Amsterdam.

5
Wear and surface damage

5.1 Introduction

5.1.1 *The definition and measurement of wear*

Wear is the progressive damage, involving material loss, which occurs on the surface of a component as a result of its motion relative to the adjacent working parts; it is the almost inevitable companion of friction. Most tribological pairs are supplied with a lubricant as much to avoid the excessive wear and damage which would be present if the two surfaces were allowed to rub together dry as it is to reduce their frictional resistance to motion. The economic consequences of wear are widespread and pervasive; they involve not only the costs of replacement parts, but also the expenses involved in machine downtime, lost production, and the consequent loss of business opportunites. A further significant factor can be the decreased efficiency of worn plant and equipment which can lead to both inferior performance and increased energy consumption.

The wear rate w of a rolling or sliding contact is conventionally defined as the volume lost from the wearing surface per unit sliding distance; its dimensions are thus those of $[\text{length}]^2$. For a particular dry or unlubricated sliding situation the wear rate depends on the normal load, the relative sliding speed, the initial temperature, and the thermal, mechanical, and chemical properties of the materials in contact. There are many physical mechanisms that can contribute to wear and certainly no simple and universal model is applicable to all situations. If the interface is contaminated by solid third bodies (for example, by entrained dirt or even just the retained debris from previous wear events) the situation can be much more complex; see Section 5.3.

5.1.2 *Classification of wear: wear maps*

The simplest classification of surface interactions is into those involving either *mild* or *severe* wear. This classification is not really based on any particular numerical value of w but rather on the general observation that, for any pair of materials, increasing the severity of the loading (e.g. by increasing either the normal load, sliding speed, or bulk temperature) leads at some stage to a comparatively sudden jump in the wear rate. The generally observable differences between the two regimes of mild and severe wear in a varitey of sliding metallic systems are summarized in Table 5.1.

It is important to note that this distinction between mild and severe wear makes no prejudgements about the actual physical mechanisms which are operating in each regime of which a plethora have been identified. Wear can be caused by adhesion, by abrasion, by oxidation, by delamination, by melting, or corrosion, as well as a variety of other phenomena. A common starting point in the investigation of such a complex field is often the *Archard wear equation* which asserts that w is directly proportional to the load W on the contact, but inversely proportional to the surface hardness $\mathcal{H}$ of the wearing material, so that $w \propto W/\mathcal{H}$, that is,

$$w = K \times \frac{W}{\mathcal{H}}. \qquad (5.1)$$

The dimensionless constant K is known as the *wear coefficient*, and a knowledge of its value is obviously vital in any attempt to apply eqn (5.1) in a predictive fashion. In practical engineering situations the hardness $\mathcal{H}$ of the uppermost layer of material in the contact may not be known with any certainty and consequently a rather more useful quantity than the value of K alone is the ratio $K/\mathcal{H}$. This is known as the *dimensional wear coefficient k* and is usually quoted in units of $mm^3 \, N^{-1} \, m^{-1}$; it represents the wear volume (in mm^3) per unit sliding distance (in metres) per unit normal load (in Newtons).

The behaviour of leaded brass sliding against a much harder smooth surface has been studied in some detail and demonstrates the transition from mild to severe conditions characteristic of many other material combinations in a clear and repeatable fashion. Figure 5.1 shows the wear rate of a pin of such an alloy sliding against a hard stellite ring as a function of the normal

Table 5.1   Distinction between mild and severe wear.

Mild wear	Severe wear
Results in extremely smooth surfaces — often smoother than original	Results in rough, deeply torn surfaces — much rougher than the original
Debris extremely small, typically only 100 nm diameter	Large metallic wear debris, typically up to 0.01 mm diameter
High electrical contact resistance, little true metallic contact	Low contact resistance, true metallic junctions formed

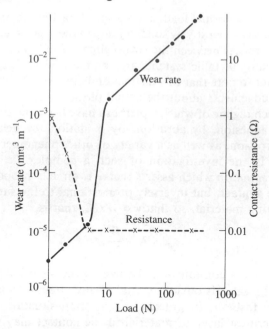

Fig. 5.1 Wear rate and electrical contact resistance for a leaded brass pin sliding against a hard stellite ring as a function of the normal load (from Hirst and Lancaster 1956).

load. At low loads the wear rate increases with load in agreement with eqn (5.1) with a dimensionless wear coefficient of approximately 2×10^{-6}. However, at a load of between 5 and 10 N there is a sharp increase in wear rate by a factor of about 100. Beyond this point, the behaviour of the contact is still reasonably well described by the Archard wear equation although now the value of the wear coefficient is approximately 10^{-4}. Also shown in Fig. 5.1 is the electrical resistance between the pin and the ring from which the extent of metallic contact at the junction can be estimated. The transition in wear rate, from mild to severe, is associated with a significant drop in the resistance indicating a much higher degree of true contact between the two materials.

The difficulties of modelling and predicting wear rates in practice can be anticipated from Table 5.2 which provides an indication of the observed values of K for a number of simple, dry sliding tribological pairs.

Several points emerge from this table. Firstly, the numerical value of K is always less than unity, usually very much smaller. Secondly, in the tests of Table 5.2 all the values of the coefficients of friction lay between 0.18 and 0.8 whereas the range of wear coefficients was very much greater. Dry

Table 5.2 Typical values of dimensionless wear coefficients in dry sliding. Materials were slid against tool steel in unlubricated pin-on-disc tests in air. In all the tests the coefficients of friction fell in the range $0.18 < \mu < 0.8$ (from Archard and Hirst (1956)).

Material	K
Mild steel (on mild steel)	7×10^{-3}
α/β brass	6×10^{-4}
PTFE	2.5×10^{-5}
Copper-beryllium	3.7×10^{-5}
Hard tool steel	1.3×10^{-4}
Ferritic stainless steel	1.7×10^{-5}
Polythene	1.3×10^{-7}
PMMA	7×10^{-6}

wear coefficients can vary by as much as four or five orders of magnitude; friction coefficients for *dry* sliding contacts vary over less than one. Combinations of similar metals fall at one end of this spectrum whereas low-friction pairs, characteristically two dissimilar materials, lie at the other end. Lastly, there is no simple correlation between friction and wear. Although in a qualitative way we might reasonably expect situations in which there are higher frictional forces to be also those involving relatively high wear, it is quite possible for material combinations to produce very similar frictional forces but very different wear behaviour.

The figures in Table 5.2 are characteristic of slow sliding under comparatively low nominal surface pressures, in other words conditions in which the increase in surface temperature is likely to be modest. If the conditions are made more severe (for example, by increasing load or sliding speed) then, although the changes in μ may not be great, there are likely to be very significant changes in the wear coefficient K. These changes often occur as one dominant wear mechanism replaces another. One way of exploring this broader pattern, and of indicating the way in which changes in the service conditions can influence the material wear response, is to construct an appropriate *map* of wear behaviour. Two approaches are possible. One is empirical: mechanism maps are built up by plotting experimental data for wear rates on suitable axes, identifying at each point the mechanism by direct experimental observation. The other uses physical modelling: model-based equations describing the wear rate for each mechanism are combined to give a map showing the total rate and the field of operation within which each is dominant. An immediate difficulty is the choice of the most appropriate coordinate axes on which to display the chosen data and two, to some extent complementary, approaches have been suggested. One plots the applied normal pressure p (i.e. the normal load W divided by the nominal area A_{nom}) versus sliding speed U. Such a map, which is specific to a particular material (grain size, hardness, surface topography, and so on) has the merit that it can also incorporate regimes of different chemical or surface reaction behaviour which are associated with temperature effects, although this is at the expense of detail on the *mechanical* forms of wear associated with different states of surface topography and asperity interaction. The general form of such a map is illustrated in Fig. 5.2, and the various regimes of wear identified upon it are discussed in more detail below.

In Fig. 5.2 pressure, speed, and wear rate have each been expressed non-dimensionally by using the relations

$$\tilde{p} = \frac{p}{\mathcal{H}}, \qquad \tilde{U} = \frac{U}{\kappa}\sqrt{\frac{A_{nom}}{\pi}} \qquad \text{and} \qquad \tilde{w} = \frac{w}{A_{nom}} = K\tilde{p}, \qquad (5.2)$$

where κ is the *thermal diffusivity* of the wearing material and is equal to the value of the thermal conductivity divided by the product of the density and specific heat (see eqn 3.117)). Regions of the map associated with different wear mechanisms are traversed by contours of equal normalized wear rate $\tilde{w}$. It can be seen that the total area is roughly divided into two regions

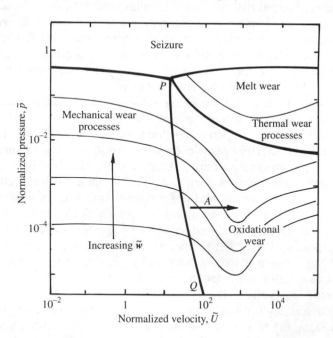

Fig. 5.2 The essential features of a load–speed wear mechanism map for steel using a pin-on-disc configuration. Load and speed have both been normalized as described in the text. Thick lines delineate different wear mechanisms and thin lines are contours of equal wear rate $\tilde{w}$ (from Lim *et al.* 1987).

by the line PQ. To the left of this divide, wear is controlled essentially by mechanical processes; here the wear rate depends on the normal pressure (or load) but is not greatly dependent on sliding velocity. On the other hand, to the right of PQ, thermal (and chemical, particularly oxidational) effects become the dominant influence and contours of $\tilde{w}$ become functions of both load and velocity. The transition from *mild* to *severe* wear occurs when the contours of $\tilde{w}$ have a steep slope, for example at arrow A. Here a comparatively small change in the sliding velocity can move the operating point from one contour to the next of greater magnitude relatively easily. Notice that the 'valley' in the wear rate contours which would be traversed by raising the sliding velocity still further in the direction of the arrow indicates that, within this region, increasing the velocity initially brings about an increase in the wear rate but that this is then followed by a *decrease* at still higher velocities. This, at first sight paradoxical, behaviour is well documented for some dry sliding pairs (including steels) and is associated with the development of protective, coherent oxide films; see Section 5.2.3.

An alternative presentation of wear data which gives more emphasis to the mechanical aspects of surface damage (but less to the thermal, since velocity is not considered as an independent variable) plots the regimes of wear on axes representing the shear strength of the interface τ between the two materials (usually normalized by the shear strength of the weaker material k) versus some surface roughness parameter such as the average surface slope ϑ (see Section 3.7). The general form of such a plot, taken from Childs (1988), is shown in Fig. 5.3; note that in order to include the region of *elastic* response the ϑ-axis is calibrated logarithmically. This axis could equally well be calibrated in terms of the plasticity index Ψ whose value indicates the gradual transition from elastic to plastic contact conditions (see Section 3.7). At steeper values of surface slope, that is, higher values of ϑ, the worn surface deforms plastically and the area of the map which represents this response is split into a number of regions depending on the nature of the resultant plastic deformation; see Section 5.2.4. Once again contours of equal wear rate $\tilde{w}$ traverse the map.

When the surfaces are comparatively smooth, so that the value of ϑ is small, elastic deformations cannot be neglected—indeed they may be sufficient to accommodate the applied loads alone—so that wear in repeated traversals, which is the usual situation in most tribological devices, depends on some form of fatigue or damage accumulation mechanism. The boundary between elastic and plastic zones corresponds to a value of Ψ close to unity. Plastic deformation of the substrate starts to become significant when ϑ is somewhat larger than the value consistent with this transition. Wear may then be due to a combination of elastic and plastic effects, as in *delamination* wear. If the surface is made rougher still *abrasion* is initiated; this always involves severe plastic deformation and can take the form of a combination of *ploughing* (in which, although the surface topography is much modified,

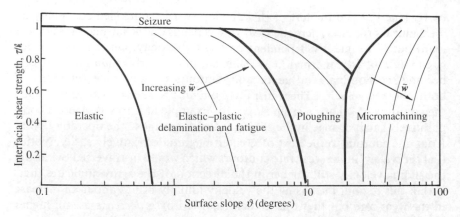

Fig. 5.3   Wear mechanism map for a rough surface; τ/k represents the relative strength of the interface and ϑ the mean slope of the rough surface or the attack angle of an individual asperity. As in Fig. 5.2 thick lines delineate different wear mechanisms and thin lines are contours of equal wear rates (from Childs 1988).

only a small proportion of the displaced material is actually detached from the surface) and *micromachining* (when a much higher proportion of the plastically deforming material is lost as wear debris). Once significant volumes are lost by machining then reducing the interfacial shear stress τ has the effect of *increasing* the volumetric loss because the efficiency of the microcutting operation is improved; we have a situation in which reducing friction (perhaps by providing or improving lubrication) enhances wear. This is in contrast to the circumstances over the rest of the map in which reducing the surface shear stress lowers the wear rate.

5.2   Mechanisms of wear

5.2.1   *Seizure*

We have seen, in Chapter 4, that when metal surfaces are brought into contact the real area over which they touch is a comparatively small fraction of the nominal contact area. The high normal pressures generated at these asperity contacts forge metallic junctions which, when they are sheared by the application of a load tangential to the interface, can grow until the actual area of metallic contact approaches the nominal area; this is the phenomenon of 'junction growth' and has been explored in Section 4.1.3. The behaviour of an individual asperity contact can be described by a relationship of the form (see eqn 4.8))

$$p^2 + \alpha_1 \tau^2 = \alpha_2 \mathcal{k}^2. \tag{5.3}$$

In this equation p is the normal pressure at the asperity, $\mathcal{k}$ the bulk shear strength of the solid, and τ the interfacial shear strength; α_1 and α_2 are constants. When $\tau/\mathcal{k} = 0$, the pressure p is equal to the material hardness $\mathcal{H}$, so that, as in eqn (3.93)

$$\mathcal{H} = \sqrt{\alpha_2} \, \mathcal{k}. \tag{5.4}$$

Using the nomenclature of the current chapter and looking at the limit when $\tau/\mathcal{k}$, that is, the friction factor m, approaches the value unity, the normalized pressure $\tilde{p}$ $(=p/\mathcal{H})$ is given by

$$\tilde{p} = \sqrt{1 - m^2 \alpha_1/\alpha_2}. \tag{5.5}$$

Setting both the value of m and the ratio α_1/α_2 equal to 0.9 (values consistent with observation) suggests a value of $\tilde{p}$ of between 0.5 and 0.6, which will be *independent* of the sliding speed parameter $\tilde{U}$. This argument is consistent with the simple straight boundary between sliding and seizure conditions on a wear map such as Fig. 5.2. In practice the local hardness depends both on the surface temperature and the surface strain rate: as the sliding velocity increases so the consequent increase in temperature tends to reduce the hardness, while the resulting increase in strain rate has the opposite effect and tends to increase local resistance to deformation. It turns out that for carbon steels these two effects more or less cancel each other out so that the experimental seizure line in Fig. 5.2 is indeed effectively horizontal.

5.2.2 *Melt wear*

Localized melting of the uppermost layer of the wearing solid is always a possibility; evidence for its occurrence has been found in wear tests on steels even at sliding speeds as low as 1 m s^{-1}. At higher velocities the coefficient of friction can drop, eventually to very low values, as a film of liquid metal forms at the interface which acts in the same way as a hydrodynamic lubricating film. The heat generated by viscous work in such a 'melt lubrica- tion' situation continues to melt more solid so that the wear rate can be high despite the fact that the coefficient of friction is low. The metal removed from the surface can be ejected as sparks or incandescent debris, or even in extreme cases, squirted out as a molten stream. It is well known that when solids slide on ice or snow, friction-induced melting generates a lubricating layer of water and the analysis of such an analogous situation can form the basis of a model of this melt wear zone. This leads to an equation of the form

$$\tilde{w} = \alpha \tilde{p} - \frac{\beta}{\tilde{U}}, \tag{5.6}$$

where α and β are terms that involve both material properties (such as hardness, thermal diffusivity, and latent heat of fusion) and process parameters (such as the local coefficient of friction and the division of thermal energy between the contacting surfaces); see Ashby and Frost (1982) and Lim and Ashby (1987). When the velocity of sliding is low then the second term of eqn (5.6) becomes large, so that wear is small: at high sliding velocities the first term of eqn (5.6) becomes of greater significance and the wear characteristics fan out as indicated in the top right-hand section of Fig. 5.2.

5.2.3   Oxidation-dominated wear

Figure 5.2 includes a large region at higher speeds designated 'oxidational wear'. The existence and extent of this regime, and the wear rates within it, are dependent on the ability of the wearing material to undergo oxidation and, equally of course, on the availability of oxygen in the immediate vicinty of the sliding contact.

When dry steel surfaces slide at speeds below about $1 \, \mathrm{m \, s^{-1}}$ the wear debris is largely metallic; at higher speeds it consists mainly of iron oxides. A velocity of $1 \, \mathrm{m \, s^{-1}}$ (which translates to a normalized velocity value of about 100) is just sufficient to give flash temperatures Θ_{max} (see Section 3.8.2) sufficient to cause oxidation, that is, of about 700°C. The value of Θ_{max} is strongly dependent on velocity but varies hardly at all with load, hence the near verticality of the boundary PQ in Fig. 5.2. The presence of an oxide film at the interface may be sufficient to reduce the wear rate merely by its role of suppressing, or at least reducing, the degree of the mechanical interaction. Close to the line PQ the oxide film is thin, patchy, and brittle, and wear is caused largely by its splitting off from the surface. This spalling away exposes more metal which rapidly oxidizes once again. At higher velocities (and to a rather lesser extent, at higher loads) the oxide film becomes thicker and more continuous—frictional heating is considerable although now the underlying metal may be partially thermally insulated by the oxide itself (which may get so hot that it can flow plastically or even melt). The first of these regimes where the oxide is thin and brittle is sometimes known as that of *mild oxidational wear*, while the second, involving much thicker films, as *severe oxidational wear*. The prefixes 'mild' and 'severe' refer here to the extent of the oxidation, and *not* the wear rate (which can be, and often is, lower during severe oxidational wear).

Mild oxidational wear has been the subject of active research for some years (see, for example, Quinn (1984)) and is much concerned with the rate kinetics of the oxidation process; this can be described by an Arrhenius relationship leading to an equation relating the wear rate $\tilde{w}$ to the other service parameters which is of the form

$$\tilde{w} = C \exp\left\{-\frac{Q_0}{R\Theta_{max}}\right\} \times \frac{\tilde{p}}{\tilde{U}}. \tag{5.7}$$

Q_0 can be taken as the activation energy for the static oxidation of iron and R is the universal molar gas constant, but the constant C must be established empirically. In the case of steels the behaviour in this regime can be complicated by the effects of martensitic formation. When a hot asperity emerges from the contact it is effectively quenched by the conduction of heat into the surrounding solid. The quench rate is often high enough to form martensite which leads to the surface becoming suddenly much harder; in addition it is likely to be left with residual compressive stresses (because of the volumetric expansion involved in the martensitic transformation). These effects combine to bring about a sudden local reduction in wear rate which shows up on the map as the valley at the tip of arrow A in Fig. 5.2.

Higher sliding velocities generate higher temperatures; not only do oxidation rates consequently increase but also now the resulting oxide film may begin to soften and deform locally, absorbing latent heat as it does so. The film in this region of severe oxidational wear subsequently flows and spreads out over cooler regions of the surface so effectively distributing this energy as the oxide solidifies. Asperity melting is thus a way of redistributing the heat input to the surface in a more uniform way. By making sensible idealizations about the nature and severity of asperity contact it has proved possible (Lim and Ashby 1987) to tentatively outline a model which has the general form

$$\tilde{w} = f_m \times \left\{ \mathcal{C} \, \tilde{p} + \mathcal{B} \frac{\sqrt{\tilde{p}}}{\tilde{U}} \right\}. \tag{5.8}$$

Here $\mathcal{C}$ and $\mathcal{B}$ are appropriate material parametric groups while f_m is a factor which accounts for the proportion of the molten oxide which is lost from (rather than being distributed over) the surface.

5.2.4 Mechanical wear processes

At normalized sliding velocities below about 10 (equivalent in the case of steels to actual velocities less than $0.1 \, \text{m s}^{-1}$) surface heating is negligible; the effect of the frictional force is principally to deform the metal surface, shearing it in the sliding direction and ultimately causing the removal of material usually in the form of small particles of wear debris. In this regime the wear behaviour often follows the Archard equation (5.1) and the aim of any model of the process is to predict the wear coefficient K (which is equal to $\tilde{w} \div \bar{p}$) in terms of material and process parameters. Since now velocity plays only a minor part in the argument, wear regimes are more readily displayed on maps such as Fig. 5.3 which include some geometric parameter desribing the roughness of the surface topography.

Running-in

When mass produced lubricated machine components are run together for the first time their ultimate load-carrying capacity is often very much less than would be the case if they had been preconditioned by running together for an initial period at a comparatively light load. This process, during which they improve in conformity, topography, and frictional compatibility, is known as *running-in* (or in the US as *breaking-in*). During this regime the wear rate is often initially quite high, but as the surfaces become smoother and the more prominent asperities are lost or flattened, the wear rate falls. After a suitable period, whose duration is invariably established empirically (and usually provided that the wear debris that has been generated is removed) the full service conditions can be applied without any sudden increase in wear rate; this sort of component history is illustrated in Fig. 5.4 in which wear (measured either as volume lost or change in dimension) is plotted as a function of either sliding distance or test duration.

During running-in a number of mechanical wear processes, especially those that depend on adhesion or abrasion, are likely to be operating simultaneously. Once running-in is complete the steady low-wear-rate regime is maintained for the operational life of the component; the wear rate rises again once the operating time becomes sufficiently long for fatigue processes in the upper layers of the loaded surface to start making a significant contribution to material loss driven by the cyclic nature of the component loading.

Adhesive wear

Models of adhesive wear have a long pedigree—they are all based on the notion that touching asperities adhere together and that plastic shearing of the junctions so formed 'plucks' off the tips of the softer asperities leaving them adhering to the harder surface. Subsequently these tips can become detached giving rise to wear particles or fragments. Severe damage of this type can result in the tearing of macroscopic chunks of material from the surface and this situation is sometimes known as *galling*. It can be a particular problem when both members of the tribological contact are made of the same sort of materials (e.g. both of ferrous alloys) or when there is poor lubrication and temperatures or sliding speeds are high. The term *scuffing* is used specifically to describe the onset of adhesive wear between lubricated surfaces which has arisen from the break down or failure of the lubricant film for whatever reason; see Section 9.4.

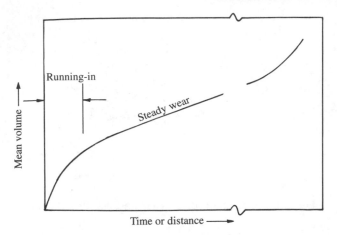

Fig. 5.4 Typical wear behaviour over the life of a component. An initial period of running-in (in which the wear rate is comparatively high) is followed by a steady low-wear regime which is terminated when fatigue mechanisms come into play.

The simplest situation, and that considered by Archard (1953), is illustrated in Fig. 5.5, which shows a single asperity junction of diameter $2a$. For a wear particle to be formed, shearing of the junction AB must occur along some path such as path 2, rather than along the original interface path 1. Even at first sight this would seem to be a relatively rare event, since the original contacting interface is likely to have both the smallest cross-sectional area and the highest density of defects, and so represents the path of least resistance. However, on occasion (particularly if the asperities interact so that the plane AB is inclined to the sliding direction) a transferred particle can be formed. In such a case, in order to make some sort of prediction of the overall wear rate, we need an estimate of the volume of the wear particle formed. If, for the sake of simplicity, this is assumed to be a hemisphere of radius a, then in sliding a distance $2a$ a wear volume of $\frac{2}{3}\pi a^3$ has been generated. Thus per unit sliding distance the wear volume is equal to $\frac{2}{3}\pi a^3$ divided by $2a$, that is, $\frac{1}{3}\pi a^2$. Since the asperity is simply deforming plastically, we can relate the dimension a to the load W_i on the asperity through the material hardness $\mathcal{H}$, so that W_i is simply equal to $\mathcal{H} \times \pi a^2$.

Thus by summing up all such wear events, the total wear rate is

$$w = \frac{\pi}{3} \times \sum a^2 = \frac{1}{3\pi} \times \sum \frac{\pi W_i}{\mathcal{H}} = \frac{W}{3\mathcal{H}}, \tag{5.9}$$

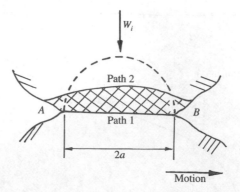

Fig. 5.5   Production of a wear particle by adhesion.

where W is now the total load carried by the two surfaces, that is, $W = \Sigma W_i$. Our predicted value of the wear coefficient K for the junction as a whole is thus $\frac{1}{3}$.

Although equation (5.9) is of the right form, with W in the numerator and hardness $\mathcal{H}$ in the denominator, the assumption that *every* junction contributes to the wear volume leads to totally unrealistic values of K (as is clear from an examination of the values in Table 5.2). To some extent we can get round this problem by supposing that the wear coefficient K represents the *probability* of generating a wear fragment: K must therefore be less than unity (this is always the case) though we have seen in practice that its range of numerical values is such that this elementary model provides very limited guidance towards its prediction.

Support for such simple models of adhesive wear, which suggest that material is removed or plucked from asperities on the wearing surface, is provided by those cases in which the wear debris is found to have an irregular, or 'blocky' shape of dimensions which seem appropriate to the surfaces in question. Although examination of the debris can give valuable clues to the mechanisms of its production it is always important to question whether the debris particles are in the same state when they are collected as when they were first produced or detached. They may, for example, have been further plastically deformed by rolling between the surfaces or have been changed chemically by oxidation.

Abrasive wear

Abrasive wear is damage to a component surface which arises because of the motion relative to that surface of either harder asperities or perhaps hard particles trapped at the interface. Such hard particles may have been introduced between the two softer surfaces as a contaminant from the outside environment, or they may have been formed *in situ* by oxidation or some other chemical or mechanical process. On the other hand, abrasion may take place simply because the counterface is both rough and intrinsically harder than the wearing component. If wear depends on the presence of free particles (albeit those entrained within the confines of the contact) the situation is known as *three-body abrasion*: if the wear-producing agent is the hard counterface itself we have *two-body abrasion*. Abrasive wear gives a characteristic surface topography consisting of long parallel grooves running in the rubbing direction. The volume and size of the grooves varies considerably from light 'scratching' at one extreme to severe 'gouging' at the other and industrial surveys invariably indicate that abrasive wear accounts for up to about 50 per cent of wear problems. The rate of damage to a surface in three-body abrasion is relatively insensitive to the precise value of the hardness of the particles involved provided they are at least 20 per cent harder than the surface itself. By far the most commonly occurring contamination in industrial machinery is that from quartz or silica particles (these minerals constitute about 60 per cent of the earth's crust). These are likely to have a hardness in excess of 8 GPa and consequently can do significant damage to even hardened bearing or martensitic steels (typically of hardness 7–8 GPa). The hardness values of some further abrasive particulates are given in Table 5.3.

Abrasive wear in practice can often be described by an equation of the same form as the Archard wear equation (5.1), that is, the volumetric loss of material in a given situation is proportional to the distance slid and the intensity of the loading. If we define the abrasive wear resistance of a material as simply the inverse of the wear volume then we should expect this quantity to be proportional to the hardness $\mathcal{H}$. The classic experimental confirmation of this effect was carried out in Russia in the early 1950s (Kruschov 1957) and the trend has been verified in many experimental observations since. Some typical results are shown in Fig. 5.6.

In experiments of this sort, specimens of each material, often in the form of cylindrical pins, are rubbed against a 'standard' abrasive; this is usually an abrasive paper or cloth carrying silicon carbide or quartz particles. The results, Fig. 5.6(a), show that the relative wear resistance of a wide range of pure metals is proportional to hardness (the relative wear resistance is the wear volume of the sample material divided by that of some standard material tested under the same experimental conditions). However, the relationship becomes more complicated for alloys: for example, steels of different alloy compositions which have been heat treated to give a range of hardness values still give linear plots but of different slopes from that of pure single-phase metals. Metallurgical investigations have suggested that while the abrasive wear resistance of steels increases with the carbon content, its actual numerical value is very dependent on details of the microstructure.

Table 5.3   Hardness values for common abrasive particles.

Material	Hardness (GPa)
Diamond	60–100
Boron carbide, B_4C	27–37
Silicon carbide	21–26
Alumina (corundum)	18–29
Quartz (silica)	7.5–12
Garnet	6–10
Magnetite, Fe_3O_4	3.7–6
Soda-lime glass	ca. 5
Fluorite CaF_2	1.8–1.9

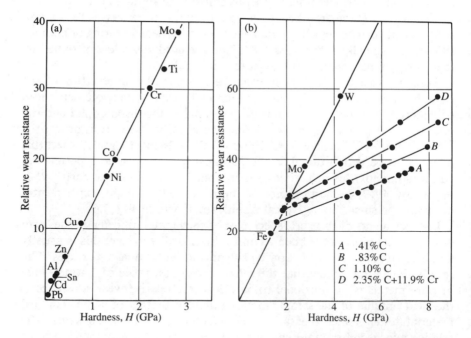

Fig. 5.6 Effect of hardness of metals on relative abrasion resistance (proportional to the reciprocal of wear rate) in two-body abrasion: (a) pure metals; and (b) steels with varying compositions and heat treatments (from Krushov 1957).

In pearlitic steels, considerable improvements in abrasion resistance can be achieved by refining the structure of the pearlite. Martensite is very resistant to wear, but even in a martensitic steel the volumes and precise compositions of the carbide phases can be extremely important in determining the value of the relative wear resistance, and careful control of the metallurgy is essential to optimize material performance.

Attempts to develop a model of abrasive wear nearly always assume that the deformations of the harder surface are negligibly small compared to those of the softer member and have concentrated on the effect of a single isolated hard asperity moving across a softer previously undeformed surface. Consider, for example, the very simple situation illustrated in Fig. 5.7 which shows a conical asperity with semi-angle $90° - \vartheta$ carrying a normal load W so that it indents the soft surface to depth h. The angle ϑ represents, by extension, the 'average' surface slope or roughness of the abrading surface which might carry an array of such asperities.

The asperity now slides through a unit distance so that the volume of displaced material or wear rate w will be given by

$$w = h^2 \times \cot \vartheta. \tag{5.10}$$

However, the depth of penetration h can be related to the hardness of the material $\mathcal{H}$. Since it is only the front half of the asperity that supports the load W, we might write

$$W = \frac{\pi}{2} \times (h \cot \vartheta)^2 \times \mathcal{H},$$

so that

$$w = \frac{2\tan\vartheta}{\pi} \times \frac{W}{\mathcal{H}}. \tag{5.11}$$

Again we have an Archard equation with a wear coefficient K this time equal to $2\tan\vartheta/\pi$. Rougher surfaces which have larger values of the angle ϑ will thus be expected to produce more wear damage, which is intuitively the right way round, but once again our simple equation (5.11) overestimates the wear observed in real life. For example, if ϑ is 1° then the predicted value of K rom eqn (5.1) is approximately 0.01 whereas typical measured values of K in the two-body abrasion of metals are in the range 5×10^{-3} to 50×10^{-3}, while for three-body abrasion values are about an order of magnitude lower. One immediate explanation for this discrepency is our assumption that *all* the material from the groove is lost from the surface: observations show that in general only some of the displaced material is actually detached (in the form of curly slivers or chips by a microcutting process) while the remainder is simply piled up at the edges of the resultant groove as ploughed ridges.

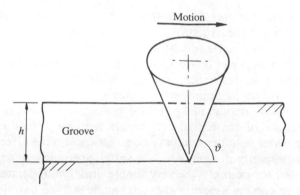

Fig. 5.7   A single conical asperity moving through a soft surface.

If we think of the abrading asperity as having the shape of a simple symmetrical pyramid rather than a cone (both are probably equally valid approximations to the shape of a real asperity) with one of its sloping edges leading, then the general appearance of the deformation is as indicated in Fig. 5.8(a). Only some of the displaced material is lost from the surface. The extent of the subsurface deformation can, to some degree, be judged from the extent of the wedge or prow of material that extends in front of the leading edge.

The material in the ridges does not strictly speaking contribute to wear at all because it is not actually detached from the surface, although the surface is certainly damaged by the production of these features. The relative volumes appearing in the ridges and the machined chips depend on the geometry of the asperity which can be conveniently described by the two angles shown in Fig. 5.8(b). The angle between the leading edge and the direction of motion is usually designated the *attack angle* ψ, and the angle at the base of the pyramidal asperity, 2ϕ, is known as the *dihedral angle*. There is a significant contribution to wear by micromachining only when the attack angle exceeds some critical value ψ_c. If the angle of attack is less than this then the process is essentially one of ploughing or rubbing. This transition in wear mode from ploughing to micromachining is evident in the wear map of Fig. 5.3 since both asperity attack angle ψ and the mean slope ϑ are angular measures of the abrasive aggresiveness or roughness of the hard surface.

The dihedral angle of an asperity can also influence the mode of material response. If 2ϕ is small then the asperity is rather like a knife moving through the surface; as ϕ increases so the asperity gets 'bluffer'. In the limit 2ϕ is equal to 180° when the pyramid would be moving with one of its flat faces forward rather than one of its edges. The effect that both these angles

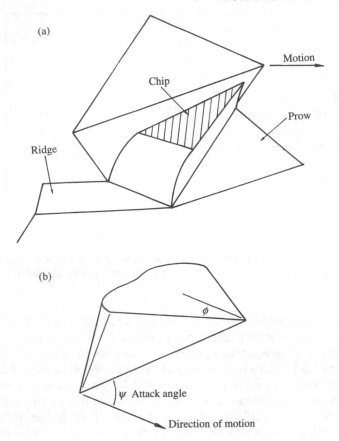

(a)

Motion

Chip

Prow

Ridge

(b)

ϕ

ψ Attack angle

Direction of motion

Fig. 5.8   (a) As a pyramidal indenter moves through a previously underformed surface material is displaced by both ploughing and micromachining. (b) The geometry of the indenter can be described by the two angles ψ and 2ϕ.

have on the form of the resultant deformation can be usefully displayed on a *wear-mode diagram* whose general form is shown in Fig. 5.9.

When a square-based pyramid or asperity moves edge forward ($2\phi = 90°$) there is a fairly clear transition from cutting or machining to ploughing or rubbing. However, if the indenter moves face forward ($2\phi = 180°$) then, as the angle of attack increases, there is a region of unstable prow or wedge formation before pure ploughing is established.

No complete plastic analysis of even a simple conical or pyramidal indenter moving through a flat surface has yet been produced. To facilitate the analysis one of two alternative simplifications is usually made. The first has been to consider the analogous two-dimensional problem which is then

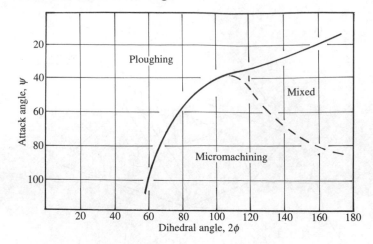

Fig. 5.9   Wear-mode diagram for abrasive wear. ψ is the attack angle and 2ϕ the dihedral angle of the asperity (from Kato *et al.* 1986).

analytically tractable by either the upper bound or slip-line field techniques. In a two-dimensional or plane strain problem, where the asperity becomes an infinitely wide wedge, there is clearly no way for the material into which the asperity is advancing to flow around the sides of the indenter. For steady state conditions to be achieved, some material must actually be removed as a chip, or else the indenter must ride up vertically during the initial stage of the deformation until its apex is at the same level as the undeformed surface; deformation is now associated with a plastic wave which is pushed ahead of the indenter, like a ruck in a carpet, Challen and Oxley (1979). A slip-line field for this deformation mode is illustrated in Fig. 5.10(a). If there is high friction between the asperity and the deforming material then the wave may be so retarded that it becomes energetically favourable for it to shear off from the surface along the path AB in Fig. 5.10(b). This mode of deformation is equivalent to the prow or wedge formation shown on Fig. 5.10. When one prow has been removed as a wear fragment the indenter digs in once more and the process restarts. As we have seen, larger values of the attack angle ψ favour the cutting mode, and a slip-line field for this process is illustrated in Fig. 5.10(c). The dimension h represents the depth of penetration of the asperity into the softer surface.

The alternative form of analysis that has been employed involves maintaining the three-dimensional aspect of the problem but relating forces and geometry through an upper bound approach (see Section 3.5.1). The indenter is taken to be rigid and the abraded material to be rigid-perfectly plastic, flowing at a constant shear stress k. Although there are difficulties

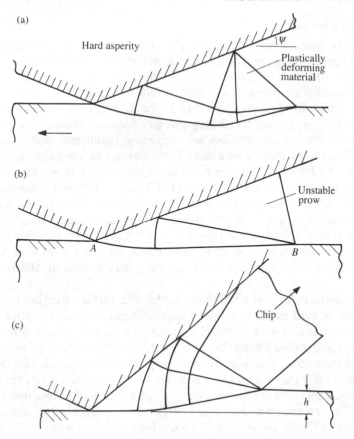

Fig. 5.10 Slip-line fields for the interaction of a hard asperity with a soft surface: (a) steady state plastic wave; (b) unsteady prow or wedge formation, a wear particle is formed by shear along *AB*; and (c) chip formation or micro-machining at larger values of attack angle ψ (from Petryk 1987).

with this approach it has proved possible to obtain reasonable agreement between theory and observation both in regard to the transition from ploughing to micromachining (in terms of the asperity geometry as indicated on maps such as Fig. 5.9) and on the magnitude of the tangential force. A real wear surface is formed as a result of many wear events which follow one another during the life of the component. The interaction of individual wear events or tracks is not well understood and there is to date no fully satisfactory way of predicting abrasive wear rates, or component lives, in terms of their initial surface geometry or topography and material properties.

Delamination wear

During the initial contact or running-in period of a typical tribological pair the surfaces in contact will be losing material in such a way that they become smoother, that is, of a lower mean surface slope: on the map of Fig. 5.3 contact conditions move from right to left. Providing the sliding velocity is insufficient for frictional heating to be an important effect (for steel surfaces this means that the sliding velocity does not exceed approximately $0.1 \, \text{m s}^{-1}$) then, as the surfaces become more conformal, the intensity of the loading on the individual areas of true contact becomes less severe. The effect of the frictional force is to deform the surface layer, shearing it in the sliding direction. Loss of material still takes place but examination of such surfaces suggests that this wear is not principally due to scratching and gouging (i.e. abrasion) but is rather due to some form of mechanism in which material is lost in the form of thin flakes or platelets. One possible sequence of events in this regime has been summarized by Suh (1986) who has been responsible for much of the work on the phenomenon of 'delamination' wear.

A key element of this mechanism is that the surface traction exerted by the harder asperities induces plastic shear deformation of the softer surface which accumulates with repeated loading. As the subsurface deformation continues, cracks are formed below the surface in the vicinity of pre-existing voids or inclusions. Crack nucleation very near the surface is not favoured because of the highly compressive stresses which exists just below the contact regions. Once cracks are present, however, further loading and deformation causes them to extend and to propagate, joining neighbouring ones. When these cracks finally shear to the surface long and thin wear sheets are lost or delaminate. The thickness of a wear sheet is controlled by the location of subsurface crack growth and this is probably determined by a balance between the hydrostatic compression beneath a contacting asperity (which tends to suppress void or crack nucleation) and the plastic shearing (which tends to cause it). The characteristic depth is typically of the same order as the size of an asperity, perhaps 10 micron. It is possible to model this situation assuming that the near-surface material contains some initial distribution of voids or inclusions (say of volume fraction f_v) from which delamination initiates and that we know γ_0, the net plastic shear strain accumulated per asperity pass. Once again we can generate an Archard wear law and in this case the non-dimensional wear coefficient K is such that

$$K = \frac{\gamma_0 \times f_v}{f_A^*} \tag{5.12}$$

where f_A^* is the critical area fraction of voids to generate a wear particle. For a given material f_v and f_A^* might be considered constants so that the wear rate depends very much on the shear strain accumulated per pass, γ_0. In Section 3.5.2 the transition from elastic to plastic conditions under a loaded single asperity was considered and, in particular, the role of the friction or traction coefficient μ noted. If μ is less than about 0.3 then the zone of plastic deformation beneath the asperity does not break through to the surface but is contained within the body of the elastically deforming material. Consequently, the magnitude of the plastic strain 'picked up' by the material per pass must be small. However, if $\mu > 0.3$ then the flow field does extend to the surface and the elastic constraint of the hinterland is much reduced: the plastic strain per pass γ_0, can become much larger, and eqn (5.12) suggests that this can lead to a rapid increase in the wear coefficient. This then offers an explanation of the transition from mild to severe wear which can be brought about by increasing the load on a tribological pair. If this increase has the effect of disrupting any protective surface or boundary layers, so locally producing a change in friction coefficient from a low value to one in excess of 0.3, then there can be a further associated significant increase in wear rate. This increase can be by as much as two orders of magnitude.

In well-lubricated, and so low-friction, contacts the rates of material loss can be very low indeed and although the wear debris is still lamellar in form it is much thinner (perhaps less than a micron) than in the type of delamination wear described above. The production of this form of 'filmy' wear appears not to depend on a fracture process as such but rather on the exhaustion of the ductility of the softer material in the uppermost layer of the wearing component and is the subject of current research as are a number of other topics in the general area of wear.

5.2.5 *Fatigue wear in rolling contacts*

The races and rolling elements in a rolling-element bearing are invariably made of high-strength material and given a very good surface finish: in normal operating conditions the loading is entirely elastic and there is little slip at the interface (the slide:roll ratio is typically less than 0.01). Operating temperatures are usually modest, less than 200°C, and clean oil or grease lubrication is specified. Under these circumstances there is likely to be very little adhesive, abrasive, or oxidative wear and many millions of cycles of operation may be carried out with no visible degradation of the surfaces of the bearing elements. However, eventually, perhaps after many thousands of trouble-free hours of operation, a particle spalls away from one of the surfaces. Once this has occurred further deterioration usually follows rapidly; for some graphic examples see Tallian (1991). The particles from such a fatigue wear process are characteristically very much larger than the very small fragments associated with adhesive or abrasive wear (perhaps of the order of a large fraction of a millimetre rather than a few tens of microns). This form of wear, which generates a characteristic 'pitted' surface, is closely related to the general phenomenon of metal fatigue. Changes to the material (or its environment) which alter its general fatigue performance will usually have a similar effect on pitting or fatigue wear. For example, surface active lubricants which would be expected to extend the life of sliding components generally *lower* the life of a rolling contact which is limited by pitting. Similarly, increases in the cleanliness of bearing steels (i.e. a reduction in the inclusion content) brought about by developments in process metallurgy have improved both fatigue and rolling-bearing performance. However, despite these similarities, there can be a significant difference between conventional fatigue and surface failure involving fatigue. In conventional fatigue tests, ferrous materials often exhibit a fatigue limit, that is, stress amplitude below which the fatigue life of the specimen is effectively inifinite. This does not generally apply to rolling contacts: in other words there is no rolling load below which there is a 100 per cent chance of infinite bearing life. Bearing lives limited by fatigue mechanisms are generally viewed statistically — conventionally on the basis of Weibull statistics (see Section 11.2.3).

Gear tooth contacts, in which there is a combination of rolling and sliding contact conditions, can also on occasion develop the pitted surface typical of surface fatigue. This form of failure is quite distinct from that of scuffing (see Section 8.3) which is primarily associated with the ultimate collapse of the lubricating film between a pair of sliding surfaces. The four-ball machine (Section 5.1) can be used to test materials (and lubricants) for their resistance to this form of pitting failure provided that each of the nest of three balls is free to rotate and so give rolling, rather than sliding, conditions. The disc machine when set up for rolling rather than sliding, can likewise be used.

5.2.6   *Fretting and corrosion wear*

Fretting wear is a phenomenon that can occur between two surfaces which have a relative oscillatory motion of small amplitude, usually only a few tens of microns (Waterhouse 1972). The main characteristic of a fretting contact in ferrous material pairs is the appearance of reddish-brown debris: this powder is predominantly made up of particles of the hard oxides of iron and can act as a grinding paste or lap producing highly polished patches on the fretted contact. The sequence of events appears usually to follow a typical pattern: mechanical action disrupts the protective oxide film so exposing clean and highly reactive metal. In the presence of an oxygen-containing atmosphere this rapidly oxidizes during the half-cycle during which it is exposed; the film is then again disrupted on the return path of the counterface and the cycle repeats. Particles of the disrupted film become trapped in the contact and because of their hardness can bring about an additional contribution by abrasive wear. If areas of the surfaces which are incompletely oxidized come into direct contact there can also be an element of adhesive wear.

There are many practical situations in which fretting can be troublesome and many static joints — flanges, couplings, keyways, and so on — which are subject to vibration can suffer from this problem. It can occur between the undersides of screw or bolt heads and the components being joined and can lead to the loosening of such joints, resulting in increased vibration and thus an acceleration in the rate of further wear and damage. On the other hand, if the predominantly oxide debris from a fretting joint is unable to escape from the interface it may, since it occupies a greater volume than the metal from which it was formed, actually lead to the seizure of parts which were originally designed to slide or rotate with small dimensional clearances. Whether fretting leads to sloppiness or seizure depends very much on the ease with which the wear debris can escape from the contact zone.

Slip at an interface occurs because the shear force applied is greater than the maximum friction force that can be generated in the opposite direction; even if there is no grosss sliding in an elastic contact subjected to an alternating tangential or shear force there can be an annular zone of microslip around the periphery of the contact (see Section 3.4.1). In an essentially static contact, subject to small-scale vibrations, the damage is often initiated at, or very close to, the edge of the Hertzian contact patch where the slip amplitude between the two surfaces is greatest. An analogous situation develops if the contact is subject to a cyclic torque about a normal axis. As the amplitude of the torque increases so the central area, within which there is no relative movement, decreases in size until eventually there is slip over the whole of the contact area. This suggests two possible approaches to the prevention or control of fretting: either reducing the applied stress at the joint or increasing the potential friction stress that can be generated at the interface. Reducing the applied stress usually means a change in the design of the component or assembly. The potential maximum friction stress can be increased by maintaining the normal load at the same value but redesigning the joint so that the apparent area of contact is reduced. Changes in surface condition by plating, coating, or chemical treatments have also been successfully employed to reduce the incidence of fretting damage as has the introduction of non-metallic inserts between the loaded surfaces.

Metals and alloys other than those based on iron can show a similar phenomenon to those described above. In the case of aluminium and its alloys, fretting produces a black powder containing a mixture of metallic aluminium and alumina. Fretting has also been observed with alloys based on nickel, cobalt, and molybdenum. The microslip present within the contact patch of a rolling contact, although very small, may be sufficient to initiate fretting wear, especially if the surrounding atmosphere is particularly chemically active, for example in sea-water environments. In the cases of such non-conformal contacts the phenomenon of fretting is sometimes called *false Brinelling*.

5.2.7 *Erosive wear*

Erosive wear is the process of material removal by the impingement of particles, usually at high velocity, on component surfaces. Erosion can be exploited deliberately as in the case of cleaning by iron shot or sand particles but, when encountered unintentionally in service, is often extremely dele-terious. Severe erosion problems have been encountered in such diverse situations as the transport of powders and slurries, the impact of dust particles on the blades of turbo-machinery, and the operation of fluid bed combusters. The particles do not have to be solid; rotating machinery operating in the presence of water droplets (for example in wet steam) can suffer similar damage.

Erosion is usually quantified rather differently from other forms of wear; it is usually described by the specific weight loss ε, that is, the loss of mass from the surface in question resulting from a unit mass of the impinging particles. A variety of tests have been developed both to assess materials for their erosion resistance and to examine a single erosion event in more detail. The most common procedure involves blasting a stream of abrasive particles at velocity V against a target specimen. Such investigations all show a rela-tionship of the form

$$\varepsilon = \text{constant} \times V^n. \tag{5.13}$$

The value of the index n varies with the target material but for most materials is in the range of 2–2.5.

In order to study the mechanisms of material removal in more detail it is necessary to examine the behaviour of an individual particle. In most studies of this sort, a gas gun is used to fire a single particle at a smooth target and high-speed photography is used to record the details of the impact. An important parameter is the angle at which the particle meets the target; for ductile metals the rate of erosion is worst, that is, a maximum, when this angle is about 20° as indicated in Fig. 5.11. This implies that the front face of the particle exhibits a large angle of attack which, by analogy with abrasion, corresponds to the conditions under which we should expect the greatest loss of material by machining or cutting. Photo-graphy confirms that there are clear similarities between the behaviour of ductile metals in these two situations which both involve considerable amounts of plastic straining. Some material is also always extruded into side ridges or lips which are subsequently detached if there are repeated

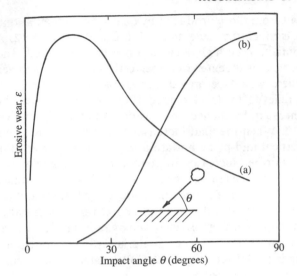

Fig. 5.11 Effect of impact angle on rates of erosion: (a) ductile metal; and (b) brittle solid.

impacts on the same area of the specimen. Studies of the impact of single particles on to metals at an impact angle of 30° show that rounded particles deform the surface predominantly by ploughing, displacing material to the front and sides. Further impacts on neighbouring areas lead to the detachment of heavily strained material from the rim of the crater or from the terminal lip. The deformation caused by an angular particle depends on the orientation of the particle when it strikes the surface and on whether the particle rolls forwards or backwards during the contact. If it rolls forwards then it indents the surface, raising material into a prominent lip, which is vulnerable to removal by subsequent nearby impacts. If the particle rolls backwards then a true machining action can occur in which the sharp corner of the particle cuts a microchip from the surface. Rounded particles are thus associated with more diffuse and less localized deformation and more impacts are required to remove each unit mass of material; an increased impact angle has the same effect. In the extreme case of spherical particles striking with normal incidence, material is removed only after neighbouring impacts have imposed many cycles of plastic deformation and the resulting surface looks very different from one eroded by angular particles.

In the case of brittle materials, such as glass or ceramics, material is removed from the surface by the formation and intersection of cracks. Although the detailed crack patterns caused by impact by angular particles differ in detail from those produced by quasi-static indentation, the general morphology is much the same and is illustrated in Fig. 5.12. At the point of initial contact, very high local stresses are generated. Indeed, if the tip of the particle or indenter were perfectly sharp (i.e. had zero radius of curvature) there would be an infinite stress at this point. These intense stresses are relieved by local plastic flow or even, under the very high pressures generated, by changes in density around the tip, indicated by zone D in Fig. 5.12. When the load increases to a critical value, tensile stresses across the vertical mid-plane initiate a *median vent crack*, indicated by M in Fig. 5.12. Further increases in load are accompanied by progressive extension of the median crack. On reducing the load the median crack closes (Fig. 5.12(d)). However, further unloading (Figs 5.12(e) and (f)) is accompanied by the formation and growth of *lateral vent cracks* (labelled L). The formation of these lateral cracks is driven by residual elastic stresses, caused by the relaxation of the deformed material around the region of contact. As unloading is completed (Fig. 5.12(f)) the lateral cracks curve upwards, terminating at the free surface.

The median cracks propagate down into the bulk of the solid with increasing load on the indenter, and do not grow further on unloading. They are not associated in the first instance with the removal of material. In contrast, lateral cracks can lead directly to wear. One model for the abrasive wear of brittle materials is based on the removal of material by lateral cracking (Wiederhorn and Hockey 1983). As a sharp particle slides over the surface forming a plastic groove, lateral cracks grow upwards to the free surface from the base of the subsurface deformed region driven by the residual stresses associated with the deformed material. It is assumed that material is removed as chips from the region bounded by the lateral cracks and the free surface, and the volume wear rate is then estimated from the volume of this region. The particle loads and thus the extent of cracking is most severe when the impact velocity is normal to the surface and thus wear of brittle materials by erosion is also at its most rapid under these conditions, as indicated in Fig. 5.11.

5.3   Third bodies and wear

5.3.1   *Wear by abrasive contaminants*

As a general rule the abrasive wear of a softer surface by a harder counter-face is adequately described by the Archard wear equation (subject to the difficulties in predicting the wear coefficient K) provided that the detached, worn material or debris is swept out of the contact. In the case of three-body conditions, whether lubricated or dry, the position is less clear-cut. A not

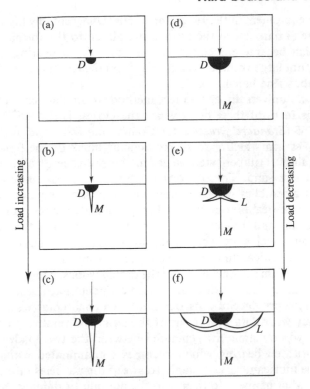

Fig. 5.12 Diagram showing crack formation in a brittle material due to a point indentation. The normal load increases from (a) to (c) and then is progressively reduced from (d) to (f) (from Lawn and Swain 1975).

uncommon situation is one in which the two surfaces in relative motion are of different hardnesses with the gap between them contaminated by abrasive particles with a hardness greater than either. Loss of material depends not only on the hardness of the wearing surface but also of those of the counter-face and the contaminant. In plain bearings it is the usual practice to make the shaft as hard as practicable (perhaps using a surface hardening treatment) and to use a much softer material on the surface of the opposing component. A shaft will be typically at least three times as hard as the bearing supporting it. However, increasing this hardness differential will not always reduce the wear rate of the shaft: if the abrasive particles become only partially embedded in the softer material then they can act as particularly aggressive asperities, gouging and machining an increased volume of material from the harder surface. This type of mechanism was responsible for the notorious 'wire wool' failures in the plain bearings of power generation equipment in the 1950s and 1960s. Unacceptably high concentrations of metal particles in the lubricating oil led to the formation of hard 'scabs' which became embedded in the white metal bearings which then 'machined' out large volumes of material from the rotating shaft in the form of continuous fine hair-like chips.

There is no universally agreed test method to simulate or evaluate materials in this form of three-body wear; the closest is the ASTM Standard Practice G65 (*Standard Practice for Conducting Dry Sand/Rubber Wheel Abrasion Test*) in which a small rectangular block of the test material is pressed against a rubber-tyred wheel in the presence of a free stream of a standard test sand. For obvious economic reasons, running complete mechanical assemblies (such as internal combustion engines or hydraulic pumps) with lubricants contaminated by particulate matter is expensive and time-consuming and, in any case, the results are often not easy to interpret. As components such as conventional journal bearings or seals wear, so their clearances and critical dimensions change and this, in turn, can modify the hydrodynamic film thicknesses from the design values. Most two-body wear is carried out on pin-on-disc or pin-on-cylinder machines. If these are operated in the presence of a viscous fluid then unacceptable scatter is often found in any wear data as a result of the stray and relatively uncontrolled degree of hydrodynamic lift generated between the two loaded surfaces.

In a lubricated bearing which becomes contaminated with particulate abrasive the lubricant serves mainly to sweep the particles into the bearing gap. Here, what happens to them and the amount of damage that they do, depends principally on their size and shape and on their hardness relative to the two solids. If the particles are no more than a little larger than the minimum film thickness then they may roll or tumble through the gap producing relatively little damage to the two solid surfaces. Such wear as does occur is associated with the indentations formed as the particle rolls through the gap and the worn surface has a characteristic pitted appearance rather similar to one subjected to random erosion. However, at some critical ratio of particle dimension to film thickness the mode of surface damage changes and becomes characterized by the machining of long wear grooves in one or both of the solid surfaces. These changes are illustrated by Fig. 5.13. Once there is an element of sliding between a particle and one of the surfaces there must be damage by grooving, though not all the displaced material is necessarily lost from the surface: some is detached by cutting or micromachining while the rest may be simply displaced into long side ridges. If the surfaces are of comparable hardnesses then on average any particular particle will spend a proportion of its time sliding relative to each, so that both will show evidence of grooving or abrasive wear. However, if one of the surfaces is softer than the other then the particles may become embedded within it: if the particles are completely 'absorbed' and so fail to break

(a)
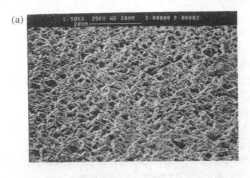

(b)

Fig. 5.13 Metallic bearing surfaces suffering from abrasive wear because of particulate contamination in the lubricating oil. (a) When the particles are small compared to the oil film thickness the wear rate is comparatively low and the mechanism one of indentation as particles tumble through the bearing gap. (b) Larger particles become wedged between the loaded surfaces and produce the characteristic parallel grooving and scoring of abrasive wear.

through the lubricant film then we should expect the wear rate to decrease. On the other hand, if the particles remain proud of the surface they may act like particularly damaging permanent asperities increasing considerably the damage done to the harder counterface. Adequate filtration of the lubricant to remove all the hard particulates above some critical size is essential.

5.3.2 *Interfacial 'third bodies'*

The loads and speeds in tribological contacts can vary from a few milligrams to hundreds of tonnes and from zero to many tens of metres per second. The difference in the velocities of the two loaded components is actually rarely accommodated directly at the interface between them – invariably, a third body is either intentionally introduced or formed within the contact.

The 'best' third body is a lubricant film and this is the situation that has been extensively analysed in classical hydrodynamics, as well as in the newer study of elasto-hydrodynamics. In such bearings velocity accommodation takes place by shear within the fluid film. However, when a pair of *dry* rubbing surfaces are examined after a period of testing or service there is very often evidence of an adherant transfer layer on one or both of them. This much less well defined form of third body is formed principally of compacted wear debris, perhaps together with solid contaminant material that has been ingested into the contact. It also usually contains many voids and defects but, like the lubricant film, must accommodate, either across its thickness or at its boundaries, the difference in the linear velocities of the two solid substrates. While there is some information available on the relevant properties of solid layers when these are formed or applied as boundary or solid lubricants (see Chapter 8), the mechanical and rheological properties of layers of compacted wear debris, although they have often been observed in practice, have very much less often been studied in any detail. The generation and stability of such debris or third-body layers can also depend on whether the overall geometry and motion of the contact encourage the retention of debris between the surfaces or tend to displace it from the interface (Godet 1990). Considerations of this sort can be important when attempting to simulate real-life engineering situations in laboratory test rigs. Even when the correct materials, specific loads, speeds, and temperatures are used, if the debris is retained in one case and ejected from the contact in another, the test results may be far from representative of service conditions.

5.3.3 *Debris analysis*

The material that is detached from the wearing surface has a form and a shape which is characteristic of the process or processes that lead to its production. For example, the debris produced during the early stages of running-in tends to have a high proportion of material in the form of finely machined chips—typical of abrasive microcutting. On the other hand, mild lubricated wear tends to produce thin flake-like debris and fatigue wear more equi-axed particles. Near-spherical particles are a characteristic feature associated with fatigue crack propagation in rolling contacts, the concentration of spheres indicating the extent of the crack propagation. These observations suggests that it should be possible to infer something about the state of a pair of wearing surfaces by monitoring the amount and form of the debris produced during their period of service provided it can be removed before it has undergone further mechanical or chemical changes. Dry debris can be swept on to a glass slide and examined in an optical microscope. Finer particles, beyond the resolving power of an optical instrument, can be examined by electron microscopy; they can be brushed directly on to a microscope grid, though shadowing *in vacuo* with a heavy metal improves the definition and eases the interpretation of the observations. The dispersion of the particles can be improved by agitating the debris in a suitable liquid and placing a single drop on the supporting film or stub; the solvent evaporates leaving the debris well dispersed. When particles are contained in a lubricant, this can be removed by centrifuging a sample with an appropriate solvent. The solvent is progressively replaced until a specimen in a volatile medium, such as ether, is produced. Once again a single drop can be placed on a supporting surface prior to SEM examination.

To obtain quantitative data various techniques are available (Scott 1976). A known quantity of the service fluid can be filtered through a very fine membrane filter. These are available with very well-controlled pore sizes in the range of less than a micron up to a few tens of microns. They are characterized by a very smooth structure-free surface between the pores which introduces a minimum of background detail in the subsequent microscopic examination. After filtering, the deposited debris is washed by clean solvent and then the surface of the filter examined in the SEM and the particles counted and their morphology noted. The counting process can be automated to reduce the time and potential error involved. Commercial particle counters are also available using the 'Coulter' principle. In such a counter the debris must be supported in an electrically conductive fluid. The suspension is then forced to flow through a small aperture with simultaneous flow of an electric current; this results in a series of pulses, each being proportional to the volume of the particle causing it. These pulses are counted and so provide direct data that can be plotted in terms of cumulative particle frequency against particle size.

'Ferrography' (Bowen and Westcott 1976; Scott *et al.* 1975) is a technique developed specifically for the quick and convenient separation of wear debris from a lubricant and its arrangement according to size on a transparent substrate. The Ferrograph analyser consists of a pump to deliver a lubricant sample at a low flow rate, a magnet to provide a high-gradient magnetic field near its poles, and a treated transparent substrate on which the particles are deposited. The lubricant sample, diluted with a solvent designed to promote the precipitation of the wear particles, is pumped across the substrate, usually a glass microscope slide, which is mounted at a slight angle to the horizontal. The magnetic particles adhere to the substrate, distributed approximately according to their size. After about 2 ml of the suspension has been pumped across the slide, a washing and fixing cycle removes the residual lubricant and causes the wear particles to adhere permanently. The quantity of wear particles and their size distribution can be determined by optical density measurements at particular locations along the substrate. When successive oil samples yield Ferrograms with essentially constant density readings it may be concluded that the machine from which the samples were taken is operating normally and wearing at a steady rate. A rapid increase in the number of particles, in particular the ratio of large to small particles, is generally indicative of a new severe wear mechanism baving been initiated and may well presage catastrophic component failure. These are examples of techniques of 'condition monitoring' which are becoming increasingly widely used in an effort to forestall the serious economic consequences of unexpected machinery or plant breakdowns.

5.4   Further reading

Archard, J. F. (1953) Contact and rubbing of flat surfaces. *Journal of Applied Physics*, 24, 981–8.

Archard, J. F. and Hirst, W. (1956) The wears of metals under unlubricated conditions. *Proceedings of the Royal Society*, A236, 397–410.

Ashby, M. F. and Frost, H. J. (1982) *Deformation mechanism maps: the plasticity and creep of metals and ceramics*. Pergamon Press, Oxford.

Bowen, E. R. and Westcott, V. C. (1976) *Wear particle atlas*. Predict Technologies .

Childs, T. H. C. (1988) The mapping of metallic sliding wear. *Proceedings of the Institution of Mechanical Engineers*, 202, c6, 379–95.

Challen, J. M. and Oxley, P. L.B. (1979) An explanation of the different regimes of friction and wear using asparity deformation models wear, 53, 229–243.

Godet, M. (1990) Third bodies in tribology. *Wear*, 136, 29–45.

Hirst, W. and Lancaster, J. K. (1956) Surface film foundation and metallic wear. *Journal of Applied Physics*, 27, 1057–65.

Kato, K., Hokkirigawa, K., Kayaba, T., and Endo, Y. (1986) Three dimensional shape effect on abrasive wear. *Journal of Tribology*, 108, 346–51.

Kruschov, M. M. (1957) Resistance of metals to wear by abrasion, as related to hardness. *Proceedings of Conference on Lubrication and Wear*. Institution of Mechanical Engineers, London, pp. 655–59.

Lawn, B. R. and Swain, M. V. (1975) Microfacture beneath point indentations in brittle solids. *Journal of Materials Science*, 10, 113–22.

Lim, S. C. and Ashby, M. F. (1987) Wear mechanism maps. *Acta Metallurgica*, 35(1), 1–24.

Lim, S. C., Ashby, M. F., and Brunton, J. H. (1987) Wear rate transitions and their relationship to wear mechanisms. *Acta Metallurgica*, 35, 1343–8.

Lipson, C. (1967) *Wear considerations in design*. Prentice-Hall, New Jersey.

Petryk, H. (1987) Slip line field solutions for sliding contact. *Proceedings of the Institution of Mechanical Engineers' Conference on Friction, Lubrication and Wear*, London, pp. 987–94.

Quinn, T. F. J. (1984). The role of oxide films in the friction and wear of metals. In *Microscopic aspects of adhesion and lubrication* (ed. J. M. Georges). Elsevier, Amsterdam.

Scott, D. (1976) Particle tribology, *Proceedings of the Institution of Mechanical Engineers*, 189, 623–33.

Scott, D. (ed.) (1979) *Treatise on material science and technology, Volume 13, Wear*, Academic Press, New York.

Scott, D., Seifert, W. W., and Westcott, V. C. (1974) *Scientific American*, **230** (5), 88–97.

Scott, D., Seifert, W. W., and Westcott, V. C. (1975) Ferrography–an advanced design aid for the 80's. *Wear*, **34**, 251–60.

Suh, N. P. (1986) *Tribophysics*, Prentice-Hall, Englewood Cliffs, New Jersey.

Tallian, T. E. (1991) *Failure atlas for Hertz contact machine elements*. ASME Press, New York.

Waterhouse, R. B. (1972) *Fretting corrosion*. Pergamon, Oxford.

Wiederhorn, S. M. and Hockey, B. J. (1983) Effect of material parameters on the erosion resistance of brittle materials. *Journal of Materials Science*, **18**, 766–80.

Winer, W. O. and Peterson, M. B. (1978) *Wear control handbook*. ASME.

Zum-Gahr, K-H. (1987) *Microstructure and wear of materials*. Elsevier, Amsterdam.

6
Hydrostatic bearings

6.1 Introduction

A hydrostatic bearing is one in which the loaded surfaces are separated by a fluid film which is forced between them by an externally generated pressure. Formation of the film, and so successful operation of the bearing, requires the supply pump to operate continuously, but it does not depend on the relative motion of the surfaces (hence the term 'hydrostatic'). Such bearings have a great attraction to engineers; machine elements supported in this way move with incomparable smoothness and the only restriction to motion arises from the small viscous losses in the fluid. A mass supported on a hydrostatic bearing will glide silently down the slightest incline.

The essential features of a typical hydrostatic single-pad thrust bearing are shown in Fig. 6.1(a). The bearing is supplied with fluid under pressure p_s which, before entering the central pocket or recess, passes through some form of *restrictor* or *compensator* in which its pressure is dropped to some lower value p_r. The fluid then passes out of the bearing through the narrow gap, shown of thickness h, between the bearing *land* and the opposing bearing surface or *slider* which is also often known as the *bearing runner*. The depth of the pocket is very much greater than the gap h. The restrictor is an essential feature of the bearing since it allows the pocket pressure p_r to be different from the supply pressure; this difference, between p_r and p_s, depends on the load applied W. At one extreme, if W is very large then the gap h may fall effectively to zero, so that there is no flow through the bearing and then p_r is equal to p_s; on the other hand, if the bearing is unloaded, then the gap h becomes very large so that the only resistance to flow is that offered by the restrictor. The lubricant flow rate will increase until the pressure losses across the restrictor are just sufficient to bring p_r down to the local external or ambient value.

The general arrangement of the lubricant supply system of a typical installation is shown in Fig. 6.1(b). The pump draws fluid from a reservoir through a coarse filter, or strainer, and supplies it to the bearing compensator through a line filter at a pressure p_s whose value is determined by the setting of the pressure relief valve. A second filter in the return line from this valve allows debris to be flushed from the system by operating the pump with the on–off valve in the line to the bearing closed.

Hydrostatic bearings are particularly attractive where normal loads are high and the advantage of low friction at zero speed is at a premium, for example, in heavily loaded devices requiring very precise positional control. In a correctly designed hydrostatic bearing the surfaces never come into contact, and so it follows that wear and damage are absent or, at least very low, and accuracy can be maintained throughout the working life of the bearing. By appropriate design, the stiffness (and thus the vibration characteristics) of the bearing can be controlled and significant amounts of mechanical damping introduced. The disadvantages of bearings of this type may be those of size and cost (hydrostatic designs for a given load tend to be larger than competing hydrodynamic or rolling-element bearings) and the fact that they are not inherently fail-safe since the external pumps they require must run continuously. In order to keep pump requirements within reasonable bounds, clearances and restrictor orifices are very small, and thus the fluid must be kept scrupulously clean if serious damage through the blockage of critical elements by contaminant particles is to be avoided. Supply pressures are generally in the range of 1–10 MPa (150–1500 lbf/in^2) and Fig. 6.2 can be used to estimate the approximate overall bearing size as a function of the load to be supported for this range of supply values.

The fluid supplied to the bearing can be either a liquid, typically a hydro-carbon oil, or a gas, usually air. Since the viscosity of a vapour is so much less than that of a typical liquid then, other things being equal, film thicknesses in air bearings will be very much smaller than those in oil-lubricated designs. Aerostatic, that is, air bearings are more expensive because of the extra cost of providing components of the required accuracy and tolerances and this factor may well outweigh the advantages that air has of abundant availability and ease of supply. In what follows we treat the fluid as incompressible. This can be taken as a universally satisfactory idealization for liquids but is clearly less so for gases, although initial calculations may be done on this basis. However, several of the characteristic features of air bearings depend on compressibility effects and these are explored more fully in Chapter 8.

6.2 Analysis of parallel pocket thrust bearings

The aim of our analysis is to relate both the load-carrying capacity of the bearing, and various other features of its performance (stiffness, flow requirements, and so on), to its dimensions and the physical properties of the pressurized fluid.

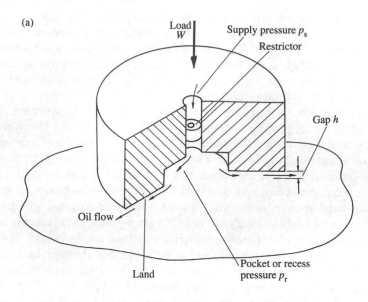

(a)

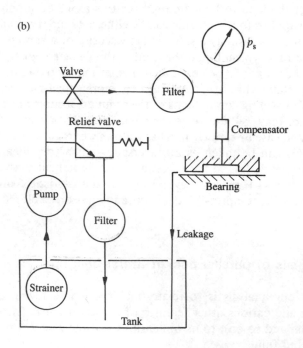

(b)

Fig. 6.1 Essential features of (a) a hydrostatic bearing and (b) the lubricant supply system.

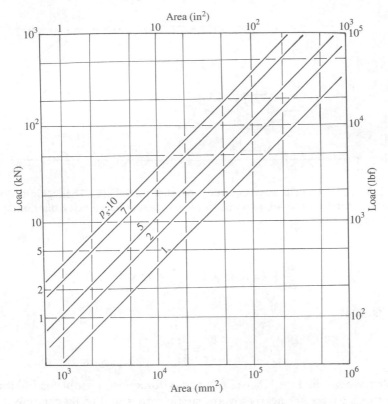

Fig. 6.2 Approximate dimensions of hydrostatic bearings. Units of supply pressure p_s are MPa for SI scales and $\times$ 100 lb/in^2 for those using conventional US units.

6.2.1 One-dimensional pressure-induced flow

We begin by considering the fluid flow between two stationary parallel plates separated by a viscous film of thickness h as shown in Fig. 6.3. The plates are 'infinitely' long in the Oy-direction, so that flow is parallel to the Ox-axis, and is maintained by the application of a pressure p at the left-hand end. The gap h is small compared to the length of the bearing in the Ox-direction and fluid flow is laminar. In small bearings there is little difficulty in satisfying this requirement; in very large hydrostatic bearings (or particularly hybrid designs which involve both hydrostatic and hydrodynamic mechanisms) there can be significant effects of turbulence within the film (see Section 7.11).

Considering the equilibrium of the shaded element of fluid shown, and taking unit depth normal to the page, we can write that

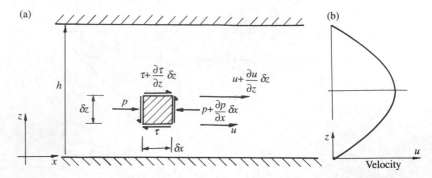

Fig. 6.3 (a) Poiseuille flow between parallel plates. (b) The volume flow is represented by the area enclosed by the parabolic velocity distribution.

$$\frac{\partial p}{\partial x} \times \delta x \times \delta z \times 1 = \frac{\partial \tau}{\partial z} \times \delta z \times \delta x \times 1$$

that is, that

$$\frac{\partial p}{\partial x} = \frac{\partial \tau}{\partial z}. \tag{6.1}$$

We have not included inertia or gravity forces as in problems in lubrication these effects are nearly always sufficiently small to be neglected. The shear stress τ occurs as a result of the different shear rates across the width of the film. If the fluid is Newtonian (see Section 1.3.2) and of viscosity η then at any section the shear stress τ is proportional to the local velocity gradient, that is,

$$\tau = \eta \frac{\partial u}{\partial z}, \tag{6.2}$$

where u is the fluid velocity in the Ox-direction. Hence, substituting eqn (6.2) in (6.1),

$$\frac{\partial p}{\partial x} = \frac{\partial}{\partial z} \left\{ \eta \frac{\partial u}{\partial z} \right\}.$$

Assuming that η is independent of z, the position of the element in the film, it follows that

$$\frac{\partial p}{\partial x} = \eta \frac{\partial^2 u}{\partial z^2}. \tag{6.3}$$

If the x-dimension of the bearing is very much greater than the film thickness h we can take the pressure to be invariant with z, so that p is a function of x only; thus

$$\frac{dp}{dx} = \eta \frac{\partial^2 u}{\partial z^2}.$$

Integrating twice then yields

$$\eta u = \frac{dp}{dx} \frac{z^2}{2} + \mathcal{A}z + \mathcal{B} \tag{6.4}$$

where $\mathcal{A}$ and $\mathcal{B}$ are constants.

Now we can apply the boundary conditions that the fluid velocity $u = 0$, at both $z = 0$ and $z = h$, that is, that elements of the fluid actually in contact with the boundary are not slipping relative to it. These conditions, virtually always assumed in problems in fluid mechanics, lead to the relation

$$u = \frac{1}{2\eta} \frac{dp}{dx} z(z - h). \tag{6.5}$$

This equation shows that, subject to the idealizations we have made, the velocities across the gap will be distributed parabolically, as indicated in Fig. 6.3(b); this is characteristic of *fully developed pressure* or *Poiseuille flow*. The volumetric flow rate, q per unit width of the bearing, assuming the fluid to be incompressible, is represented by the area under the velocity distribution; thus

$$q = \int_0^h u \, dz = -\frac{h^3}{12\eta} \frac{dp}{dx}. \tag{6.6}$$

The negative sign shows that fluid flows *down* the pressure gradient. If we were to apply the same argument to flow along a narrow circular passage of radius b (as opposed to a film bounded by parallel plates) then the corresponding equation for *fully developed pipe flow* is

$$Q = -\frac{\pi b^4}{8\eta} \frac{dp}{dx} \tag{6.7}$$

where Q is the total volume flow rate. We shall use this result in considering bearing compensation in Section 6.2.3.

6.2.2 *Single circular pad bearing*

As an example of the application of these equations consider the case of the circular pad illustrated in Fig. 6.4. The bearing has an outer radius r_o

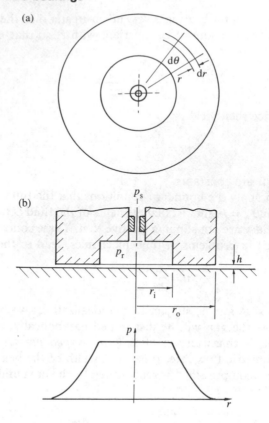

Fig. 6.4 (a) Geometry and (b) pressure distribution within a circular hydrostatic pad bearing.

while the central recess is of radius r_i. At a radius r within the land, that is, $r_i < r < r_o$, cylindrical symmetry leads to eqn (6.6), which describes the volumetric flow rate, becoming

$$Q = -\frac{h^3}{12\eta} \times \frac{dp}{dr} \times 2\pi r,$$

where Q is the *total* flow rate through the bearing. Rearranging and integrating from the inner to the outer edges of the land,

$$\frac{\pi h^3}{6\eta} \int_{p_r}^{0} dp = -Q \int_{r_i}^{r_o} \frac{dr}{r}$$

so that

$$Q = \frac{\pi h^3 p_r}{6\eta \ln(r_o/r_i)} \tag{6.8}$$

where p_r is the fluid pressure in the central pocket or recess, and the ambient pressure at the outer edge of the land is taken as zero. We can now note that since the flow rate at all annular sections must be the same, we could equally well write that the conditions at some intermediate radius r must satisfy the relation

$$\frac{\pi h^3}{6\eta} \int_{p_r}^{p} dp = -Q \int_{r_i}^{r} \frac{dr}{r}$$

so that

$$Q = \frac{\pi h^3 (p_r - p)}{6\eta \ln(r/r_i)}. \tag{6.9}$$

Thus, equating our two expressions for Q from (6.8) and (6.9), the variation of pressure p with radius r is given by the expression

$$\frac{p}{p_r} = \frac{\ln(r_o/r)}{\ln(r_o/r_i)}. \tag{6.10}$$

This fall in fluid pressure across the land is illustrated in Fig. 6.4(b). The total load W carried by the bearing is given by the integral

$$W = \int_{0}^{r_o} p \times 2\pi r \, dr.$$

Integrating by parts we can thus write this as

$$W = [p \times \pi r^2]_0^{r_o} - \int_{0}^{r_o} \pi r^2 \frac{dp}{dr} \, dr.$$

Since we have taken $p = 0$ at radius r_o it follows that the first term on the right-hand side of this equation is zero. Over the central pocket, $r_i < r < r_o$, the fluid pressure is constant at p_r and so the pressure gradient dp/dr is zero; thus the lower limit of the definite integral can be moved to r_i. In addition we can substitute for the pressure gradient dp/dr leading to

$$W = \frac{6\eta Q}{h^3} \int_{r_i}^{r_o} r \, dr$$

or

$$W = \frac{3\eta Q}{h^3} \left(r_o^2 - r_i^2 \right).$$

Substituting now the expression for the flow rate Q from eqn (6.9) gives

$$\frac{W}{\pi r_o^2 p_r} = \frac{1 - (r_i/r_o)^2}{2\ln(r_o/r_i)}. \tag{6.11}$$

If the width of the land $\Delta r = r_o - r_i$ is comparatively small compared to the radius of the bearing, then eqn (6.11) can be simplified, without great loss of accuracy, to

$$\frac{W}{\pi r_i^2 p_r} \approx 1 + \frac{\Delta r}{r_o}, \tag{6.12}$$

demonstrating, as suggested by Fig. 6.3(b), that most of the load in this case is carried by the pocket or recess.

The load-carrying capacity and associated flow rate for a hydrostatic bearing can often be conveniently expressed in non-dimensional terms by defining a normalized or non-dimensional load W^* and the non-dimensional flow rate Q^* as, respectively

$$W^* = W \div (\text{area of pad } A \times \text{pocket pressure } p_r) \tag{6.13}$$

and

$$Q^* = Q \div (h^3 p_r/\eta). \tag{6.14}$$

W^* and Q^* depend only on shape of the bearing: for a single circular pad eqns (611) and (6.8) show that

$$W^* = \frac{1 - (r_i/r_o)^2}{2\ln(r_o/r_i)} \quad \text{and} \quad Q^* = \frac{1}{6\eta \ln(r_o/r_i)}.$$

The variations in these quantities with the ratio of the outer to inner land radii r_o/r_i are plotted in Fig. 6.5.

In the more general case of a two-dimensional pocket bearing, that is, one without the essential degree of symmetry of the circular pad, there will be pressure gradients, and so fluid flow, in both the x- and y-directions, as indicated in Fig. 6.6. Referring to Fig. 6.6, continuity of flow requires that

$$\frac{\partial q_x}{\partial x} \times \delta x \times \delta y \times h + \frac{\partial q_y}{\partial y} \times \delta y \times \delta x \times h = 0, \tag{6.15}$$

where q_x and q_y are the volumetric flow rates per unit width in the Ox- and Oy-directions. These flows are still pressure driven and so in each case are proportional to the pressure gradient according to the equations

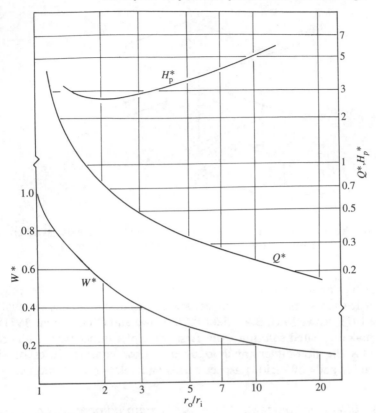

Fig. 6.5 Circular pad hydrostatic bearing non-dimensional characteristics. W^* is the non-dimensional load and Q^* the corresponding flow rate. H_p^* is the non-dimensional pumping power, see Section 6.2.4.

$$q_x = -\frac{h^3}{12\eta}\frac{\partial p}{\partial x} \quad \text{and} \quad q_y = -\frac{h^3}{12\eta}\frac{\partial p}{\partial y}. \tag{6.16}$$

Substituting into eqn (6.15) and cancelling the common factors yields the governing equation

$$\frac{\partial^2 p}{\partial x^2} + \frac{\partial^2 p}{\partial y^2} = 0. \tag{6.17}$$

This is Laplace's equation in two dimensions which, in principle, must be solved for the geometry of the particular bearing under consideration. Closed-form solutions are possible for some shapes while for others computer-assisted numerical analyses can be obtained without much diffi-

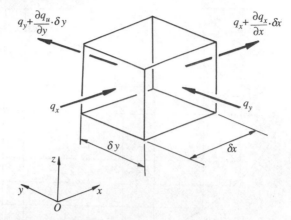

Fig. 6.6 Pressure-induced two-dimensional flow. The z-axis is taken to be normal to the two parallel faces of the bearing.

culty using standard routines. However, for the majority of simple, practical geometric shapes (annuli, rectangles, and so on) design guides are available (Neale 1973; Rowe 1983; Stansfield 1970; Rowe and O'Donoghue 1971) and these make repeated solution from first principles unnecessary. Figure 6.7 illustrates the non-dimensional loads and flow characteristics of single rectangular pads of various aspect ratios and relative land widths.

6.2.3 *Effect of non-central loading: bearing compensation*

In our analysis of a single hydrostatic pad we have made the assumption that the line of action of the applied load W passes through the geometric centre of the bearing. If this is not the case then the bearing will tilt as illustrated in Fig. 6.8(a). The increasing gap at the right-hand side means that the resistance to flow progressively decreases across the land on this side. Continuity of flow therefore requires that the pressure gradient becomes less steep, or more *concave*, as the fluid flows from the pocket. By the same token the narrowing gap at the left-hand side of the bearing leads to a *convex* pressure distribution as shown in Fig. 6.8(b). The bearing will continue to rotate or tilt until the centre of pressure of the modified profile is once again on the line of action of the load W. However, it is possible that by this stage the minimum film thickness $h_{\min}$ may have fallen to a dangerously small value, that is, a dimension which is comparable or even less than the size of surface asperities or lubricant contaminants, so leading to solid contact with its associated likely damage and wear.

   The practical solution in cases in which non-central loads may be encountered and which avoids this potentially damaging situation is to use a *pair*

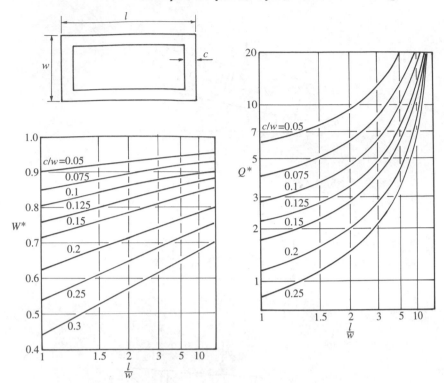

Fig. 6.7 Characteristics of a single rectangular hydrostatic pad bearing: the bearing has dimensions $l \times w$ and the land is of width c. W^* and Q^* are respectively the non-dimensional load and lubricant flow rate as defined in the text (from Rowe 1983).

of bearings as indicated in Fig. 6.8(c) and so arrange that the recess pressures are different, since if $p_2 > p_1$, a restoring couple, tending to push the surface back parallel, is provided. One way of achieving this is to arrange for *each* pocket to be supplied by a separate constant flow pump. To maintain the same flow through the narrower gap h_2 requires a greater pressure gradient, that is, a higher pocket pressure, than is associated with the greater gap h_1. For large and expensive pieces of machinery it may be economically feasible to provide as many flow pumps as bearings but this is unlikely to be realistic in the majority of applications. Note that merely supplying the two separate pockets from a common source or manifold will not produce an effective solution.

The alternative, and more usual, solution is to provide each bearing with some form of compensator which will allow the pocket pressures to vary in

(a)

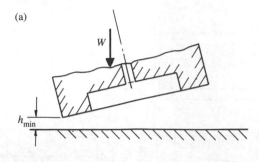

(b)

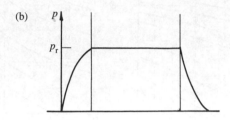

(c)

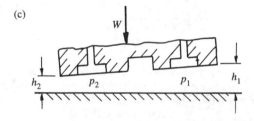

Fig. 6.8   Tilted pad: (a) geometry; (b) resultant pressure distribution; and (c) design with two pockets at differing pressures.

response to off-centre loads. Three such methods are commonly employed. In the first, each pad is provided with a constant-flow valve; this has a variable orifice controlled in some way by the flow itself and a number of such valves can be fed from the main supply manifold. Each pocket pressure can vary from a relatively small value up to a pressure approaching that of the supply pump. If the flow rate Q for each pocket is held constant then clearly from eqns (6.13) and (6.14) the load-carrying capacity increases as the gap h falls according to

$$W = W^*Ap_r = \left\{\frac{W^*A\eta}{Q^*}\right\}\frac{Q}{h^3}. \tag{6.18}$$

Changes in the load W give rise to variations in the film thickness h so that the bearing behaves as a stiff but (because of the inverse cube relation) very non-linear spring. A bearing stiffness k can be defined by differentiating this expression with respect to h, so that

$$k = -\frac{\partial W}{\partial h} = 3\left\{\frac{W^*A\eta}{Q^*}\right\}\frac{Q}{h^4}. \tag{6.19}$$

This expression implies that bearings with this form of *constant-flow compensation* exhibit greater stiffness with decreasing film thickness since k is proportional to h^{-4} (the minus sign in eqn (6.19) indicates that as W increases so h decreases).

A second, simple method of bearing compensation employs a restrictor constructed out of a length, say L, of small-bore capillary tube placed between the supply manifold, at pressure p_s and the recess where the pressure is p_r. The flow rate through a circular capillary of radius b is given by eqn (6.7) which, when integrated, is of the form

$$Q = \frac{C}{\eta}(p_s - p_r), \tag{6.20}$$

where the value of the constant C is equal to $\pi b^4/8L$.

Since the oil flow to the pad enters via the capillary (which should be as close to the pad as possible), we can equate the expression for Q given in (6.20) to that for the bearing given in (6.14), so that

$$\frac{h^3 Q^* p_r}{\eta} = \frac{C}{\eta}(p_s - p_r).$$

Provided there is no change in the viscosity between the restrictor and the pad then η can be cancelled through, so that after rearranging,

$$p_r = \frac{1}{1 + Kh^3}p_s \tag{6.21}$$

where $K = Q^*/C$ and so is a factor which depends on both the bearing shape and the type of restrictor. Since the load on the bearing is given by $W = W^*Ap_r$ it follows that the bearing stiffness is

$$k = \left|\frac{\partial W}{\partial h}\right| = \frac{3W^*Ap_s Kh^2}{(1 + Kh^3)^2}. \tag{6.22}$$

It is usually considered advantageous to maximize the stiffness of machine elements and, for a bearing of given size and shape, this can be achieved by an appropriate choice of the compensator characteristic C. Differentiating eqn (6.22) with respect to K and setting to zero shows that stiffness has a maximum when

$$Kh^3 = 1. \tag{6.23}$$

Equations (6.21) and (6.22) show that under this condition the recess pressure p_r is one-half of the supply pressure p_s. The bearing stiffness k_o at this design operating point is given by

$$k_o = \frac{3W^*Ap_s}{4h_o} = \frac{3W^*Ap_s}{4}\left\{\frac{Q^*}{C}\right\}^{1/3}. \tag{6.24}$$

The corresponding volumetric flow rate Q_o is given by

$$Q_o = \frac{p_s Q^* h_o^3}{2\eta}. \tag{6.25}$$

If the load is changed from the nominal design value then the film thickness, bearing stiffness, and volumetric flow rate will also change. These variations are shown, each non-dimensionalized by the design values W_o, h_o, k_o and Q_o respectively, in Fig. 6.9. In this plot the nominal operating point of the bearing has coordinates (1, 1).

The third form of compensator is a simple orifice plate placed in the supply line. The analysis of this form of compensation is rather more algebraically involved, largely as a result of the absence of laminar flow through a short, sharp-edged constriction. The flow through such a device of radius b can be written as

$$Q = \frac{\pi b^2}{2}\sqrt{\frac{p_s - p_r}{2\rho}} \times C_d, \tag{6.26}$$

where C_d is the discharge coefficient and ρ the fluid density; the value of C_d depends on the Reynolds number of the flow but is typically about 0.6. Equating the flow rates from (6.26) and (6.14) gives a quadratic for p_r whose value in this case is no longer independent of the viscosity η. This fact can be significant in the case of oil lubrication, since it means that as the bearing warms up and the viscosity falls, there will be a reduction in the bearing clearance h. In general an orifice plate compensator operating at its most favourable conditions will give a rather stiffer bearing than the corresponding best capillary-controlled design albeit at the cost of an increased flow rate.

6.2.4 Bearing optimization

Various criteria of optimization can be adopted in the design of hydrostatic bearings — these might include the maximization of either the load carried for a given flow rate or the bearing stiffness for a given supply pressure. We have seen that this latter condition requires that the pocket pressure is one-half of the supply value. The cost of running a bearing will depend

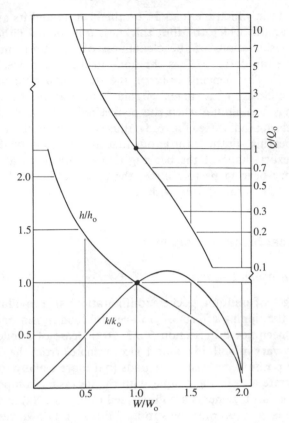

Fig. 6.9 Variations in non-dimensional film thickness, volumetric flow rate, and bearing stiffness, as a function of load for an optimized hydrostatic pad bearing (from Stansfield 1970).

greatly on the pumping power required to supply it, and consequently one reasonable design criterion is that of the power required by the bearing per unit load which it carries. The power supplied to the bearing, H_p, is related to the supply pressure and the flow rate simply by

$$H_p = p_s \times Q. \tag{6.27}$$

Viewed from this stand-point, an efficient design should minimize the power required per unit load supported, that is, H_p/W. For a bearing in which the stiffness is optimized, so that $p_r = 0.5 p_s$, this ratio can be expressed as

$$\frac{H_p}{W} = \frac{2h^3 W}{\eta A^2} \left\{ \frac{Q^*}{W^{*2}} \right\}. \tag{6.28}$$

In a situation in which the load W on the bearing and its available area A are fixed, and in which the film thickness h is limited either by manufacture or considerations of the cleanliness of the lubricant (which has viscosity η) we can define H_p^* as the non-dimensional power, that is, the group Q^*/W^{*2}. This depends only on the shape and proportions of the bearing. For the case of a single, circular pad bearing the curve of H_p^* versus r_o/r_i has a minimum when the pocket radius is approximately one-half of the bearing radius (see Fig. 6.5): thus each land is about one-quarter of the total bearing width. It turns out that in the region of the minimum in H_p^*, the exact shape of the bearing is not critical; as a general rule, for satisfactory bearing performance, the land width should be close to 25 per cent of the total bearing width.

6.3   Practical bearing designs

6.3.1   Multi-recess bearings

The advantages of multiple pocket configurations in providing improved resistance to the tipping effects of non-central loadings in one dimension have already been noted in Section 6.2.3. In the case of a load whose line of action may vary in both the x- and y-coordinates from the central, symmetric design position, at least three pads (but more usually four) are provided to generate a restoring couple with the necessary components. Two commonly used arrangements are illustrated in Figs 6.10(a) and 6.11(a). In these designs each recess must be supplied through its own restrictor. The non-dimensional characteristics of these bearings are plotted in Figs 6.10(b) and 6.11(b). The recommended design range is also indicated in the figures.

6.3.2   Opposed-pad designs

When a single hydrostatic pad is unloaded the volumetric flow rate will be limited solely by the characteristics of the restrictor. This is normally chosen to give close to optimum conditions within the bearing at the normal operating load so that if this is removed, or falls substantially, the fluid flow rate increases by a large factor. This can be seen from the data in Fig. 6.9; reducing the load by 50 per cent increases the flow rate by a factor of approximately 5 over that at the design conditions. Such a large increase may be unacceptable either from the demands that this implies on the supply pump or because of considerations on sealing, handling, and draining such an increased volume of fluid. The conventional arrangement used to overcome some of this difficulty utilizes a pair of pads operating in opposition to one another, as illustrated in Fig. 6.12.

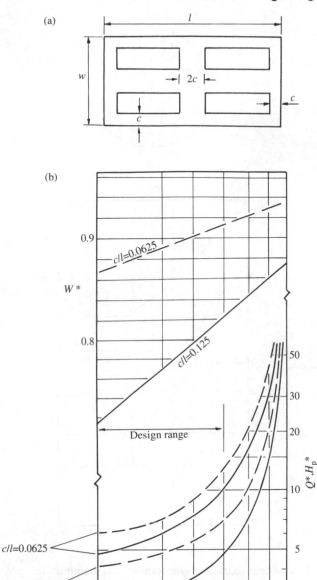

Fig. 6.10 Four-recess rectangular hydrostatic pad bearing: (a) geometry; and (b) non-dimensional characteristics. Full lines denote values of Q^* and dotted lines those of H_p^* (from Rowe 1983).

(a)

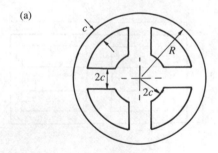

(b)

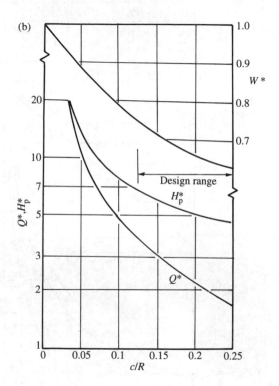

Fig. 6.11 Four-recess circular hydrostatic pad bearing: (a) geometry; and (b) non-dimensional characteristics (Rowe 1983).

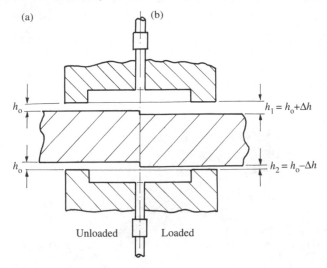

Fig. 6.12 Equal-area opposed-pad hydrostatic bearing: (a) left-hand side, unloaded, gaps symmetrical top and bottom; and (b) right-hand side loaded, upper gap increases while lower gap falls.

As the load on the centrally supported member is increased so the gap h_1 grows, while gap h_2 falls from the unloaded value of h_o. If the working load is as likely to be 'up' as 'down' then the pockets can be designed to have the same areas, and so the bearing, and its behaviour, are symmetrical. On the other hand, if the load is predominantly in one direction, then the pockets can have unequal areas to allow for this. The characteristics of a symmetrical, opposed design are plotted in Fig. 6.13. The bearing stiffness k has been non-dimensionalized by the value at the mid-position when $\Delta h = 0$; and load W non-dimensionalized by the load W_o which is that needed to cause Δh to grow to h_o so that one of the bearing clearances has fallen to zero. A practical maximum limit to the range of $\Delta h/h$ during the working life of the bearing, to avoid surface interactions and damage, might be of the order of 0.6 and over this range of loads ($W < 0.6\ W_o$) Fig. 6.13 shows that for an opposed-pad design the stiffness varies by only about 25 per cent and the flow rate by a factor of 2.

6.4   Velocity effects in hydrostatic thrust bearings

6.4.1   *Squeeze film effects*

In all the examples of hydrostatic bearings so far discussed the two opposed surfaces have had no component of relative velocity. In practical situations

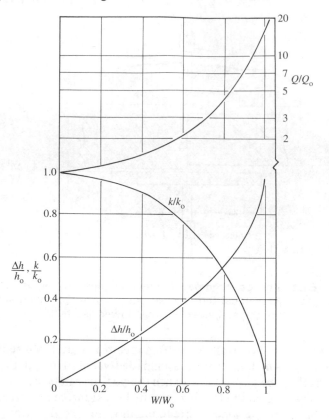

Fig. 6.13 Bearing characteristics of an equal-area opposed-pad hydrostatic bearing such as that illustrated in Fig. 6.12. W_o is the load required to reduce one of the bearing gaps to zero.

bearings may well be required to operate under dynamic as well as static conditions and we now consider some of these effects.

The general relative motion of the bearing can be resolved into two components, one being normal and the other parallel to the bearing surfaces. Normal velocities occur when the load on the bearing varies with time and the subsequent effects are known as *squeeze film* phenomena (see Section 7.9.1). As the surfaces move together fluid must be displaced at a volumetric rate equal to the product of the total bearing area and the velocity of approach. Some of this fluid adds to the leakage through the bearing gap while the remainder attempts to flow back through the control restrictor so increasing the pocket pressure. The squeeze force generated is thus directly proportional to the squeeze velocity and in a direction opposed to it; in other words the bearing possesses an inherent degree of viscous damping which effectively acts in parallel with the bearing stiffness.

The simplest dynamic model of a mechanical structure supported on such a hydrostatic bearing is shown in Fig. 6.14. k_s and λ_s represent the inherent stiffness and damping of the structure, and k_b and λ_b those of the bearing. The damping coefficient of the bearing is likely to be much greater than that of the structure and so, provided that the displacement x_b across the bearing is at least comparable with x_s (the displacement of the structure) significant amounts of beneficial damping can be introduced into the system via the bearing. Note, however, that if the bearing is designed to be very much stiffer than the rest of the device, that is, $k_b \gg k_s$, then the bearing will act more or less as a rigid member and damping of the overall system will depend solely on λ_s: the potential benefit of squeeze film damping will have been lost.

6.4.2 Sliding effects

Returning to the case of a steadily loaded bearing we can now consider the case in which there is a component of relative sliding velocity between the surfaces. Suppose that this is such that the lower slider is translating with a speed U relative to the upper surface which carries the pocket, as shown in Fig. 6.15. Analysis of the fluid flow through the gap will follow the same lines as for the static case outlined in Section 6.2.1 but with the appropriate changes in the velocity boundary conditions, namely that $u = U$ at $z = 0$ and $u = 0$ at $z = h$.

Applying these conditions to eqn (6.4) it is clear that the distribution of velocity across the fluid film is given by

$$u = \frac{1}{2\eta} \frac{\mathrm{d}p}{\mathrm{d}x} z(z - h) - \frac{U}{h}(z - h), \qquad (6.29)$$

a combination of the parabolic distribution of Poiseuille flow (first term) and the *linear* variation in velocity of *Couette* flow (second term).

The volumetric flow rate per unit width q is now given by

$$q = \int_0^h u \, \mathrm{d}x = -\frac{h^3}{12\eta} \frac{\mathrm{d}p}{\mathrm{d}x} + \frac{Uh}{2}. \qquad (6.30)$$

Designating the magnitude of the flow *out* of the bearing at *zero* velocity as q_0 we can write q_2, the corresponding flow out of the bearing at the right-hand edge when the lower surface is in motion, as

$$q_2 = q_0 + \frac{Uh}{2}. \qquad (6.31)$$

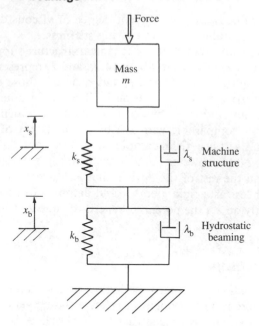

Fig. 6.14 Simple dynamic model of a machine structure supported on a hydrostatic pad bearing.

This distribution of velocity across the gap is shown in Fig. 6.15. The pocket pressure p_r is acting to pump fluid out of the gap and the effect of the sliding velocity U is to enhance this effect. At the left-hand edge of the bearing, the motion of the lower surface tends to draw fluid in so that the flow rate q_1 out of the bearing is given by

$$q_1 = q_0 - \frac{Uh}{2}.$$
(6.32)

The movement of the lower slider has thus not changed the *net* flow of fluid through the bearing since the sum of q_1 and q_2 is the same as in the stationary case. However, it is clearly possible to have a situation in which $Uh/2$ is of a greater magnitude than q_0, so that q_1 will become negative. This implies that fluid is being drawn into the bearing gap over part of the periphery of the land; this can only happen if the sliding surface approaching the bearing is provided with a sufficiently large volume of lubricant. If the approaching surface is dry, or carries an insufficient reservoir of fluid, then the fluid film on the upstream land must be incomplete and no longer extends across the full width of the bearing. The consequence of this state of affairs is that the load-carrying capacity will be much reduced. Even more

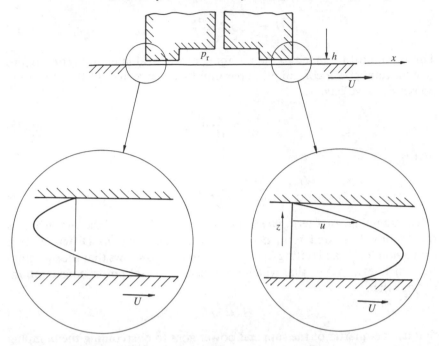

Fig. 6.15 Hydrostatic pad bearing in which the lower slider is moving with velocity U. Insets show the velocity profiles in the fluid film at the edges of the bearing.

serious may be the entrainment of air in an oil-lubricated bearing which can lead to the sudden collapse of the bearing through this phenomenon of *oil recession*. The practical remedy to this problem is to provide a source of low-pressure lubricant at the upstream edge of the bearing.

The maintenance of the motion of the lower slider in Fig. 6.15 requires the application of a horizontal force F whose value is related to the viscosity of the fluid and the velocity gradient across the film. If the pocket has a depth very much greater than the film thickness h (perhaps at least twenty times as great), then the shear rates across the land will be very much greater than across the pocket and we may write that

$$F \approx \frac{\eta U A_1}{h},$$

(6.33)

where A_1 is the area of the land. The frictional power then absorbed by the bearing is then

$$H_f = F \times U = \frac{\eta U^2 A_1}{h}. \tag{6.34}$$

This term must be added to the pumping power evaluated in Section 6.3 to give the total power consumption per unit load supported. Calling this total power H_{tot}, we have

$$\frac{H_{tot}}{W} = \frac{H_p}{W} + \frac{H_f}{W} \tag{6.35}$$

that is

$$\frac{H_{tot}}{W} = \frac{2h^3 W}{\eta A^2} \left\{ \frac{Q^*}{W^{*2}} \right\} + \frac{\eta U^2 A_1}{hW}. \tag{6.36}$$

The pumping, first term *increases* with the thickness of the film whereas the frictional, second term *decreases* with increasing h. Differentiating expression (6.36) and setting $\partial\{H_{tot}/W\}/\partial h$ to zero shows that the optimum film thickness, that is, that which minimizes the total power consumption, occurs when

$$H_f = 3H_p, \tag{6.37}$$

that is, one-quarter of the supplied power goes to overcoming the pumping losses. In the more general case when the bearing is not operating under these optimal conditions we might write that $H_f = S \times H_p$. Then

$$H_{tot} = (1 + S)H_p. \tag{6.38}$$

$S\ (>3)$ is a numerical factor which expresses the ratio between the frictional and the pumping losses.

Oil is a relatively poor conductor of heat and so it follows that in many applications it can be assumed that *all* the mechanical energy adsorbed by the bearing is convected away by heating the oil pumped through it: this oil is subsequently cooled elsewhere in the lubrication circuit. An estimate of the temperature rise of the oil on this basis will be an upper bound on that likely in practice. Such an estimate is of very real practical importance because the viscosity of many fluids, especially hydrocarbon oils, is very dependent on their temperature. Equating the mechanical power H_{tot} to the thermal energy absorbed by the oil per unit time we have

$$H_{tot} = (1 + S)H_p = Q\rho c \Delta\Theta \tag{6.39}$$

where ρ and c are the density and specific heat of the oil and $\Delta\Theta$ is the mean temperature rise. Since $H_p = Q \times p_s$ we can write that from eqn (6.39)

$$\Delta\Theta = \frac{p_s(1 + S)}{\rho c}. \tag{6.40}$$

If S is of the order of 3 and p_s a few MPa then eqn (6.40) indicates that using hydrocarbon oils (see Appendix 8) hydrostatic bearings generate only modest temperature increases of the order of 10–20°C.

Friction in hydrostatic thrust bearings

The simple circular thrust pad examined in Section 6.2.2 is an appropriate geometry for supporting the axial or thrust load in a shaft. If the shaft is at rest then the lubricant flows radially outwards from the recess but, once the shaft starts to rotate, say at a rate of Ω, a circumferential velocity will be imposed on this pattern. At any radius across the land r, the circumferential velocity will vary linearly through the thickness of the film from zero at the stationary component to $r\Omega$ at the surface of the rotating element. For a Newtonian lubricant the shear stress τ will again be given by eqn (6.2) so that

$$\tau = \eta \frac{\partial u}{\partial z} = \eta \frac{r\Omega}{h}. \tag{6.41}$$

This shear stress acts in opposition to the imposed motion, in other words it represents a frictional stress. Normally the recess depth is very much greater than the bearing clearance and so shear stresses over the recess can be neglected in comparison with those generated over the bearing land. To evaluate the total friction torque we consider a circumferential element of width δr on which the elemental torque δM will be equal to $\tau \times 2\pi r^2 \times \delta r$, so that the total torque M is given by

$$M = \int_{r_i}^{r_o} 2\pi \tau r^2 \, dr = \int_{r_i}^{r_o} \frac{2\pi \eta \Omega r^3}{h} \, dr = \frac{\pi \eta \Omega (r_o^4 - r_i^4)}{2h}. \tag{6.42}$$

6.5   Hydrostatic journal bearings

An obvious and important extension of the technology of hydrostatic thrust bearings is the hydrostatic support of loaded, circular shafts or *journals*, both stationary and rotating. Such hydrostatic journal bearings rely on an external pump, in just the same way as hydrostatic thrust bearings, but their behaviour and analysis is complicated by two additional factors. Firstly, except in the unloaded, concentric position the film thickness around the bearing is not uniform, and secondly, shaft rotation often results in comparatively high relative sliding velocities which can have significant effects on the pressure distribution.

Nevertheless the major advantages of hydrostatic designs remain, namely a high load-carrying capacity and very low friction—even at zero speed—

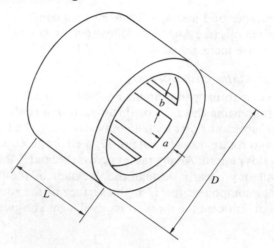

Fig. 6.16 General arrangement of a multi-recess hydrostatic journal bearing of diameter D and length L.

often coupled with a high bearing stiffness. The essential features of such a hydrostatic journal bearing are illustrated in Fig. 6.16. Typically there will be between four and eight pockets around the circumference of the bearing, each controlled by its own restictor, so that each recess and the surrounding land acts rather like an individual thrust pad. When the journal rotates, viscous drag will lead to additional circumferential flow from one recess to another which helps build up the supporting pressure under the load. Although these hydrodynamic pressures can be considerable, for the sake of safety such recessed hydrostatic bearings are usually designed on the basis that the maximum load must be carried entirely hydrostatically.

The shape of the fluid film in a stationary but loaded hydrostatic journal is shown in Fig. 6.17. O is the centre of the bearing and C the centre of the shaft or journal which is displaced from O by a small *eccentricity e*. Measuring the angle θ as shown the film thickness h can be written as

$$h = c(1 + \varepsilon\cos\theta), \qquad (6.43)$$

where c is the radial clearance, that is, the difference between the radii of the bearing and the journal; this is shown very much exaggerated in the diagram. ε is the *eccentricity ratio* defined as

$$\varepsilon = e/c. \qquad (6.44)$$

The minimum film thickness is thus $c = 1 - \varepsilon$. A typical design might have an $a{:}L$ ratio of 0.25 and four pockets with the compensation such

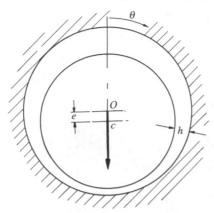

Fig. 6.17 Stationary hydrostatic journal bearing carrying a load *W*. *O* is the centre of the bearing and *C* the centre of the shaft or journal.

that the pocket pressure is one-half of the supply pressure. Defining a non-dimensional load W^*, in this case, by

$$W^* = \frac{W}{p_r L D} \tag{6.45}$$

the increase in the eccentricity of the journal with load is shown in Fig. 6.18 for designs with three different, but typical, length-to-diameter ratios.

Oil flow out of the bearing must escape from the two end lands each of width *a*. When the shaft is concentric a good estimate of the leakage flow rate can be made by assuming that each axial flow land is a slot of width πD and length *a*. It thus follows that the non-dimensional flow rate Q^* depends simply on the L/D ratio, thus

$$Q^* = \frac{Q}{p_r h_o^3 / \eta} = \frac{2\pi/3}{L/D}. \tag{6.46}$$

This relation is plotted in Fig. 6.19.

A hydrostatic journal bearing will exhibit a stiffness *k*, now in the radial direction, entirely analogous to that of a thrust bearing, so that

$$k = -\frac{\partial W}{\partial e}. \tag{6.47}$$

As in the case of a thrust bearing the value of *k* will depend on the ratio of the recess pressure to the supply pressure and the type of compensator employed: for a journal bearing the number of pads also has an influence. Defining a non-dimensional stiffness k^* by the relation

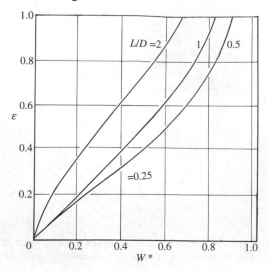

Fig. 6.18 Eccentricity ratios of four pocket hydrostatic journal bearings as functions of load for various L/D or length: diameter ratios. $a/L = 0.25$ in each case (from Stansfield 1970).

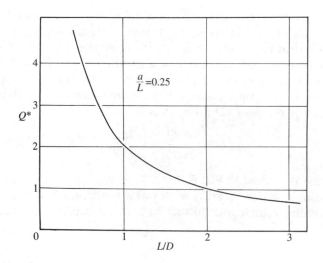

Fig. 6.19 Non-dimensional flow rate Q^* through a hydrostatic journal bearing as a function of its length/diameter, L/D, ratio.

Table 6.1 Dimensionless stiffness k^* of hydrostatic journal bearings of n pads: effect of different compensating elements (data adapted from Neale (1973)).

Number of pads, n	k^*		
	Constant flow	Orifice	Capillary
4	0.85	0.60	0.45
5	0.69	0.55	0.47
6	0.68	0.56	0.48

$$k^* = \frac{k}{p_r LD/e} \qquad (6.48)$$

the values of k^* for bearings with four, five, and six pads and the three forms of restrictors discussed in the previous section are given in Table 6.1.

The majority of loaded shafts will carry end thrusts as well as transverse loads and economies in flow rate and power consumption can be made by using a conical journal bearing to carry both components of load in place of separate thrust and parallel bearings. The basic arrangement of a recessed-pad conical bearing is shown in Fig. 6.20. Again here, each pad must be separately compensated. The choice of the most appropriate length-to-diameter ratio, cone angle, and so on, depend largely on the ratio of the thrust force to the maximum radial load. Stansfield (1970) and Rowe and O'Donoghue (1971) contain more details of appropriate design procedures.

6.6 Material selection and other design considerations

The choice of materials in hydrostatic bearings is generally less critical than in many other types of tribological components since, in normal operation, there should be no direct contact between the bearings faces. However, this does not mean that careful thought should not be given to material selection and the following considerations may well be of significance.

Movement with power disconnected

If the device is likely to be moved, adjusted, loaded, or unloaded, with the pressure supply disconnected then materials should be chosen to minimize the danger of damage to the surfaces. If the bearing is likely to be left unpressurized but carrying some non-negligible load for any length of time then the bearing surface materials should be selected to be an inherently low

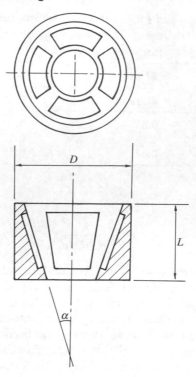

Fig. 6.20   General arrangement of a recessed-pad conical bearing.

friction or adhesion pair: the general rule of not using 'like' on 'like' materials should be observed (see Chapter 4).

High pressures or temperatures

Soft bearing materials may extrude, or creep, under the influence of high temperatures and pressures. Differential thermal expansion of the bearing components can sometimes be important, particularly with bearings operating with hot gas lubrication.

Corrosive lubricants

The bearing surfaces should be chemically stable to the lubricant or to any additives or contaminants that may find their way into the lubrication system.

An important aspect of the design of any hydrostatic bearing system is the collection and return of the leakage fluid. This not only requires the provision of a return flow system but also involves consideration of fluid sealing. It is vital to prevent, or at the very to least to control, the ingress of dirt or other foreign solid contaminants which could be damaging to the bearing surfaces. It may also be important, for both economic and environmental reasons, to prevent the loss of the lubricant to the outside world.

In a few situations it may be possible to totally discard leakage fluid. Obviously this is most often the case in air-lubricated bearings, but there are virtually no examples in which this is acceptable when using liquid lubricants. Many systems are totally enclosed so that the whole of the bearing area is shielded by a moving apron or bellows which not only traps the lubricant but also completely seals against contamination. Such an arrangement may be almost essential in arduous service conditions or in aggressive environments.

Conventional seals are either of the contacting or non-contacting designs. Contact seals rely on the rubbing seal formed between a metal surface and a flexible, usually elastomeric, element such as an 'O' ring or lip seal. Carefully designed these components can produce very low sealing forces and virtually negligible wear for long periods, although there is always the danger of sudden, serious damage to one of the sealing elements leading to a very rapid increase in the leakage rate. Non-contacting or *labyrinth* seals are those which allow some small leakage through a very long, tenuous, and thus high-resistance, flow path, such as a helical spiral on a circular shaft. These designs have the advantage that they are unlikely to suffer sudden breakdown, but they require careful and accurate manufacture and so are expensive. More details on the practical technology of sealing can be found in Section 10.5 and in the various tribology handbooks now available.

6.7 Further reading

Neale, M. J. (ed.) (1973) *Tribology handbook*. Butterworth, London.

Rowe, W. B. (1983) *Hydrostatic and hybrid bearing design*. Butterworth, London.

Rowe, W. B. and O'Donoghue, J. P. (1971) *Design procedures for hydrostatic bearings*. Machinery Publishing Co., Brighton.

Stansfield, F. M. (1970) *Hydrostatic bearings for machine tools*, Machinery Publishing Co., London.

7
Hydrodynamic bearings

7.1 Introduction

With a few important exceptions, engineering devices which involve the contact of loaded, sliding surfaces will only operate satisfactorily, that is, without giving rise to unacceptable amounts of surface damage or wear, when they are provided with adequate lubrication. The lubricant can act in two distinct, but not necessarily mutually exclusive, ways. The first of its functions may be to physically separate the surfaces by interposing between them a coherent, viscous film which is relatively thick (i.e. significantly larger than the size of likely surface asperities). In hydrostatic bearings this film is provided by an external pump and so its presence depends on the continuous operation of an external source of energy. In hydrodynamic bearings its generation relies only on the geometry and motion of the surfaces (hence the term dynamic) together with the viscous nature of the fluid. The second role of the lubricant may be to generate an additional thin, protective coating on one or both of the solid surfaces, preventing, or at least limiting, the formation of strong, adhesive and so potentially damaging friction junctions between the underlying solids at locations of particularly acute loading. If this protective coating has a comparatively low shear strength then the ultimate tangential force of friction can be much reduced: this mechanism of friction limitation is generally known as *boundary lubrication*. Such boundary films are generally very thin, perhaps only a few (albeit very large) molecules thick, and their formation and survival depends very much on the physical and chemical interactions between components of the lubricant and the solid surfaces. In this chapter we shall examine the development of hydrodynamic lubricant films: boundary lubrication is considered in Chapter 9.

All self-acting hydrodynamic bearings depend for their successful operation on the presence of a converging, wedge-shaped gap into which the viscous fluid is dragged by the relative motion of the solids. A pressure is generated which tends to push the faces of the wedge apart and it is the integrated effect of this pressure distribution within the fluid that balances the normal load on the bearing. The classical problem in hydrodynamic theory is the establishment of a relation between the velocity of sliding, the geometry of the surfaces, the properties of the lubricant and the magnitude of the normal load the bearing can support.

Figure 7.1 shows the rudiments of a two-dimensional hydrodynamic pad bearing: the bearing is 'long' in the y-direction so that there is no fluid flow in a direction normal to the plane of the paper. The upper, inclined fixed member is of length B, while the lower flat slider moves from left to right with velocity U. The continuous fluid film is of thickness h_i at the left-hand or inlet end of the bearing, and h_o at the outlet. Figure 7.1(b) shows diagrammatically the pressure distribution in the viscous fluid. The load supported by the bearing per unit length in the Oy-direction is W/L and its line of action is located at a distance χB from the leading edge; note that in general the pressure distribution is not symmetrical and so χ is not equal to 0.5.

Before investigating some aspects of this problem analytically we need to remember that diagrams such as Fig. 7.1(a) always, for clarity, greatly exaggerate the angle of the wedge or taper; in reality this is typically only about a quarter of a degree, equivalent to about 40 microns in 10 mm, or 1 inch in 20 feet. It is also worth noting that, as we shall see, the load-carrying capacity of a bearing of this sort is not greatly dependent on the precise *shape* of the wedge profile between the entry and exit regions, but is very sensitive to the *ratio* of the film thickness at entry to that at exit. In the analysis we shall assume, as in the previous chapter on hydrostatic bearings, that fluid flow is everywhere laminar. Conditions in large, rapidly rotating bearings can (although this is not common) be such as to generate superlaminar or turbulent flow and the implications of this change of regime are briefly discussed in Section 7.10.

7.2 Reynolds' equation in one dimension

The analysis of the type of situation illustrated in Fig. 7.1 (in particular that of a circular shaft supported in a journal bearing) was first carried out by Osborne Reynolds in 1886 and the general governing equation is now universally known by his name. Below we derive a simplified version of Reynolds' equation which, despite its simple form, is nevertheless applicable to a wide range of technologically important problems. The argument employed follows closely that already presented in Section 6.2.1 and 6.4.2 on hydrostatic bearings so that eqn (7.5) for example is identical to equation (6.4).

We start by considering the equilibrium of some small element of the fluid within the gap which has a local thickness h, as shown in Fig. 7.1. The magnitude of h varies in a known way from h_i at the entry to the con-vergent wedge, to h_o at the exit. Oxy is a stationary set of axes in which we

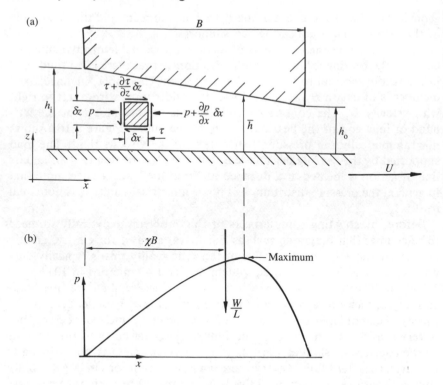

Fig. 7.1 (a) Essential features of a hydrodynamic bearing. (b) The resulting pressure profile, W/L, is the load per unit length (into the page).

observe the motion of the surfaces and that of the oil film. Clearly, for oil to be drawn into the gap it is necessary that $U > 0$.

Taking unit depth normal to the paper we have, by balancing forces on the incremental element (and neglecting the effects of gravity and inertia),

$$\frac{\partial p}{\partial x} \times \delta x \times \delta z = \frac{\partial \tau}{\partial z} \times \delta z \times \delta x$$

where p is the pressure in the fluid and τ the shear stress acting on the faces of the element. Thus

$$\frac{\partial p}{\partial x} = \frac{\partial \tau}{\partial z}. \tag{7.1}$$

On the basis that the fluid exhibits a constant Newtonian viscosity η throughout the bearing, we can relate the value of the shear stress to the local velocity gradient in the z-direction, so that

$$\tau = \eta \frac{\partial u}{\partial z},$$ (7.2)

where u is the local fluid velocity in the x-direction. Substituting into eqn (7.1) we obtain

$$\frac{\partial p}{\partial x} = \eta \frac{\partial^2 u}{\partial z^2}.$$ (7.3)

Now since h, the film thickness, is very much smaller than the dimensions of the wedge in the Ox- and Oy-directions, we can reasonably take the pressure p to be constant across the film thickness; in other words, p can be taken as a function of x only—its value is not dependent on the values of either y or z. Equation (7.3) then becomes

$$\frac{dp}{dx} = \eta \frac{\partial^2 u}{\partial z^2}.$$ (7.4)

This can be integrated twice with respect to x to yield

$$\eta u = \frac{dp}{dx} \frac{z^2}{2} + \alpha z + \beta$$ (7.5)

where α and β are constants.

We can now apply the boundary conditions that at the solid interfaces the fluid velocity is equal to that of the solid with which it is in contact: this is the usual no-slip boundary assumption of fluid mechanics. Setting $u = U$ at $z = 0$ and $u = 0$ at $z = h$ enables the constants α and β to be evaluated, and eqn (7.5) becomes and

$$u = \frac{1}{2\eta} \frac{dp}{dx} z(z - h) + \left\{ 1 - \frac{z}{h} \right\} U.$$ (7.6)

The velocity distribution within the fluid film is a combination of the parabolic distribution (characteristic of pressure-driven Poiseuille flow) represented by the first term of eqn (7.6), and the linear variation in velocity of shear (or Couette flow) represented by the second.

7.2.1   Volumetric flow rate

The volumetric flow rate q through a unit width (in the Oy-direction) of a section of the bearing can be directly obtained by integration, thus

$$q = \int_0^h u \, dz = -\frac{h^3}{12\eta} \frac{dp}{dx} + \frac{Uh}{2}.$$ (7.7)

For an incompressible fluid, the volume flow rate q must have the same value at all sections, or values of x, through the bearing gap. Since h_i, the film thickness at entry, is greater than h_o, that at exit, it follows that the pressure gradien dp/dx must be *positive* in the entry region of the bearing and change in sign to become *negative* as the exit is approached. This is consistent with a pressure distribution of the form shown diagramatically in Fig. 7.1(b). Designating the film thickness $\bar{h}$ at the point at which dp/dx is equal to zero, that is, where p has a maximum value, we can immediately write, from eqn (7.7), that

$$q = \left\{ \frac{U\bar{h}}{2} \right\}. \tag{7.8a}$$

This equation can also be written as

$$q = \bar{U}\bar{h}, \tag{7.8b}$$

by introducing $\bar{U}$, which is known as the *entraining velocity*; in this simple case $\bar{U}$ is simply equal to $U/2$. When both surfaces are in motion (and in particular when there is some rolling as well as sliding contact) some care must be taken to evaluate $\bar{U}$ correctly; see Section 7.6.1.

By equating the right-hand side of eqns (7.7) and (7.8b) and rearranging we have

$$\frac{dp}{dx} = 12\eta\bar{U}\frac{h - \bar{h}}{h^3}. \tag{7.9}$$

This is an elementary, but useful and often-quoted, form of Reynolds' equation in one dimension. The essential problem in classical hydrodynamics involves the integration of eqn (7.9) for a given film shape, that is, a knowledge of h as a function of x, to give a pressure distribution. A second integration, over the bearing area, then provides the value of the normal load that the bearing can carry.

Before embarking on an examination of some applications of Reynolds' equation, it is worthwhile looking at some of the essential physical idealizations on which it is based; these are briefly summarized in Table 7.1. In the analytical design of a component utilizing bydrodynamic action it is usual to start from this simplest form of Reynolds' equation to arrive at a first estimate of likely performance. This may be later refined by taking into account more realistic fluid or material properties such as non-Newtonian viscous behaviour or the elastic deformation of the solid surfaces.

Table 7.1 Simplifying idealizations in the one-dimensional form of Reynolds' equation.

1. *Fluid incompressibility*: generally satisfactory for all single-phase liquids, but may not be so for gases or liquids containing gas bubbles.
2. *Newtonian viscosity*: satisfactory for mineral oils at low pressures, but less so at higher pressures. Many other fluids, polymers, organic chemicals, biological fluids, and so on, are definitely non-Newtonian; see Chapter 8.
3. *Viscosity constant*: only even approximately true in mild conditions. Viscosity is very dependent on both temperature and pressure variations, see Chapters 1 and 8.
4. *Negligible inertia and turbulence*: generally satisfactory except at very high speeds or in very large bearings; see Section 7.11.
5. *Rigid solid surfaces*: at high pressures, or with compliant solids (e.g. rubbers and polymers) elastic deformation can make significant changes to geometry; see Chapters 3 and 8.
6. *Pressure constant through thickness of film*: satisfactory in virtually all cases.
7. *Smooth solid surfaces*: possibility of microasperity hydrodynamic lubrication where surface asperities are not small compared to hydrodynamic film thickness; see Chapter 8.

7.3 Plane slider pad bearings with infinite width

A pad bearing is one in which the load vector is perpendicular to the direction of relative sliding between the bearing faces. Typically, a single pad bearing, or an array of such bearings, is used to take the gravity loads between heavy sliding machine components, or to carry the axial component of load, or thrust, in a rotating shaft which passes through the casing of a machine or prime mover.

7.3.1 The exponential film profile

As an example of the application of Reynolds' equation consider the case of the converging wedge formed between a moving flat slider and a fixed component which has the exponential profile illustrated in Fig. 7.2. Although it would not be easy to manufacture, this shape does demonstrate all the essential features of pad bearings and has the advantage of being comparatively easy to analyse.

The fluid film extends over the region $-B > x > 0$ and is of thickness h_i at entry and h_o at exit. Suppose the bearing extends over the length L in the Oy-direction but that $L \gg B$ so that, to a first approximation, we can consider it to be 'infinitley wide' in the direction perpendicular to the page.

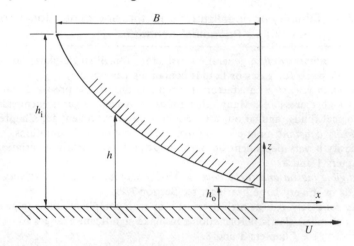

Fig. 7.2   Geometry of an exponential fluid film. The upper surface which carries the exponential profile is at rest while the lower plane surface has a velocity of magnitude U.

Through the bearing gap the film thickness h depends on the position x according to the equation

$$h = h_o \exp(-\alpha x) \tag{7.10}$$

where

$$\alpha = \frac{1}{B} \ln \frac{h_i}{h_o}. \tag{7.11}$$

If we suppose that $\dfrac{\mathrm{d}p}{\mathrm{d}x} = 0$ at $h = \bar{h}$ at which the x-coordinate has the value $\bar{x}$ then, since $\bar{U} = U/2$, Reynolds' equation, that is, eqn (7.9), becomes

$$\frac{h_o^2}{6U\eta} \frac{\mathrm{d}p}{\mathrm{d}x} = \exp(2\alpha x) - \exp(-\alpha\bar{x}) \exp(3\alpha x), \tag{7.12}$$

which can be integrated immediately to give

$$\frac{h_o^2}{6U\eta} p = \frac{\exp(2\alpha x)}{2\alpha} - \frac{\exp(-\alpha\bar{x}) \exp(3\alpha x)}{3\alpha} + C. \tag{7.13}$$

Although this equation only contains one constant of integration C we have two gets of boundary conditions, namely that the pressure is zero at both the entry to the bearing and at its exit. From these conditions we can evaluate both C and $\bar{x}$, the location of the peak pressure (or equally well, $\bar{h}$, the value of the film thickness at this point).

At the entry section of the bearing $x = -B$ and $p = 0$, while at the exit

$$x = 0 \quad \text{and} \quad p = 0. \tag{7.14}$$

If the dimension B is large then we might simplify the first of these conditions by setting the pressure equal to zero when $x = -\infty$. Together with the second condition of (7.14) this leads immediately to the values of C and $\bar{h}$ being given by

$$C = 0 \quad \text{and} \quad \bar{h} = \frac{3}{2} h_{\mathrm{o}}, \tag{7.15}$$

so that

$$p = \frac{3U\eta}{h_{\mathrm{o}}^2 \alpha} \left\{ \exp(2\alpha x) - \exp(3\alpha x) \right\}. \tag{7.16}$$

It is convenient to write this equation in a normalized or non-dimensional form by defining a non-dimensional pressure p^*, and a film thickness ratio H, by

$$p^* = \frac{ph_{\mathrm{o}}^2}{6U\eta B} \quad \text{and} \quad H = \frac{h_{\mathrm{i}}}{h_{\mathrm{o}}}, \tag{7.17}$$

so that

$$p^* = \frac{1}{2 \ln H} \left\{ \exp(2\alpha x) - \exp(3\alpha x) \right\}. \tag{7.18}$$

The load carried per unit width of the bearing, W/L, can be found from the straightforward integration

$$\frac{W}{L} = \int_{-\infty}^{0} p \, \mathrm{d}x = \frac{U\eta}{2h_{\mathrm{o}}^2 \alpha^2}. \tag{7.19}$$

In the case of a bearing in which B cannot reasonably be considered to be large, the corresponding equation to (7.18) for the non-dimensional pressure p^* is

$$p^* = \frac{1}{2 \ln H} \left\{ \exp(2\alpha x) - \frac{1}{H^2 + H + 1} \left[H(H+1) \exp(3\alpha x) + 1 \right] \right\}. \tag{7.20}$$

Integrating this expression gives the load or thrust on the bearing; this can be made non-dimensional by defining W^* as

$$W^* = \frac{Wh_{\mathrm{o}}^2}{U\eta LB^2}, \tag{7.21}$$

and in this case

$$W^* = \frac{3}{(H\ln H)^2} \left\{ \frac{H^2 - 1}{6} - \frac{H^2 \ln H}{H^2 + H + 1} \right\}. \tag{7.22}$$

The form of eqn (7.22) suggests that the load-carrying capacity W^* will be maximized by some particular value of H. Setting the value of dW^*/dH to zero shows that this occurs when H is approximately 2.3, and at this condition W^*_{max} is equal to 0.165.

We can also obtain an estimate of the *tangential* force necessary to maintain the faces of the bearing in relative motion. Since we have taken the fluid to exhibit constant Newtonian viscosity, the shear stress on the lower, sliding element will be given by

$$\tau_{z=0} = \eta \left(\frac{\partial u}{\partial z} \right)_{z=0}, \tag{7.23}$$

where u, the fluid velocity, is taken from eqn (7.6); it follows that

$$\tau_{z=0} = -\frac{h}{2} \left(\frac{dp}{dx} \right)_{z=0} - \frac{U\eta}{h}. \tag{7.24}$$

Using expression (7.16) for the variation in pressure along the bearing and considering the exponential expression for the film thickness (7.10) leads to

$$\frac{h_o \tau_{z=0}}{U\eta} = \frac{9}{2} \exp(2\alpha x) - 4\exp(\alpha x). \tag{7.25}$$

F/L, the tangential force per unit width of the lower element of the bearing, and thus the external force required to maintain the surfaces in relative motion, is given by integrating the surface shear stress over the total length of the bearing:

$$\frac{F}{L} = \int_{-\infty}^{0} \tau_{z=0} \, dx$$

whence

$$\frac{F}{L} = \frac{7U\eta}{4h_o\alpha}. \tag{7.26}$$

The ratio of this tangential force to the normal load carried is μ, the coefficient of friction of the bearing; it is thus given by

$$\mu = \frac{F}{W} = \frac{7}{2} h_o \alpha$$

which can be written as

$$\frac{7}{2\sqrt{2}}\sqrt{\frac{LU\eta}{W}}. \tag{7.27}$$

It is generally characteristic of hydrodynamic bearings that their coefficient of friction is dependent on the square root of a group which has the form of $(LU\eta/W)$, the value of the numerical factor depending on the shape of the surfaces.

7.3.2 Plane inclined bearings

Although easy to analyse, an exponentially shaped film is not the most practicable of film shapes, and indeed, it is not that profile which leads to the greatest load-carrying capacity. The simplest, and most common, geometry used in lubricated slider bearings is the plane inclined pad of fixed slope, as was illustrated in Fig. 7.1. The stationary, upper component is made with a sloping surface, where the angle of the slope is set during component manufacture. The lower plane surface moves from left to right at velocity U, so drawing the viscous lubricant into the convergent gap which, as before, decreases in thickness from h_i at entry to h_o at exit. The values of h_i and h_o will depend on the operating conditions but the difference between them, $h_i - h_i$, is a constant Δh. We shall again assume, initially at least, that the bearing extends infinitely in the Oy-direction that is, that there is no flow of lubricant perpendicular to the xz-plane.

Setting up an origin of coordinates at the entry to the bearing, h, the film thickness at position x, is given by

$$h = h_i - \frac{\Delta h}{B}x$$

or

$$\frac{h}{h_o} = H - (H-1)\frac{x}{B} \tag{7.28}$$

where again $H = h_i/h_o$.

To obtain estimates of the pressure distribution, and hence of the normal load-carrying capacity of a bearing of this configuration, eqn (7.28) must be substituted into Reynolds' equation, which from (7.9) with $\bar{U}$ again equal to $U/2$, can be written as

$$\frac{dp}{dx} = 6U\eta\frac{h - \bar{h}}{h^3}. \tag{7.29}$$

The integration is most easily carried out if it is noted that

$$\frac{dh}{dx} = -(H-1)\frac{h_o}{B}$$

so that

$$\frac{dp}{dh} = \frac{-6U\eta B}{(H-1)h_o}\frac{h-\bar{h}}{h^3}.$$

This expression can be immediately integrated between the limits $p = 0$ both at $h = h_i$ and $h = h_o$, showing that

$$\bar{h} = \frac{2h_o h_i}{h_o + h_i} \tag{7.30}$$

and

$$p = \frac{6U\eta B}{(H-1)}\frac{h_i}{h_o + h_i}\left\{\frac{1}{h} - \frac{1}{h_i}\right\}\left\{\frac{1}{h_o} - \frac{1}{h}\right\}. \tag{7.31}$$

Replacing h by the simple linear function of eqn (7.28) gives the variation of pressure with distance along the bearing. Expressed in terms of the non-dimensional pressure p^*, defined as in eqn (7.17), this becomes

$$p^* = \frac{(H-1)(B-x)x}{(H+1)\{BH - (H-1)x\}^2}. \tag{7.32}$$

The peak pressure occurs when $x = \bar{x}$ and $h = \bar{h}$ and is given by

$$p^* = \frac{H-1}{4H(H+1)}. \tag{7.33}$$

The point at which the pressure reaches this peak is rather more than half-way along the pad, since

$$\frac{\bar{x}}{B} = \frac{H}{H+1} \qquad \text{and} \qquad H > 1. \tag{7.34}$$

A second integration of eqn (7.32) with respect to x provides an expression for W/L the load per unit width of the bearing, once again this relation can be expressed in non-dimensional form,

$$W^* = \frac{Wh_o^2}{U\eta LB^2} = \frac{6}{H-1}\left\{\frac{\ln H}{H-1} - \frac{2}{H+1}\right\}. \tag{7.35}$$

The non-dimensional load W^* has a maximum numerical value of approximately 0.16 and occurs when H is close to the numerical value 2.3, so that at this configuration the film thickness at inlet is once again about 2.3 times that at exit; this can be thought of as the optimum geometry of the bearing. For a given normal load this value maximizes the smallest gap between the two solid surfaces and therefore is likely to minimize the damage to either of them by abrasive contaminant particles carried into the gap by the lubricant. If the load W on the bearing is increased above this optimum value, W_{op}, then the minimum clearance, h_o, will fall below the optimum value h_{op} according to the relation

$$\frac{W}{W_{op}} = 22.2 \left\{ \ln\left(1 + \frac{\Delta h}{h_o}\right) - \frac{2\Delta h/h_o}{2 + \Delta h/h_o} \right\}. \tag{7.36}$$

The variation in the ratio h_o/h_{op} with W/W_{op} is shown in Fig. 7.3 on which the point P represents the operating point of the bearing at optimum conditions and so has the coordinates $(1, 1)$. Increasing the load on the bearing by a factor of 2 *reduces* the minimum clearance by about 40 per cent.

Since the thickness of the oil film varies with changes in the applied load the bearing acts as a very stiff spring. The presence of this compliance may be important in the overall dynamic performance of the the machine of which the bearing is a part. An expression for its stiffness k, that is, the change in clearance per unit increase in normal load, can be obtained by differentiating eqn (7.35). Designating the stiffness at the operating point P by k_{op}, then as the load is changed so the stiffness is given by

$$\frac{k}{k_{op}} = \frac{3.44 (\Delta h/h_o)^3 (4 + 3\Delta h/h_o)}{(2 + \Delta h/h_o)^2 (1 + \Delta h/h_o)^2}. \tag{7.37}$$

This variation is also shown in Fig. 7.3.

The pressure variation within the bearing is not symmetrical about the mid-point. This is clear from eqn (7.34), which indicates that the maximum pressure must be developed when $x > 0.5$, so that the peak in the pressure curve always occurs towards the exit, or trailing, edge of the bearing. The position of the centre of pressure, that is, the location of the line of action of the normal force W, can be found by taking moments about the leading edge of the bearing. Locating the line of action of W at $x = \chi B$ we have

$$W\chi B = \int_0^B px \, dx. \tag{7.38}$$

The results of this evaluation for a range of values of the geometric factor H are given in Table 7.2. For practical values of H, the value of χ is always numerically close to 0.6.

The volumetric flow rate of the lubricant per unit width q can be written down immediately from eqns (7.8), (7.28), and (7.30) as

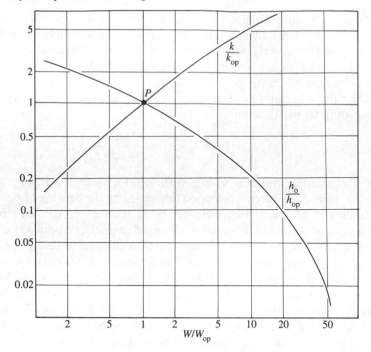

Fig. 7.3 Changes in minimum film thickness h_o and bearing stiffness k for a tilted-pad bearing as the load is changed from the optimum value W_{op}.

Table 7.2 Plane inclined-pad slider bearing: summary of bearing performance; H, W^*, F^*, and χ are defined in the text

H	1	1.5	2	2.3	2.5	3	4	5	6
W^*	0	0.13	0.159	0.160	0.158	0.148	0.124	0.104	0.087
F^*	1.0	0.844	0.773	0.745	0.729	0.697	0.648	0.609	0.576
χ	0.5	0.540	0.569	0.578	0.590	0.607	0.634	0.654	0.669

$$q = \frac{Uh_o}{2}\frac{H}{H+1}.\qquad(7.39)$$

We can also evaluate F, the friction force or tangential resistance to motion offered by the bearing, in the same way as was completed for the exponential film. Per unit width in the y-direction, the non-dimensional tangential force F^* required to maintain motion is given by

$$F^* = \frac{Fh_o}{B\eta UL} = \left\{ 4\frac{\ln H}{H-1} - \frac{6}{H+1} \right\}. \tag{7.40}$$

Values of F^* are also given in Table 7.2. The coefficient of friction exhibited by the bearing is given by

$$\mu = \frac{F}{W} = \frac{h_o}{B}\frac{F^*}{W^*}. \tag{7.41}$$

We can note that close to the optimum ($H \approx 2.3$), Table 7.2 shows that the ratio F^*/W^* is close to 4.7. In a pad bearing h_o/B might be typically 1 in 2500, so it follows that

$$\mu = \frac{1}{2500} \times 4.7 \approx 0.002. \tag{7.42}$$

Although this coefficient is small compared to dry sliding values, there is still a finite amount of mechanical work dissipated within the fluid film. At least to a first approximation, it can be assumed that all this energy is removed from the bearing by the oil flow, in other words, that only a negligible quantity of thermal energy is conducted or radiated away from the pad and the slider. It is then a straightforward matter to estimate $\Delta\Theta$, the temperature rise of the fluid, provided its appropriate physical properties, namely the density ρ, and specific heat c, are known. Equating the rate of mechanical working $\dot{H}$ to the rate of energy removal by the oil

$$\dot{H} = F \times U = Lq\rho c\Delta\Theta, \tag{7.43}$$

and by substitution for q, F, and h_o, it follows that

$$\Delta\Theta = \left\{ \frac{F^*}{W^*}\frac{H+1}{H} \right\} \frac{W}{LB}\frac{1}{\rho c}. \tag{7.44}$$

This is clearly just about the simplest idealization we could make. In practice more effort is likely to be dedicated to the estimation of bearing and lubricant temperatures; see for example the techniques of Section 7.8.

As soon as the relative sliding motion of the two elements of the bearing ceases then all the hydrodynamic load-carrying capacity falls to zero and the bearing collapses. For the simple geometry of Fig. 7.1 this is clearly a serious state of affairs as it means that any normal load applied to the bearing, even if this is only that of self-weight, will be applied to the sharp edge at one end of the upper surface with potentially damaging results. The obvious practical solution is to provide a parallel 'land' at the end of the taper, as shown in Fig. 7.4(a), which can carry any residual static load safely. The configuration is then known as a 'tapered-land'

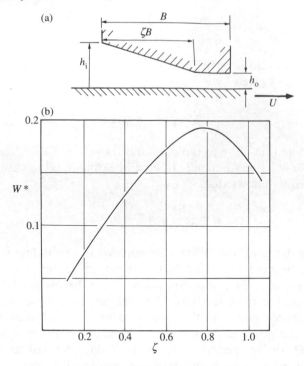

Fig. 7.4 Plane tapered land pad bearing: (a) geometry; and (b) variation of load W^* with the extent of the bearing land measured by the geometric parameter ζ.

bearing and can be analysed in the same straightforward manner, if with rather more algebraic labour, as the simple tapered bearing. Fig. 7.4(b) shows the variation of the non-dimensional load W^* with the extent of the land ζ, defined in the figure. An optimum configuration for the infinitely wide bearing occurs at $\zeta = 0.8$ and $H = 2.25$ when $W^* = 0.192$.

7.3.3 Rayleigh step bearing

In one of the classic papers in the development of hydrodynamics Lord Rayleigh, in 1918, used the calculus of variations to show that the film shape in an infinitely wide pad bearing which gives the greatest load-carrying capacity is one consisting of two parallel sections, as shown in Fig. 7.5. This design has become known as the Rayleigh step bearing.

To begin the analysis we can note that as the films in both the input and output sections are parallel it follows from Reynolds' equation that

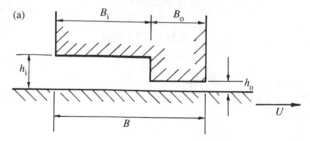

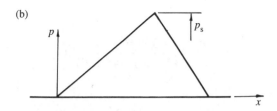

Fig. 7.5 Rayleigh step bearing: (a) geometry consisting of two parallel sections; and (b) the pressure distribution is linear in both regions of the bearing.

the pressure gradients in each must be constant. The pressure distribution along the bearing must therefore consist of two linear regions, as shown in Fig. 7.5(b); this means that we can write

$$\left\{\frac{dp}{dx}\right\}_i = \frac{p_s}{B_i} \quad \text{and} \quad \left\{\frac{dp}{dx}\right\}_o = -\frac{p_s}{B_o}, \tag{7.45}$$

where p_s is the peak pressure at the step and B_i and B_o are the lengths of the input and exit regions respectively.

Since the fluid is incompressible and we are idealizing the bearing as being infinitely long, it follows that q, the volume flow rate per unit width, must be the same at every section. Thus from eqn (7.7)

$$q = -\frac{h_i^3}{12\eta}\left\{\frac{dp}{dx}\right\}_i + \frac{Uh_i}{2} = -\frac{h_o^3}{12\eta}\left\{\frac{dp}{dx}\right\}_o + \frac{Uh_o}{2}. \tag{7.46}$$

Substituting from eqn (7.45) and rearranging, this become

$$\frac{p_s}{12\eta}\left\{\frac{h_i^3}{B_i} + \frac{h_o^3}{B_o}\right\} = \frac{U}{2}(h_i - h_o), \tag{7.47}$$

whence

$$p_s = \frac{P(H-1)}{H^3+P} \frac{6U\eta B_o}{h_o^2},$$ (7.48)

where $H = h_i/h_o$ and $P = B_i/B_o$.

W/L, the load carried per unit width, is given by integrating the expression for pressure, thus

$$\frac{W}{L} = \int_0^B p\,dx = \tfrac{1}{2} p_s (B_i + B_o),$$

so that

$$\frac{W}{L} = \frac{U\eta B^2}{h_o^2} \frac{3P(H-1)}{(1+P)(H^3+P)},$$ (7.49)

or in non-dimensional form,

$$W^* = \frac{Wh_o^2}{U\eta LB^2} = \frac{3P(H-1)}{(1+P)(H^3+P)}.$$ (7.50)

In step bearings of this sort, as well as the parameter H the designer has a second variable whose value must be chosen, namely P, the ratio of the lengths of the two parallel sections of the bearing. Differentiation shows that W^* has a maximum numerical value of $\frac{4}{3}\left(\frac{2}{\sqrt{3}}-1\right) = 0.206$ when $H = 1 + \tfrac{1}{2}\sqrt{3} = 1.87$, and $P = (5+3\sqrt{3})/4 = 2.59$.

7.4 Practical pad bearings: design considerations

7.4.1 *Effects of finite pad dimensions*

In the derivations of Section 7.3 we have made the idealization that the bearing is infinitely wide in the Oy-direction, so that no fluid flows out of the bearing gap in a direction perpendicular to the xz-plane. The finite width of real pad bearings means, of course, that significant volumes of fluid can leak out via this path; in addition, the fact that at the edges of the pad the pressure falls to zero means that the load-carrying capacity of a bearing with finite width is very much lower than the same width cut from an infinitely wide bearing.

The general solution to the analysis of bearings of finite length involves setting up and solving Reynolds' equation in two dimensions. This is left until the following section, although we can here display the results of some specific solutions which are relevant to real bearing designs. The normalized load W^* on a pad of finite width L can, by analogy with eqn (7.21), be defined as

$$W^* = \frac{(W/L)h_o^2}{\eta UB^2},\tag{7.51}$$

where W is the total load on the bearing. Figure 7.6 illustrates two aspects of the behaviour of a bearing of the sort. Q^* is the proportion of the fluid entering the bearing gap which is lost from the edges. For any given value of H this relative loss increases as the bearing gets narrower, that is, as L/B is decreases. For example, in a 'square' bearing (i.e. one in which $L/B = 1$), W^*

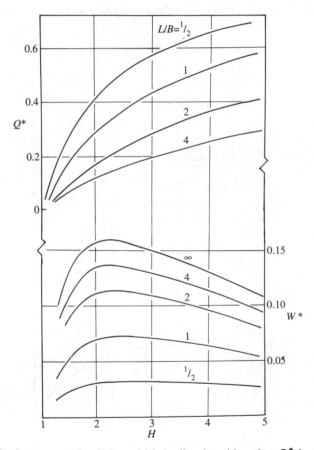

Fig. 7.6 Performance of a finite-width inclined-pad bearing Q^* is the proportion of fluid leaking from the sides and W^* the non-dimensional load. Figures on the curves are L/B ratios. H is the ratio of input to output film thickness (data from Pinkus and Sternlicht 1961).

250 Hydrodynamic bearings

has a maximum of 0.07 at a value of $H \approx 2.2$; under these conditions about one-third of the lubricant entering the bearing leaks away from the sides.

The most common application of inclined-pad bearings is to carry the thrust in a rotating shaft. Thus the pads will not be strictly square, or even rectangular, but rather sector shaped, and the effective sliding velocity will vary across the width of the pad. Defining in this case the normalized load per pad W^* as

$$W^* = \frac{W/L}{\eta R \omega} \left\{ \frac{h_0}{R\beta} \right\}^2 \tag{7.52}$$

where β is the angle subtended by each sector pad and R and L are as indicated in the inset to Fig. 7.7, then the variation of W^* with the film

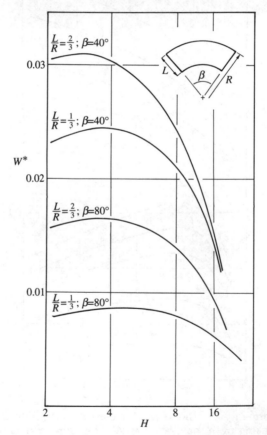

Fig. 7.7 Performance curves for inclined-pad sector bearings (data from Pinkus and Sternlicht 1961).

thickness parameter H is shown in Fig. 7.7. Values of of β of 80 and 40 degrees correspond approximately to systems having 4 or 8 pads in the thrust collar when making a reasonable allowance for the size of oil supply grooves that would be required to provide the necessary flow.

It is clear from the preceding comments that the side leakage associated with bearings of finite width has a major effect on their specific load-carrying capacity. For a square bearing there is little difference in the value of W^* between a plane inclined slider pad bearing and a Rayleigh step. The advantage of the step can, to some extent, be regained if the shape of the pocket is changed to restrict the side flow of fluid and retain the pressure; the design of Fig. 7.8 has been found to be a satisfactory optimum. The specific load-carrying capacity of this design is about twice that of a simple Rayleigh step.

7.4.2 *Tilting-pad bearings*

The angular taper (or the depth of the step) in a pad of fixed inclination is very small, and this in itself can create manufacturing difficulties in maintaining the components within the required tolerances. In addition, the taper required for efficient bearing operation may well be comparable with the changes in geometry resulting from thermal distortions of the bearing and associated machine components: there is thus always the danger of the load-carrying capacity being adversely affected by these effects. These difficulties can to a considerable extent be overcome by using a design in which the pad is pivoted and so is free to take up its own optimum angle to the slider. This form of thrust bearing is often

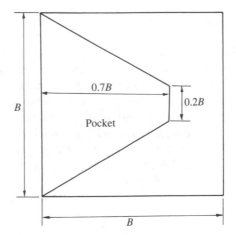

Fig. 7.8 Rayleigh step bearing with restricted side flow.

known, after its inventor, as the Michell bearing. In fact, this design of bearing was conceived simultaneously and independently in the early years of this century by two inventors, the other being Kingsbury, an American professor of mechanical engineering. In principle the correct position for the pivot will be at (or at least very close to) the centre of pressure and this has been shown, in Table 7.2, to be somewhat nearer the trailing edge of the bearing than its geometric centre. However, in practice, it has been found that in the great majority of cases placing the pivot on the centre line of the bearing produces a device which works perfectly well, so that often there is little point in calculating the theoretical optimum pivot position. Moreover a pad centrally pivoted has the advantage that it allows motion of the slider in either direction.

That such bearings actually work, when simple theory suggests that a centrally pivoted pad should be incapable of supporting a steady load, appears to violate the foundations of hydrodynamic theory. The generally accepted explanation is that in-service pressure (and perhaps also temperature) distortions of the pad, which is manufactured flat, lead to its becoming slightly convex in section so producing a geometry which is self-stable about a central pivot; however, there is still considerable debate about the full mechanism of load support in such 'parallel' bearings. It is ironic that this most subtle of tribological devices was invented so early in the history of the technology when no satisfactory explanation for its success could be offered. Tiliting-pad bearings can also use air as a lubricant but in them these distorsion effects are even smaller and some 'crowning' of the pad to create a convergent wedge geometry is essential; see Section 8.2.5.

7.5   Reynolds' equation in two dimensions

In order to facilitate the analysis of short' or 'narrow' bearings, both of inclined-pad geometry and of rotating or journal bearings it is necessary to extend our statement of Reynolds' equation. Consider the conditions illustrated in Fig. 7.9. This represents an incremental column with a cross-section having dimensions $\delta x \times \delta y$ situated between two rigid surfaces 1 and 2; these surfaces are separated by a fluid film which is of thickness h. The upper, inclined surface 2 has no *translational* velocity but may have a vertical velocity equal to $\mathcal{W}_2$. The lower surface 1 has an entraining velocity with components $\bar{U}$ and $\bar{V}$, parallel respectively to the Ox- and Oy-coordinate axes, together with a vertical component of velocity equal to $\mathcal{W}_1$.

Designating the volume flow rate per unit width in the Ox-direction as q_x, and in the Oy-direction as q_y, then the net volumetric outflow of fluid from the elemental column shown will be

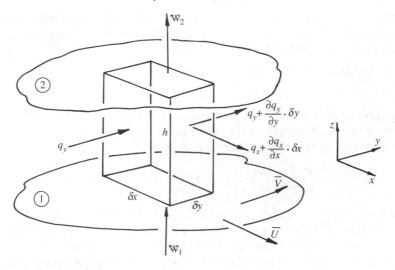

Fig. 7.9 Development of Reynolds' equation in two dimensions. The principles of continuity of flow are applied to the column element shown.

$$\frac{\partial q_x}{\partial x} \times \delta x \times \delta y + \frac{\partial q_y}{\partial y} \times \delta y \times \delta x. \qquad (7.53)$$

This must just balance the rate of *decrease* of the volume enclosed by the column as a result of differences in the velocity components in the Oz-direction, that is, $\mathcal{W}_1$ and $\mathcal{W}_2$, so that

$$\frac{\partial q_x}{\partial x} \times \delta x \times \delta y + \frac{\partial q_y}{\partial y} \times \delta y \times \delta x = (\mathcal{W}_1 - \mathcal{W}_2)\delta x \times \delta y$$

that is,

$$\frac{\partial q_x}{\partial x} + \frac{\partial q_y}{\partial y} + (\mathcal{W}_2 - \mathcal{W}_1) = 0. \qquad (7.54)$$

The arguments given in Section 7.2, leading to a relation between the volumetric flow rate and the pressure gradients in the individual coordinate axes directions still hold, so that we can write

$$q_x = -\frac{h^3}{12\eta}\frac{\partial p}{\partial x} + \bar{U}h \quad \text{and} \quad q_y = -\frac{h^3}{12\eta}\frac{\partial p}{\partial y} + \bar{V}h. \qquad (7.55)$$

Substituting eqn (7.55) into (7.54), and noting that since the solid surfaces are rigid their components of tangential velocity can be taken outside the differential operators, leads to the result

$$\frac{\partial}{\partial x}\left\{\frac{h^3}{\eta}\frac{\partial p}{\partial x}\right\} + \frac{\partial}{\partial y}\left\{\frac{h^3}{\eta}\frac{\partial p}{\partial y}\right\} = 12\left\{\bar{U}\frac{\partial h}{\partial x} + \bar{V}\frac{\partial h}{\partial x} + (\mathcal{W}_2 - \mathcal{W}_1)\right\}. \quad (7.56)$$

In the simpler forms of hydrodynamic analysis it is conventional to suppose that the viscosity of the lubricant is everywhere constant so that η can be taken outside the differentiation, hence

$$\frac{\partial}{\partial x}\left\{h^3\frac{\partial p}{\partial x}\right\} + \frac{\partial}{\partial y}\left\{h^3\frac{\partial p}{\partial y}\right\} = 12\eta\left\{\bar{U}\frac{\partial h}{\partial x} + \bar{V}\frac{\partial h}{\partial y} + (\mathcal{W}_2 - \mathcal{W}_1)\right\}. \quad (7.57)$$

The term $\mathcal{W}_2 - \mathcal{W}_1$ which expresses the rate at which the height of the incremental column changes, and so could also be written as dh/dt, is generally known as the *squeeze film* term. In steadily running bearings its value goes to zero; however it is important to note that in many practical situations, where conditions are not those of a steady state, this squeeze film term in neither zero, nor necessarily negligibly small compared to the contribution made to the hydrodynamic pressure form the convergent wedge action.

General analytic solutions to this full two-dimensional form of Reynolds' equation are difficult to obtain. In the majority of cases of practical interest, as a first approximation it is possible to simplify the equation by choosing axis directions so that the entrainment velocity $\bar{V}$ is effectively zero. In addition, it is usual to set either the remaining wedge term, $12\eta\bar{U}\dfrac{\partial h}{\partial x}$, or more commonly, the squeeze film term, to zero. Making this latter idealization eqn (7.57) becomes

$$\frac{\partial}{\partial x}\left\{h^3\frac{\partial p}{\partial x}\right\} + \frac{\partial}{\partial y}\left\{h^3\frac{\partial p}{\partial y}\right\} = 12\eta\bar{U}\frac{dh}{dx}. \quad (7.58)$$

The alternative idealization, in which the squeeze film term is dominant is considered in Section 7.9.1.

We shall take eqn (7.58) as our general statement of Reynolds' equation in two dimensions. It is clear that if the bearing is very long in the Oy-direction, perhaps in the limit approaching infinite length, then there can be no axial flow q_y or pressure gradient $\partial p/\partial y$. In such a case eqn (7.58) will become

$$\frac{\partial}{\partial x}\left\{h^3\frac{\partial p}{\partial x}\right\} = 12\eta\bar{U}\frac{dh}{dx}, \quad (7.59)$$

which can be integrated directly to give

$$\frac{dp}{dx} = 12\eta \bar{U} \frac{h - \bar{h}}{h^3}, \tag{7.9 bis}$$

where h is the value of the film thickness at which the pressure gradient becomes zero. We have now recovered the simple one-dimensional form derived in Section 7.2. As a useful rule of thumb, real bearings can usually be regarded for the purposes of analysis as being 'infinitely long' if their longer dimension perpendicular to the lubricant entraining velocity is at least four times their shorter dimension.

Infinitely narrow bearing

The converse idealization, the 'infinitely narrow' or 'infinitely short' bearing, involves neglecting the first term of eqn (7.58) in comparison with the second. Effectively this is to say that if the length L, that is, the dimension of the bearing in the Oy-direction is very much *less* than the dimension B then the corresponding pressure gradient must be very much *greater*, that is, that $\frac{\partial p}{\partial y} \gg \frac{\partial p}{\partial x}$, so that

$$\frac{\partial}{\partial y}\left\{ h^3 \frac{\partial p}{\partial y}\right\} \approx 12\eta \bar{U} \frac{dh}{dx}. \tag{7.60}$$

If h, the fluid film thickness, does not vary with y-coordinate we can further simplify this to

$$\frac{d^2 p}{dy^2} = \frac{12\eta \bar{U}}{h^3} \frac{dh}{dx}. \tag{7.61}$$

This equation, sometimes known as Ocvirk's equation, can be integrated directly, to give

$$p = \frac{12\eta \bar{U}}{h^3} \frac{dh}{dx} \frac{y^2}{2} + \mathcal{Q}y + \mathcal{B}, \tag{7.62}$$

in which $\mathcal{Q}$ and $\mathcal{B}$ are constants of integration. If the Cartesian origin is taken to lie on the mid-section of the bearing (see Fig. 7.10) then, because the pressure must be zero at $y = \pm L/2$, eqn (7.62) becomes

$$p = \frac{12\eta \bar{U}}{h^3} \frac{dh}{dx}\left\{ y^2 - \frac{L^2}{4}\right\}. \tag{7.63}$$

This model suggests that in these circumstances the fluid pressure is distributed parabolically in the axial or Oy-direction. As we shall see in the following section this simple analysis can be used with reasonable success to estimate the behaviour of narrow journal bearings (i.e. those with a length-to-width ratio of less than 1/4) although it is generally rather less satisfactory for predicting the behaviour of pad bearings.

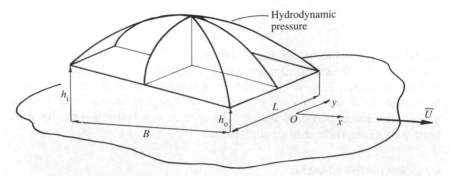

Fig. 7.10   Narrow pad bearing: the dimension L is much less than the dimension B.

7.6   Lubrication of discs

7.6.1   *Geometry*

Sliding bearings of the types described in Sections 7.3 and 7.4 are not the only machine elements within which lubrication by a fluid film plays a vital role; other examples are the contacts between the rollers, or balls, and the races in rolling-element bearings, and between individual gear teeth in geared drives. Many of these situations can be visualized or analysed (and often indeed physically modelled) as the contact between two cylinders or discs rotating at the appropriate speeds to give the correct combination of rolling and sliding contact. The first analysis of this geometry was carried as long ago as 1916 by Martin and is sometimes known by his name. In the initial treatment of this topic we shall assume, as has been implicit in the previous section, that the solid surfaces are rigid. In reality all solids must exhibit some compliance and so will deform as a result of the pressures applied to their surfaces; our idealization is really that these changes in surface profile are small in comparison with the other dimensions of the problem, in particular the thickness of the generated fluid film. This simplification makes the analysis much more straightforward although there are many practical cases in which it does not hold and where the elastic deformation of the surfaces, though small, is of vital importance to the development of a successful lubrication regime. These ideas are explored in Chapter 8.

Figure 7.11(a) shows two long circular cylinders 1 and 2 of radii R_1 and R_2 rotating about their *fixed* centres in opposite directions and in the presence of a viscous fluid. The peripheral speeds through the line of contact

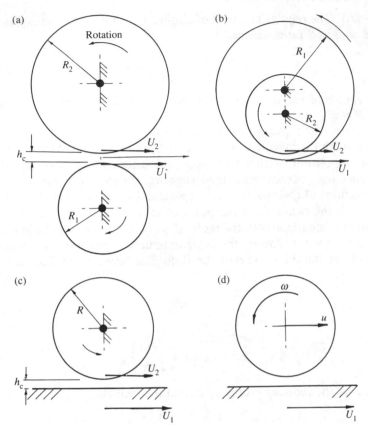

Fig. 7.11 (a) Geometry of two loaded cylinders of radii R_1 and R_2, each of which rotates about its fixed centre, in external contact. (b) Internal contact between a cylinder and a ring. (c) The cylinder and plane giving equivalent contact conditions to cases (a) and (b). (d) Cylinder 2 now translates with velocity u as well as rotating and care must be taken when evaluating the entraining velocity.

are U_1 and U_2 respectively. In this case $U_1 > U_2$, but the directions of both are such as to draw the viscous fluid into the gap between them, thus the entraining velocity $\bar{U}$ that is, that velocity which must be incorporated into Reynolds' equation, is given by

$$\bar{U} = \tfrac{1}{2}\,(U_1 + U_2) \qquad (7.64)$$

This is sometimes known as the *rolling velocity*. In our example, because U_1 does not equal U_2 there is also an element of relative sliding velocity between the two surfaces; this sliding term has a magnitude of ΔU equal to

$|U_1 - U_2|$. The relative amounts of sliding and rolling may be expressed by S, the *slide:roll* ratio, defined by

$$S = \frac{\text{sliding velocity}}{2 \times \text{rolling velocity}} = \frac{\Delta U}{2\bar{U}}.$$ (7.65)

If the speeds of rotation were such that U_1 and U_2 were equal then clearly $\Delta U = S = 0$.

Note that if the discs were to rotate in the same sense and in such a way that although the magnitudes of U_1 and U_2 remained equal their directions were in opposition, so that $U_2 = -U_1$, the entraining velocity (eqn (7.64)), and hence the hydrodynamic load support, would fall to zero.

The centres of the two discs are separated by a distance of $R_1 + R_2 + h_c$, where h_c is the clearance at the point of closest approach and we suppose is small in comparison with the radii. If we set our origin on the line joining the centres, then, following the same argument as was used in Section 3.2, we can write the thickness h of the fluid film between the disc surfaces as

$$h \approx h_c + \frac{x^2}{2R_1} + \frac{x^2}{2R_2}$$

or

$$h \approx h_c + \left\{1 + \frac{x^2}{2Rh_c}\right\},$$ (7.66)

where R is the reduced radius of curvature such that

$$\frac{1}{R} = \frac{1}{R_1} + \frac{1}{R_2}.$$ (7.67)

If the contact is between a disc and the internal surface of a ring as shown in Fig. 7.11(b) then eqn (7.67) still applies provided the radius of curvature of the *concave* surface is taken as negative. Equation (7.66) shows that the conditions in Figs 7.11(a) and (b) are geometrically equivalent to the lubricated contact of a cylinder of radius R and a plane surface as illustrated in Fig. 7.11(c) — once again the relative motion of the surfaces is such as to draw in a convergent film of viscous fluid. The contact in all these cases is 'long' in the direction perpendicular to the figure, so that there is no axial flow. Within the fluid the pressure will satisfy Reynolds' equation (7.9):

$$\frac{dp}{dx} = 12\eta\bar{U}\frac{h - \bar{h}}{h^3}.$$ (7.9 bis)

The object of our analysis is again to integrate Reynolds' equation and so to obtain estimates of the pressure p when the variation in film thickness h is given by (7.66).

Entrainment in sliding and rolling contacts

Contact between two surfaces, such as between a cylinder and a flat surface, may be somewhat more complicated than that described above when an element of the sliding velocity arises from the *translation* of the centre of the cylinder. The entraining velocity $\bar{U}$ which must be used in Reynolds' equation, depends on the velocity with which the oil or lubricant is dragged into the contact and therefore does not involve absolute velocities but those seen relative to the contact patch or zone itself. If surface 1 has an absolute velocity U_1 and surface 2 a corresponding velocity U_2, but, in addition, the contact zone itself has velocity u_c, then the entraining velocity $\bar{U}$ is given by

$$\bar{U} = \tfrac{1}{2}\{(U_1 - u_c) + (U_2 - u_c)\}. \tag{7.68}$$

In the situations of Figs 7.11(b) or (c) where u_c is equal to zero, the position of the contact patch is stationary and so there is no problem. However, consider the situation of Fig. 7.11(d) in which the lower plane surface 1 is moving with speed U_1 but the upper cylinder, which is of radius R, is not only rotating about its centre at speed ω, but is also moving from left to right with speed u: thus U_2 is equal to $u + R\omega$. The translation of the cylinder effectively *reduces* the tendency for the lubricant to be entrained in the gap to the left of the line of contact and this effect must be incorporated into Reynolds' equation. The contact line must always be below the centre of the upper cylinder and so is moving to the right with velocity u, that is, $u_c = u$. Thus the entraining velocity is given by

$$\bar{U} = \tfrac{1}{2}\{(U_1 - u) + (u + R\omega - u)\} = \tfrac{1}{2}\{(U_1 - u) + R\omega\}. \tag{7.69}$$

If it so happens that u is equal to $U_1 + R\omega$ then it is clear that $\bar{U} = 0$ and this geometry will produce no hydrodynamic load support so that the bearing collapses; this can be an important effect in *bearing whirl* (see Section 8.2.4) and in the contact between a rotary cam and a reciprocating follower when at some point in the operating cycle the lubricant entraining velocity must fall to zero.

The value of the component of sliding velocity between the solid surfaces ΔU is given by $\Delta U = U_1 - U_2$, that is

$$\Delta U = (U_1 - u_c) - (U_2 - u_c).$$

Again the relative significance of sliding and rolling may be expressed by S the slide:roll ratio. From eqn (7.65) we can write

$$S = \frac{\Delta U}{2\bar{U}} = \frac{|(U_1 - u_c) - (U_2 - u_c)|}{|(U_1 - u_c) + (U_2 - u_c)|}. \tag{7.70}$$

In the case illustrated in Fig. 7.11(d) it follows that

$$S = \frac{|U_1 - R\omega - u|}{|U_1 + R\omega - u|}.$$

In pure rolling contact (e.g. in the above example if $U_1 = R\omega$ and $u = 0$) then the slide:roll ratio S is equal to 0, while in pure sliding contact (e.g. if ω is zero) then $S = 1$.

7.6.2 Sommerfeld, half-Sommerfeld, and Reynolds boundary conditions

We now return to the contact of two discs. To evaluate the load we must integrate Reynolds' equation and specify some suitable boundary conditions, that is, values of the pressure p at two specific points on the profile. Clearly, remote from the origin the pressure in the film must fall to zero but it is not at all obvious at what values of the x-coordinate this can be taken to have occurred. Perhaps the best initial estimate is to take p as zero at very large distances from the point of closest approach of the surfaces, that is, to put $p = 0$ where $x = \pm \infty$. This choice is known as the *full Sommerfeld* condition and leads to a neat solution of Reynolds' equation which is consistent with the disc profiles, namely that

$$p = -4\frac{\bar{U}\eta x}{h^2}. \tag{7.71}$$

Once again this can be expressed conveniently in a non-dimensional form. In this case the non-dimensional pressure p^* is best defined by

$$p^* = \frac{ph_c^2}{12\bar{U}\eta\sqrt{2Rh_c}}, \tag{7.72}$$

and ξ is the non-dimensional form of the position coordinate x, sometimes known as the *Sommerfeld* variable, defined by

$$\xi = \frac{x}{\sqrt{2Rh_c}}. \tag{7.73}$$

Equation (7.71) which describes the variation of pressure through the contact then becomes

$$p^* = -\frac{\xi}{3(1 + \xi^2)^2}. \tag{7.74}$$

This distribution is plotted in Fig. 7.12 as curve A. It is immediately obvious from this, or indeed from an examination of either of eqns (7.71) and (7.74),

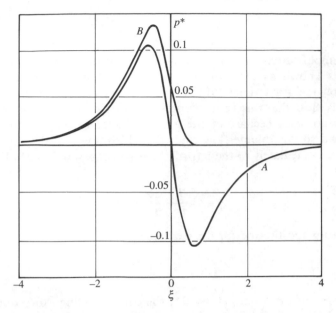

Fig. 7.12 Pressure distributions through cylindrical contact with (A) full Sommerfeld conditions, and (B) Reynolds boundary conditions.

that this full Sommerfeld pressure distribution is *antisymmetric* about the position axis, the negative pressure curve being exactly the reverse image of the positive. The maximum value of p^* occurs when $\xi = 1/\sqrt{3}$ and p^* is then of magnitude $\sqrt{3}/16 = 0.1083$. At this point the local film thickness is equal to $\frac{4}{3}h_c$.

Since the load per unit length on the discs is represented by the area under the pressure curve, the full Sommerfeld solution provides *no* net load support because of the negative pressure region. Engineering lubricants are generally incapable of withstanding large, continuous, negative pressures and this feature of their behaviour leads to the lubricant film in the exit region exhibiting the phenomenon of *cavitation*: the film ruptures into a series of fingers or streamers separated by spaces within which the pressure is very low compared to the positive pressures that are generated within the convergent film. The obvious and easiest way of treating this situation mathematically is simply just to discount the lobe of negative pressure so that W/L, the load per unit length, is given by the contribution from the positive pressure only, that is, to suppose that $W/L = \int_{-\infty}^{0} p \, dx$, whence, substituting from eqns (7.66) and (7.71) it follows that $W/L = 4\bar{U}\eta R/h_c$ or non-dimensionally,

$$W^* = \frac{Wh_c}{\eta RL\bar{U}} = 4.00. \tag{7.75}$$

The simplification used here of simply neglecting the negative pressure regime is known as the *half-Sommerfeld* condition. Although giving a reasonably realistic estimate of the load-carrying capacity in any particular case, it is clear that this form of pressure distribution cannot be *actually* realized, not least because of the abrupt change in slope in the pressure curve (and so the volume flow rate) that this implies at the origin. From eqn (7.7), immediately to the left of the origin at $x = 0^-$, the volume flow rate q is

$$q = -\frac{h_c^3}{12\eta}\frac{dp}{dx} + \bar{U}h_c$$

which, after substituting for dp/dx, becomes

$$q = -\frac{\bar{U}h_c}{3} + \bar{U}h_c. \tag{7.76}$$

To the right of the origin, at $x = 0^+$, the corresponding expression for the volumetric flow rate will not contain the first term of the right-hand side of eqn (7.76) since $dp/dx = 0$. Thus there is a discrepancy in flow implied by the half-Sommerfeld conditions which has a magnitude of 33% of the flow out of the bearing; this is hardly negligible.

This difficulty with the solution can be overcome if an alternative exit boundary condition is sought. We suppose that somewhere in the outlet region, that is, when $x > 0$, there is a point within the fluid at which the pressure p and its first derivative dp/dx or their non-dimensional counterparts) are both simultaneously zero. This idealization is known either as the Reynolds boundary condition (as it was implied, if not stated explicitly in Reynolds' paper of 1886) or the Swift–Stieber condition, after independent theoretical justification by these two authors fifty years later. The entry boundary condition remains the same, namely that the pressure in the film is zero when $x = -\infty$. In non-dimensional terms this exit boundary condition can be written as

$$p^* = \frac{dp^*}{d\xi} = 0.$$

Reynolds' equation, expressed non-dimensionally, then becomes

$$\frac{dp^*}{d\xi} = \frac{\xi^2 - \xi'^2}{(1 + \xi^2)^2}.$$

where the parameter ξ has the particular value ξ' at the exit. A solution of this equation is

$$p^* = \frac{1}{2}\left\{\gamma + \frac{\pi}{2} + \frac{\sin2\gamma}{2} - \sec^2\gamma'\left[\frac{3}{4}\left(\gamma + \frac{\pi}{2}\right) + \frac{\sin2\gamma}{2} + \frac{\sin4\gamma}{8}\right]\right\} \qquad (7.77)$$

where $\tan\gamma = \xi$ and, correspondingly, $\tan\gamma' = \xi'$.

Equation (7.77) gives the pressure distribution plotted as curve B in Fig. 7.12. The maximum value of p^* in this case is 0.127 and occurs when ξ' is equal to -0.475 (which point corresponds to values of $\gamma = 25.4°$ and $H = 1.23$). It follows from our definition of p^* in eqn (7.71) that the peak pressure p_{max} is given by

$$p_{max} = 2.15\bar{U}\eta\sqrt{\frac{R}{h_c^3}} \qquad (7.78)$$

The load W/L carried by the discs can be found by straightforward integration. In non-dimensional terms,

$$W^* = \frac{Wh_c}{\eta R\bar{U}L} = 4.89. \qquad (7.79)$$

When this is compared to the corresponding equation for the half-Sommerfeld boundary conditions (7.74) it can be seen that adopting the Reynolds boundary conditions has led to an *increase* in the estimated non-dimensionalised bearing load W^* from 4 to 4.89.

7.7 Plain journal bearings

Journal bearings are the most widely used form of machine component using hydrodynamic load support. A loaded, rotating shaft or *journal* is supported in a circular bush which has a slightly greater diameter. The geometry is shown in Fig. 7.13 (a). C is the centre of the journal and O the centre of the bush or bearing; c is the *clearance* or difference in radii between the journal and the bearing which is here shown very greatly exaggerated – in practice it is typically less than 1% of the bearing radius. Consider what happens if the shaft which carries a unidirectional load W starts to rotate with a small angular speed ω. The journal and the bearing surface are in contact at the point P and here a contact force R will be generated which is equal and opposite to the load W. This force R is often for convenience shown split into a pair of perpendicular components N and F, as shown in the figure: N is the normal contact force and F the frictional force which, since there must be sliding contact at P, is equal in magnitude to μN. The height of P above the lowest point in the bearing depends on the magnitude of the coefficient of friction μ at the contact point. To maintain the shaft in steady motion requires the application of a torque equal in magnitude to the couple formed by W and R.

(a)

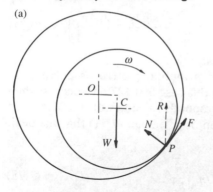

(b)

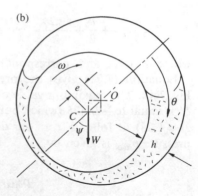

Fig. 7.13 Plain journal bearing; the journal rotates at speed ω, with (a) no lubrication present, or (b) hydrodynamic load support.

Now consider how the picture changes if both the bearing is supplied with a viscous lubricant and the speed of the journal is sufficient for hydrodynamic activity to be initiated. There is clearly a convergent wedge formed by the clearance gap on the upstream side of P: fluid is drawn into this gap pushing the solid surfaces apart. At some speed the point C will be vertically below the bearing centre O. However, this is rather a special circumstance; in general, the geometry will be as shown in Fig. 7.13 (b) with C to the left of O.

If the *eccentricity* or distance between the centres of the bearing and journal is designated by e then we can define an *eccentricity ratio* ε by the relation

$$\varepsilon = \frac{e}{c}, \tag{7.80}$$

and the minimum gap between the solid surfaces h_{min} is then given by

$$h_{min} = c - e = c(1 - \varepsilon). \tag{7.81}$$

To a satisfactory degree of accuracy for most purposes, the gap h will be related to the circumferential position θ by the equation

$$h = c(1 + \varepsilon \cos\theta) \tag{7.82}$$

where θ measures the angular position from the position of maximum film thickness. The line OC and the load vector will *not* be collinear and the angle ψ between them is known as the *attitude angle*.

7.7.1   *Narrow-bearing solution*

An important geometric parameter in the behaviour of journal bearings is the ratio of the bearing length to its diameter, L/D, in Fig. 7.14. In practice this ratio varies from about $\frac{1}{8}$ in the narrowest of journals to about 2. A 'square' bearing is one in which $L/D = 1$.

If the ratio of L/D is comparatively small, say less than $\frac{1}{4}$, then the short or narrow bearing analysis of Section 7.5 can be employed with acceptable accuracy. We have, from Section 7.5 for such a bearing, a parabolic pressure distribution, namely

$$p = \frac{12\eta}{h^3} \frac{dh}{dx} \bar{U} \left\{ y^2 - \frac{L^2}{4} \right\} \tag{7.83}$$

provided that y, the axial position coordinate, is measured from the central section of the bearing. Combining this equation with (7.82) and noting that $\bar{U} = R\omega/2$, $x = R\theta$, and $dh/dx = -c\varepsilon/R\sin\theta$, we can write that the pressure distribution around the bearing is given by

$$p = \frac{3\pi\omega\varepsilon \sin\theta}{c^2(1 + \varepsilon \cos\theta)^3} \left\{ \frac{L^2}{4} - y^2 \right\}. \tag{7.84}$$

Note that, because of the sine term, the distribution of pressure is anti-symmetric about the line OC, that is, the full Sommerfeld conditions are implied, so that we are in effect supposing that the lubricant film is able to withstand large negative pressures.

An estimate of the load carried by this narrow journal bearing can be made by adopting the half-Sommerfeld boundary conditions, that is, by setting the film pressure equal to zero at values of θ between π and 2π, the range over which they would otherwise be negative. In this case, referring to Fig. 7.15,

$$W_z = \int_0^\pi \int_{-L/2}^{+L/2} pR \sin\theta \, dy \, d\theta \tag{7.85}$$

and

$$W_x = -\int_0^\pi \int_{-L/2}^{+L/2} pR \cos\theta \, dy \, d\theta. \tag{7.86}$$

Introducing the expression (7.84) for the pressure into these integrals leads to the expressions

$$W_z = \frac{\eta R\omega L^3 \varepsilon \pi}{4c^2(1 - \varepsilon^2)^{3/2}} \quad \text{and} \quad W_x = \frac{\eta R\omega L^3 \varepsilon^2}{c^2(1 - \varepsilon^2)^2}. \tag{7.87}$$

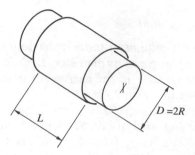

Fig. 7.14 Plain journal bearing length L and diameter D ($=2R$); in a narrow bearing $L/D < 0.25$.

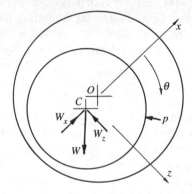

Fig. 7.15 Loads on a journal bearing. The applied load W must be carried by the sum of the component W_x and W_z.

The magnitude of the total load W on the journal is given by that of the resultant of these two components, that is,

$$W = \sqrt{W_x^2 + W_z^2}$$

so that, after a little algebra

$$W = \frac{\eta R \omega L^3}{4c^2} \frac{\pi \varepsilon}{(1 - \varepsilon^2)^2} (1 + 0.62\varepsilon^2)^{1/2}. \qquad (7.88)$$

Defining a non-dimensional load W^*, in this case, by the expression

$$W^* = \frac{W/L}{\eta R \omega} \left\{ \frac{c}{R} \right\}^2 \qquad (7.89)$$

and using the fact that the shaft diameter $D = 2R$, eqn (7.88) becomes

$$W^* = \left\{\frac{L}{D}\right\}^2 \frac{\pi\varepsilon}{(1-\varepsilon^2)^2} (1 + 0.62\varepsilon^2)^{1/2}. \tag{7.90}$$

Although we could express results in the analyses of journal bearings in terms of the non-dimensional number W^*, the more usual non-dimensional group is known as the *Sommerfeld number S*. In what follows we shall take S to be defined by

$$S = \eta\omega \frac{LD}{W} \left\{\frac{R}{c}\right\}^2 \tag{7.91}$$

so that

$$S = \frac{2}{W^*}. \tag{7.92}$$

In terms of the entraining velocity, the Sommerfeld number S can be written as

$$S = \frac{4\eta L\bar{U}}{W} \left\{\frac{R}{c}\right\}^2. \tag{7.93}$$

Equation (7.90) therefore becomes

$$\frac{1}{S} = \left\{\frac{L}{D}\right\}^2 \frac{\pi\varepsilon}{2(1-\varepsilon^2)^2} (1 + 0.62\varepsilon^2)^{1/2}. \tag{7.94}$$

Other definitions of the Sommerfeld number are possible and are sometimes used. Consequently, care must be taken to make certain of the form of this group when referring to design guides, manufacturers' data, and so on. In eqn (7.91) ω is measured in radians per second; another possible non-dimensional group S' is defined by the identity

$$\frac{1}{S'} = \frac{1}{\eta\mathfrak{N}} \frac{W}{LD} \left\{\frac{c}{R}\right\}^2 \tag{7.95}$$

wherec $\mathfrak{N}$ is the rotational speed in revolutions per second; this definition is often found in American texts. It thus follows that

$$S' = 2\pi S. \tag{7.96}$$

The *attitude angle* ψ of the bearing, that is, the angle between the load vector and the line joining the centres of the bearing and the journal, can be evaluated from the relation

$$\tan\psi = -\frac{W_z}{W_x} = \frac{\pi(1-\varepsilon^2)^{1/2}}{4\varepsilon}. \tag{7.97}$$

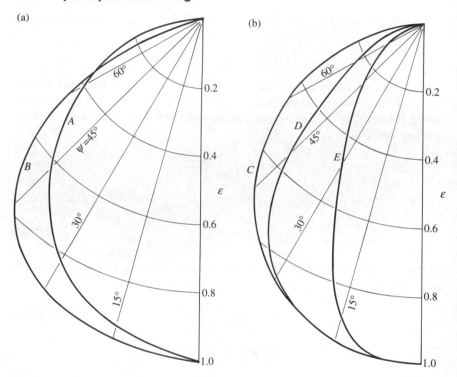

Fig. 7.16   Attitude angles, as functions of eccentricity ratio ε, for: (a) full 360°
journal bearings A: the infinitely narrow bearing, B: the infinitely long bearing;
and (b) journal bearings with $L/D = 1$ and included angles of C: 360°, D: 180°,
and E: 60° (data from Raimondi and Boyd (1958) and Pinkus (1958)).

Figure 7.16(a), curve A, shows this relationship graphically as a polar plot.
For *narrow* full bearings this curve is close to a semicircle in shape.

Figure 7.16(a), curve B, shows a corresponding plot for the infinitely long
bearing, while Figure 7.16(b) illustrates the effect of changing the included
angle of the journal for a square bearing ($L/D = 1$). Bearings with a reduced
included angle, that is, partial bearings which do not completely enclose the
shaft, can have some practical advantages; see Section 7.7.3.

An important practical matter in the design of the lubricant system is
the volumetric flow rate of fluid through the bearing. Adopting the half-
Sommerfeld boundary conditions, that is, supposing that pressure is zero
at values of the angle θ between π and 2π, the flow of lubricant per unit
length at entry Q_{in}, and exit Q_{out} of the convergent section are respectively

$$Q_{in} = \tfrac{1}{2}R\omega c(1 + \varepsilon)L \quad \text{and} \quad Q_{out} = \tfrac{1}{2}R\omega c(1 - \varepsilon)L. \quad (7.98)$$

The difference between these two flows must leak from the ends of the bearing and so represents the flow which must be supplied to prevent the bearing becoming starved of lubricant. Expressed non-dimensionally as Q^*, the proportion of the volume flow rate supplied to the bearing, this loss is given by

$$Q^* = \frac{Q_{in} - Q_{out}}{Q_{in}} = \frac{2\varepsilon}{1 + \varepsilon}. \tag{7.99}$$

Figure 7.17 illustrates this relationship for the narrow bearing. Also plotted on the same axes are the relative make-up flows for two other plain journal bearings with length-to-diameter ratios of 1 and $\frac{1}{2}$.

7.7.2 Long-bearing solution

Pressure distribution and loads carried

A journal bearing whose length is more than about four times its diameter can reasonably be considered to be infinitely long; in this case there can be little or negligible axial lubricant flow. Conditions are governed by the one-dimensional form of Reynolds' equation given in eqn (7.9):

$$\frac{dp}{dx} = 12\eta\bar{U}\frac{h - \bar{h}}{h^3}, \tag{7.9 bis}$$

where, as in the previous example, x is measured around the bearing circumference. To convert this equation to a more convenient form in cylindrical polar coordinates we can again replace x by $R\theta$ and $\bar{U}$ by $R\omega/2$. Also introducing eqn (7.82), which relates the gap h to the angular position θ, Reynolds' equation can be written in a normalized form as

$$\frac{dp^*}{d\theta} = \frac{c^2}{6R^2\eta\omega}\frac{dp}{d\theta} = \frac{1}{(1 + \varepsilon\cos\theta)^2} - \frac{1 + \varepsilon\cos\bar{\theta})}{(1 + \varepsilon\cos\theta)^3}, \tag{7.100}$$

where $\bar{\theta}$ identifies the position of the maximum pressure and p^* is a non-dimensional pressure defined by

$$p^* = \frac{c^2}{6R^2\eta\omega}p. \tag{7.101}$$

Equation (7.100) can be solved explicitly in a number of ways. For example, making use of the *Sommerfeld substitution*,

$$\cos\gamma = \frac{\varepsilon + \cos\theta}{1 + \cos\theta},$$

and using the boundary conditions that p^* falls to zero when θ takes the values 0 and 2π leads to the equation

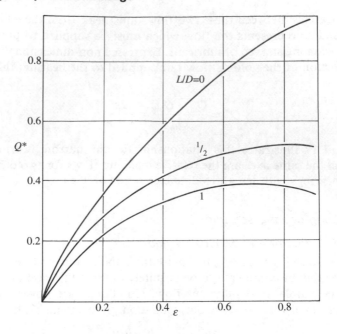

Fig. 7.17 Lubricant normalized make-up flows Q^* for full journal bearings of various L/D ratios plotted as functions of the bearing eccentricity ratio.

$$p^* = \frac{\varepsilon\sin\theta\,(2 + \varepsilon\cos\theta)}{(2 + \varepsilon^2)\,(1 + \varepsilon\cos\theta)^2}. \tag{7.102}$$

The load on this infinitely long bearing can be evaluated in much the same way as that on the infinitely short bearing. Adopting the full Sommerfeld conditions and so allowing the possibility of negative pressures within the fluid, the expressions for the components of the load in the Ox- and Oz-directions of Fig. 7.15 become, respectively,

$$W_x = 0 \quad \text{and} \quad W_z = \frac{12\eta R^3 \omega L}{c^2}\,\frac{\pi\varepsilon}{(1 - \varepsilon^2)^{1/2}(2 + \varepsilon^2)}. \tag{7.103}$$

Thus it follows from the definition of the Sommerfeld number, S, that

$$\frac{1}{S} = \frac{6\pi\varepsilon}{(1 - \varepsilon^2)^{1/2}(2 + \varepsilon^2)}. \tag{7.104}$$

Additionally, we can note that since W_x is always zero, the attitude angle ψ must always be 90°, in other words, that the angle between the line of centres of the journal and the bearing is normal to the load line. The predic-

tion that the locus of the shaft centre is a line perpendicular to the direction
of the applied load seems at first sight unlikely, but is a natural consequence
of the adoption of the *full* Sommerfeld conditions. For this to be a realistic
picture, the fluid forming the film must not cavitate, which means that the
negative pressures over the region $\pi < \theta < 2\pi$ must be small, well below one
atmosphere. In fact, under these circumstances there is good experimental
evidence that the shaft locus is close to that predicted.

The preceding analytic solutions can easily be adapted, by changing the
limits of integration, into the half-Sommerfeld conditions. These discount
the lobe of negative pressure but suffer the disadvantage of introducing a
significant discontinuity in flow at the position $\theta = \pi$ or 180°. More satisfactory is the adoption of the Reynolds' conditions, that is, to seek a pressure
distribution which is consistent with an exit point at which both p and its
derivative $dp/d\theta$ are zero. In this case the film extends beyond the point of
minimum clearance (at which $\theta = 180°$) by the angle θ'. Table 7.3 gives some
values for this angle as well as other performance parameters of such an
infinitely long bearing.

In practice the bearing designer must choose an acceptable combination
of the values of the Sommerfeld number (which is a measure of the severity
of the bearing duty) and the eccentricity ratio which will leave the minimum
gap sufficiently large both to make solid contact between the asperities on
the surfaces of the journal and the bearing unlikely, and reduce damage
arising from the entrainment of abrasive dirt or contaminant in the oil. A
typical, if rather conservative, design value of ε might be about 0.6, and
Fig. 7.18 illustrates the circumferential pressure distribution (plotted non-
dimensionally as p^*) according to both Sommerfeld (curve A) and Reynolds
(curve B) boundary conditions for an eccentricity ratio of this magnitude.
Also shown, as curve C, for comparison is the Sommerfeld curve for a value
of $\varepsilon = 0.4$.

As well as the effect of the value of the Sommerfeld number on the
eccentricity ratio (and hence the minimum film thickness) we may also be
interested in the location of the point where this occurs, relative to the load

Table 7.3 The behaviour of the infinitely long journal bearing with
Reynolds boundary conditions; variables are defined in the text.

S	2.99	1.52	0.78	0.53	0.40	0.31	0.24	0.19	0.13	0.07	0.04
ε	0.05	0.1	0.2	0.3	0.4	0.5	0.6	0.7	0.8	0.9	0.95
W^*	0.67	1.32	2.57	3.80	5.06	6.46	8.17	10.6	15.1	27.7	52.3
ψ (°)	70.0	69.0	66.9	64.5	61.6	58.3	54.2	49.1	42.2	31.7	23.2
θ' (°)	73.3	69.2	61.3	53.8	46.6	39.7	33.1	26.6	20.2	13.2	9.0
M^*	6.24	6.23	6.35	6.64	7.11	7.78	8.75	10.2	12.7	18.1	25.8

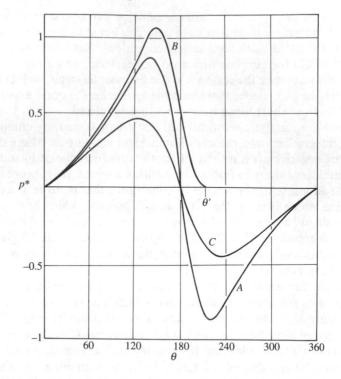

Fig. 7.18 Circumferential non-dimensional pressure distribution p^* around a 'long' journal bearing; curves A and C are for Sommerfeld boundary conditions with $\varepsilon = 0.6$ and $\varepsilon = 0.4$ respectively, and curve B is for Reynolds boundary conditions with $\varepsilon = 0.6$.

line. This information is provided by the attitude angle ψ, and the variation in this with ε for a long full (i.e. 360°) bearing is also given in Table 7.3 and displayed in Fig. 7.16 as curve B; curve A is the corresponding information for an infinitely short bearing.

Observation in the field suggests that some bearings operate satisfactorily for periods at which their eccentricity ratios are very high, perhaps in excess of 0.9, and thus with their surfaces separated by very thin lubricant films. The thickness of the predicted lubricant film under these conditions may well be less than the asperity size on the journal and bearing. Were the solid surfaces to be absolutely rigid and undeformable this would imply significant areas of true contact between them and so we should expect considerable consequential damage and wear, which is often not in fact observed. The explanation for the successful operation of such contacts lies principally in the local elastic flattening of surface roughnesses that takes place within the confines of the contact zone; these *elasto-hydrodynamic* effects are considered more fully in the following chapter.

7.7.3 Practical considerations

Slenderness ratio

The influence of the bearing Sommerfeld number on the eccentricity ratio for full journal bearings of various values of L/D or *slenderness ratio* is summarized in Fig. 7.19. Increasing the load on the bearing reduces the numerical value of the Sommerfeld number S and so the operating point must move from right to left. Thus for a bearing of a particular value of L/D the eccentricity ratio increases. Once the slenderness ratio has fallen to about $\frac{1}{4}$ the situation is effectively dominated by axial lubricant flow and so the short-bearing analysis of Section 7.7.1 can be applied; the short-bearing solution, eqn (7.94), is shown dotted in Fig. 7.19. Also on this diagram are displayed the results from the application of the half-Sommerfeld and Reynolds boundary conditions to the infinitely long bearing.

From a practical point of view there are limits on both the maximum and minimum values of the L/D ratios that can be sensibly used. If this ratio is too low, that is much less than about $\frac{1}{4}$, then the volumetric flow rate demanded by the bearing becomes excessively large for the load that can be supported. On the other hand, if L/D is much more than about 2 there can be difficulties in satisfactorily aligning the journal and bearing. Even if these can be set up with the shaft unloaded, subsequent elastic flexing under the action of the applied and dynamic loads can lead to unacceptably thin lubricant films at the edges of the bearings. The recommended design range is indicated by the area with the dashed perimeter.

Friction in journal bearings

The presence of a film of viscous lubricant between the journal and the bearing means that a torque must be continuously supplied to maintain rotation of the bearing elements. If the journal and the bearing were concentric then the velocity gradient across the fluid film would be equal to $R\omega/c$ and thus the frictional torque M/L per unit length would simply be given by

$$\frac{M}{L} = \frac{2\pi\eta\omega R^3}{c}.\tag{7.105}$$

This is known as the *Petrov equation*, and can be used to provide a quick estimate of the frictional effects in a real design, even when the actual operating value of ε is unknown. In practice, we know that the journal and the bearing will not be concentric, but will operate at some characteristic value of the eccentricity ratio ε. Nevertheless, friction effects can be accounted for: taking the case of an infinitely long bearing operating with a full fluid film between the solid surfaces, we can evaluate the total drag force F experienced by the shaft. F represents the integration around the circumference of the viscous shear force and is given by the equation

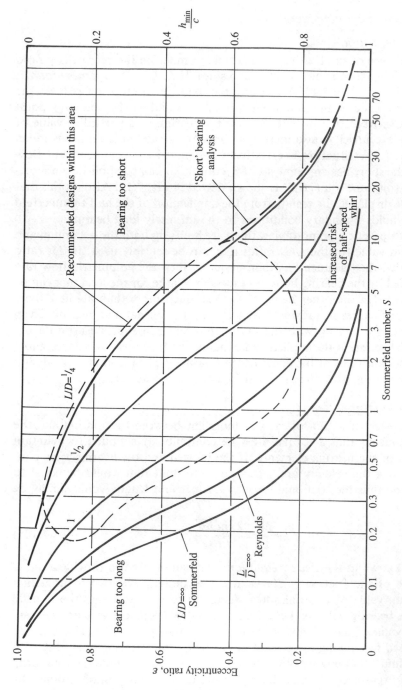

Fig. 7.19 Eccentricity ratio ε versus Sommerfeld number S for 360° journal bearings of various L/D ratios (data from Raimondi and Boyd (1958)).

$$F = \frac{4\pi\eta\omega R^2 L}{c} \frac{1 + 2\varepsilon^2}{(2 + \varepsilon^2)(1 - \varepsilon^2)^{3/2}}. \tag{7.106}$$

We can relate the value of the eccentricity ratio ε to the Sommerfeld number S and this means that the drag force F can also be expressed in terms of S. One way of displaying such a relationship in a non-dimensional form is to define a coefficient of friction μ for the bearing as F divided by the bearing load W. From eqns (7.103) and (7.106) it follows that for an infinitely long bearing operating with the full Sommerfeld conditions

$$\mu \left\{ \frac{R}{c} \right\} = \frac{1 + 2\varepsilon^2}{3\varepsilon}. \tag{7.107}$$

This equation indicates that friction will be a minimum when the eccentricity ratio ε has the numerical value $1/\sqrt{2} = 0.71$. Although the prediction of a minimum frictional resitance is perhaps rather more a result of the idealizations made in the analysis than a true representation of realtity, this value of ε is not unrepresentative of many practical steadily loaded journal bearings. Figure 7.20 shows some curves of $\mu(R/c)$ versus S for full journal

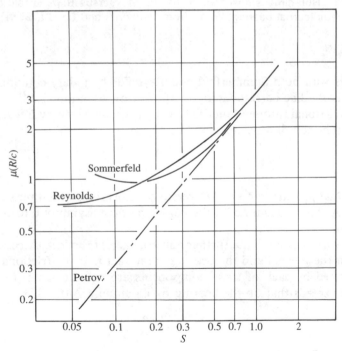

Fig. 7.20 Bearing friction plotted as $\mu(R/c)$ versus Sommerfeld number S for full journal bearings with $L/D \gg 1$ (data from Raimondi and Boyd (1958)).

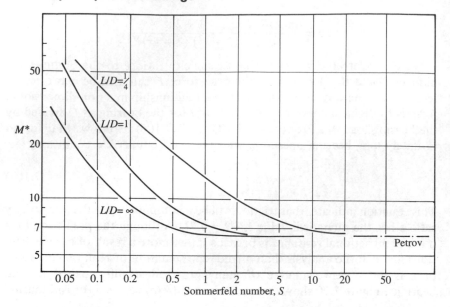

Fig. 7.21   Non-dimensional friction torque M^* versus Sommerfeld number S full journal bearings (data from Raimondi and Boyd (1958)).

bearings with both Sommerfeld and Reynolds boundary conditions. Also shown is the line corresponding to the Petrov equation (7.105).

The frictional torque M on the bearing is simply given by $F \times R$; this can be suitably non-dimensionalized to M^*, where

$$M^* = \frac{Mc}{\eta \omega R^3 L}. \tag{7.108}$$

Figure 7.21 provides plots of M^* versus S for bearings with three slenderness ratios. At higher values of S the curves become asymptotic to the Petrov solution.

It is worth noting that a further consequence of the lack of concentricity between the journal and the bearing is the fact that the frictional torques experienced by each of these components are not the same; that on the journal exceeds that on the bearing by an amount ΔM, where

$$\frac{\Delta M}{M} = \frac{2\varepsilon \sin \psi}{SM^*}. \tag{7.109}$$

This effect may not be negligible, for example, if $\varepsilon = 0.6$, $S = 0.4$, and $M^* = 12$ then $\Delta M/M$ might be as great as 25 per cent.

Partial journals

It is clear from an examination of the pressure profile of Fig. 7.18 that only a portion of the circumference of the bearing actually carries the applied load; this 'effective' arc extends to rather more than 180°. Viscous losses in the film, on the other hand, are generated all the way around the circumference. In those cases in which the load line is in a fixed and known direction it may be sensible, from an energy-saving point of view, to use a bearing with a restricted arc of contact, of perhaps 180° or sometimes even less. In this way viscous losses may be reduced by a significant factor compared to the full bearing. Figure 7.22 illustrates the relationship between the angle of the bearing arc β and the Sommerfeld number S for bearings with three different slenderness ratios, all operating at an eccentricity ratio of 0.6. In these cases it has been assumed that the load vector bisects the angle β. Figure 7.23 illustrates the effect that the extent of the bearing arc has on the relationship between the frictional torque M^* and the Sommerfeld number for three cases each with a ratio of $L/D = 1$. Once again the curves are asymptotic to the Petrov solutions at higher values of S. The effects of these restricted ares of contact on the value of the attitude angle is shown in Fig. 7.16(b).

Lubricant supply

Lubricant is usually fed to a plain journal bearing through a pocket, groove, or hole located in the unloaded arc of the bearing shell, in the manner illustrated in Fig. 7.24. In oil lubrication, the supply pressure is typically between 70 and 350 kPa (10–50 psi) which is very much less than the hydrodynamic pressures generated over the loaded area of contact. Although the bearing does not depend directly on this pressure to carry the applied load it is essential that a continuous supply of lubricant is maintained if the bearing is not to suffer from oil starvation and possible collapse. In practice, full journals are invariably split into two halves, each subtending an angle of 180°, mechanically clamped together. It is usual to assume that the joint between them is oil-tight and so does not affect the operation of the bearing. The standard position for the oil feeds will be at 90° to this parting line as indicated in Fig. 7.24(a), although this can be modified if the load on the bearing is always unidirectional. If the bearing is reasonably long then the oil hole may supply an oil feed pocket extending some way axially along the bearing, as illustrated in Fig. 7.24(b). In cases where the load vector is variable in direction the oil feed pocket may extend around the bearing circumferentially, as shown in Fig. 7.24(c). Although this arrangement will prevent oil starvation, whatever the load direction, it is at the cost of splitting the bearing into two independent halves each with a slenderness ratio of half the original value.

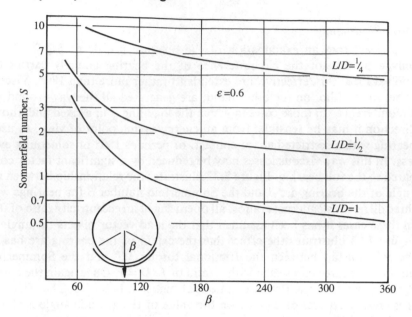

Fig. 7.22 Partial journal bearings: influence of the included angle β on the Sommerfeld number for L/D ratios of 1, $\frac{1}{2}$, and $\frac{1}{4}$. The load vector bisects the included angle (data from Raimondi and Boyd (1958)).

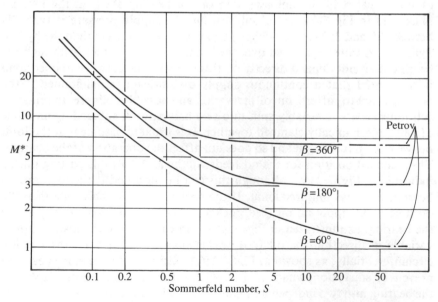

Fig. 7.23 Partial journal bearings: influence of the bearing arc angle on the relationship between M^* and S (data from Raimondi and Boyd (1958)).

(a)

(b)

(c)

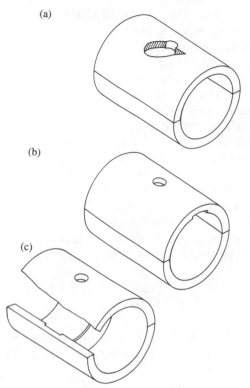

Fig. 7.24 Oil supply arrangements in plain journal bearings: (a) single oil hole; (b) axial oil feed pocket; and (c) circumferential oil feed pocket.

7.8   Thermal effects in lubricated journal bearings

The shaft size, rate of rotation, and radial applied load of a journal bearing are usually design parameters outside the decision of the bearing designer who will be asked to suggest a satisfactory tribological design involving the specification of bearing length and clearance, and the materials, oil viscosity (or grade), and flow rate. We have seen that an important decision is the choice of the steady eccentricity ratio ε, since the minimum film thickness and so the tolerance that the bearing is likely to display to contamination or surface irregularities or degradation depends on the combination of its value and that of the clearance c. Setting the bearing clearance too high will result in excessive oil flow requirements, and so unrealistically expensive supply pumps and feed arrangements; on the other hand, setting the clearance too low will bring about premature bearing failure through solid-to-solid or rubbing contact. In service, frictional shearing of the oil film will increase the local temperature of the oil and so reduce its viscosity; some of this thermal energy is carried away by convection as the lubricant passes through the bearing, and some lost by conduction through the bearing and its housing and adjacent components.

For the case of a slider pad bearing (Fig. 7.1) Cameron and Ettles (1981) have suggested the following simple argument to determine, very approximately, the relative losses by conduction and convection. Suppose that the lower moving surface is maintained at ambient temperature while that of the upper pad increases linearly from ambient at entry (where $x = 0$) to a maximum $\Delta\Theta$ above ambient at $x = B$. At a point on the surface of the pad with coordinate x, the temperature rise will be equal to $\Delta\Theta x/B$. If we also now take the temperature gradient *across* the oil film as linear, and use an average film thickness h, this gradient will be equal to $\Delta\Theta x/Bh$. The rate of local heat flow per unit area by conduction will thus be $K \times \Delta\Theta x/Bh$ where K is the thermal conductivity of the lubricant. This expression can be integrated to give the total heat flow per unit width of bearing, namely

$$\dot{H}_{\text{cond}} = \int_0^B \frac{\Delta\Theta x}{Bh}\, dx = \frac{\Delta\Theta}{h} \frac{KB}{2}. \qquad (7.110)$$

Heat flow by convection $\dot{H}_{\text{conv}}$ is simply the product of the mass flow of oil, its specific heat, and average temperature rise $\Delta\Theta/2$. Thus

$$\dot{H}_{\text{conv}} = \frac{Uh\rho c}{2} \times \frac{\Delta\Theta}{2}. \qquad (7.111)$$

We can thus write that the ratio of conducted to convected heat is equal to

$$\frac{\dot{H}_{\text{cond}}}{\dot{H}_{\text{conv}}} = \frac{K}{\rho c} \frac{2B}{Uh^2} \qquad (7.112)$$

$\Delta\Theta$ can thus be estimated from eqns (7.111) and (7.112) since $\dot{H} = \dot{H}_{\text{cond}} + \dot{H}_{\text{conv}}$ will be equal to the total energy dissipation in the bearing; see eqn (7.43). The lubricant property group $K/\rho c$ is known as its diffusivity κ (see Chapter 3), and for mineral oils has a value of approximately $0.08 \times 10^{-6}\,\text{m}^2\,\text{s}^{-1}$. This form of simple argument can be extended to journal bearings when it suggests that the ratio of conducted to convected heat is now given by

$$\frac{\dot{H}_{\text{cond}}}{\dot{H}_{\text{conv}}} = \kappa \times \frac{4\pi}{\omega R^2} \left\{ \frac{R}{c} \right\}^2. \qquad (7.113)$$

As we should expect, convection becomes increasingly important at higher bearing speeds.

The full analytical treatment of this thermal balance problem, leading to estimates of actual bearing and lubricant temperatures, is likely to be both difficult and time-consuming as it requires simultaneous solution of the lubrication equations together with the equations of heat flow both in the oil film and the metallic components. Below are suggested some much simplified but rapid procedures which can be used to obtain rough estimates of the likely temperature rises in real cases and can thus assist in the choice of design parameters and lubricant grades. In many cases, for the sake of compactness, the hope will be to specify a bearing with a length L less than its diameter D. If we choose to make $L/D = 0.5$, then as a first approximation we might adopt the short-bearing analysis of Section 7.7.1, in which we showed that

$$\frac{1}{S} = \left\{\frac{L}{D}\right\}^2 \frac{\pi \varepsilon}{2(1-\varepsilon)^2} (1 + 0.62\varepsilon^2)^{1/2} \qquad (7.94 \text{ bis})$$

where S is the Sommerfeld number defined by eqn (7.91). A sensible working compromise for the value of ε is 0.7 and a 'standard' value for the radial clearance ratio c/R (i.e. $2c/D$) might well be 0.001, so it follows, by substituting this data, that

$$\eta = \frac{WD}{\omega L^3} \times 2 \times 10^{-7} \qquad (7.114)$$

where η is the effective viscosity of the lubricant in the bearing gap.

Even though the centre of any journal is likely to be displaced relative to that of the bearing the power loss can be reasonably approximated by idealizing them as running concentrically, so that the frictional moment can be taken from the Petrov equation (7.105). The rate of dissipation of energy $\dot{H}$ in the bearing is then

$$\dot{H} = M \times \omega = \frac{2\pi R^3 L}{c} \eta \omega^2. \qquad (7.115)$$

If we make the assumption that all this thermal energy is carried away by convection via the flow of oil through the bearing then, using eqns (7.98) and (7.99), we can write that

$$\dot{H} = \frac{2\varepsilon}{1+\varepsilon} \times \rho c \Delta\Theta \times (1+\varepsilon) \frac{R\omega L c}{2}, \qquad (7.116)$$

where ρ and c are the density and specific heat of the oil respectively, and $\Delta\Theta$ is the temperature rise. Setting ε to 0.7, and taking note that for mineral oils ρ is very close to 900 kg m^{-3} and c is approximately $1.88 \text{ kJ kg}^{-1}\text{K}^{-1}$, we can say that

$$\Delta\Theta \approx \frac{2\pi}{\varepsilon} \left\{\frac{R}{c}\right\}^2 \frac{\omega\eta}{\rho c} \approx 5.30 \, \omega\eta. \qquad (7.117)$$

The problem is therefore to identify a value of η which is consistent with both eqns (7.114) and (7.117) and the physical characteristics of the oil, in particular the way in which its viscosity varies with temperature. This behaviour has been discussed for a range of practical mineral oils in Section 1.3.

For example, suppose the problem is to place a limit on the inlet temperature of SAE30 oil supplied to a journal bearing of radius $R = 0.025$ m and length $L = 0.025$ m with $c/R = 0.001$ carrying a load of 4500 N at 1000 rpm. Here $L/D = 0.5$, $\omega = 104.7$ rad s^{-1} and so from eqn (7.114)

$$\eta = \frac{4500 \times 0.05}{104.7 \times 0.025^3} \times 2 \times 10^{-7} = 0.0275 \text{ Pa s}$$

But, from eqn (7.117) the corresponding temperature increase $\Delta\Theta$ is given by

$$\Delta\Theta = 5.3 \times 104.7 \times 0.0275 = 15.3°C.$$

We can now refer to Fig. 1.10 to establish that SAE30 oil has a viscosity of 0.0275 Pa s at a temperature of approximately 65°C and hence the oil should enter the bearing at a temperature of $(65 - 15)°C$, that is at no more than 50°C.

The Engineering Sciences Data Unit publication 66023 (see Appendix 2) contains a much fuller iterative procedure for establishing the steady state running conditions of lubricated journals. The aim must be to ensure that the minimum film thickness will be adequate to prevent mechanical surface damage without the maximum bearing and oil temperature being so large as to lead to premature thermal degradation of either. The general form of these limitations can be displayed on axes representing journal speed and radial load; see Fig. 7.25. At low speeds and high loads, and so small film

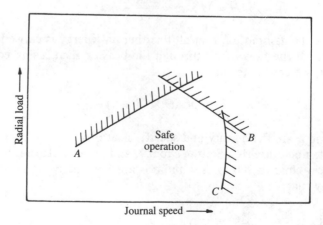

Fig. 7.25 Limits for the satisfactory operation of hydrodynamically lubricated journal bearings: (*A*) scuffing; (*B*) wiping; and (*C*) chemical degradation.

thicknesses, there is the danger of true metallic contact with its associated scuffing and scoring; this limitation gives line A on this map of bearing operation. At high speeds and high loads if metallic contact occurs then the high sliding speed may lead to the generation of surface temperature sufficient to perhaps soften or even locally melt the bearing alloy leading to smearing or wiping of the surface; this limitation is shown as line B. Finally at even higher speeds but low loads the very high rates of shear in the oil film may lead to lubricant temperatures becoming so great that there is a danger of chemical degradation, principally excessive oxidation; this effect gives the limit illustrated by line C.

Figure 7.26 summarizes the procedures set out by Martin and Garner (1974) and which enables the dimension of the minimum film thickness to be estimated in an oil-lubricated journal bearing. Basically, three 'grids' are used to define the problem and by linking these along the appropriate guide lines a point in the fourth 'solution' grid is obtained. (The lower case letters a to m allow the transfer of data between parts (a) and (b) of the figure.) In the previous design example the inputs to the design are the clearance ratio c/R, the length-to-diameter ratio L/D, the speed, oil grade, and the specific bearing load, W/LD, and these have the numerical values 0.001, 0.5, $105 \, \text{s}^{-1}$, SAE 30, and 3.6 MPa respectively. Using the arrows as indicated in Fig. 7.26 produces an estimated value of ε of approximately 0.75 so that $h_{\min} = (1 - 0.75) \times 0.001 \times R = 6.3$ microns. Figure 7.26 is based on the design of a plain journal bearing in which the bearing gap is supplied by two axial oil grooves with lubricant at a pressure of 100 kPa (i.e. about 15 psi).

The maximum oil temperature $\Theta_{\max}$ occurs within the lubricant film at a position which roughly coincides with that of the minimum film thickness. Figure 7.27 can be used to estimate its value. An example of the use of the chart is provided in the figure in which Θ_{eff} represents the 'effective' temperature of the oil, which is assumed to enter the bearing at 50°C; conditions are as for the example in Fig. 7.26.

7.9 Dynamic effects in hydrodynamic bearings

7.9.1 *Squeeze films effects*

Journal bearings

In our analyses of journal bearings we have made the assumption that the load is steady with time. Although many bearings do operate under conditions that approximate to this, there are many devices within which bearings are subject to loads which are constantly changing. As a result of these fluctuations, the lubricant is alternately squeezed out and drawn into the bearing gap; this imposed motion, combined with the viscosity of the lubricant, results in a contribution to the load-carrying capacity which is

(a)

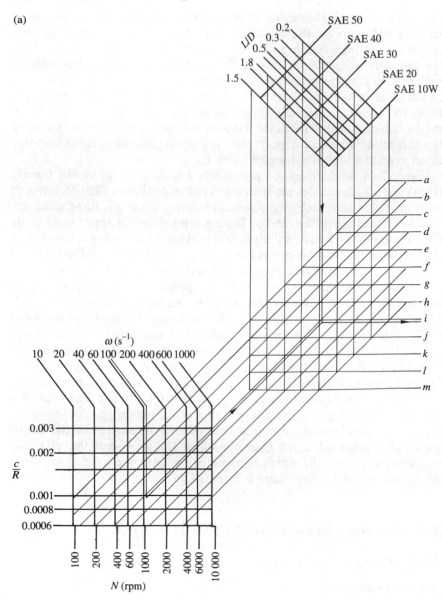

Figure 7.26(a)

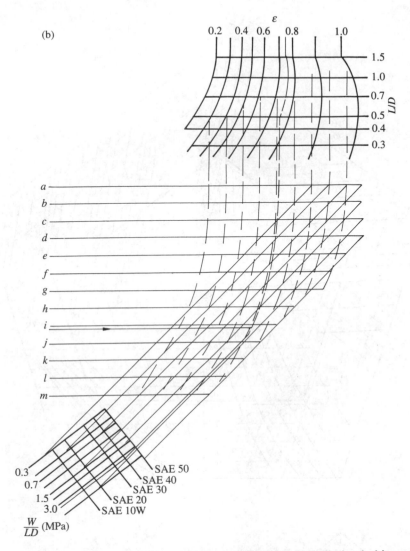

Fig. 7.26 Prediction of minimum oil film thickness in steadily loaded journals (from Martin and Garner 1974).

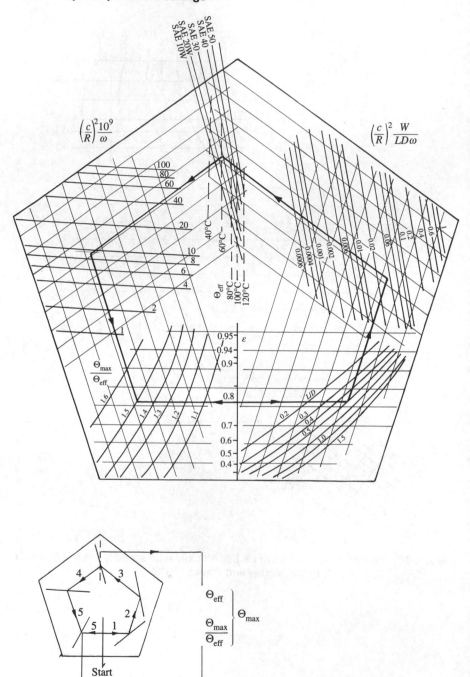

Fig. 7.27 Prediction of maximum bearing temperature (from Martin and Garner 1983). The lubricant enters the bearing at a temperature of 50°C.

independent of any sliding motion between the surfaces. This mechanism is known as the *squeeze film effect* and can be of considerable practical significance. To generate loads of the same order of magnitude, the relative vertical velocity of a pair of surfaces need only be equal to their linear sliding velocity multiplied by the film thickness-to-length ratio. In a real bearing this ratio may easily be of the order of 1:1000 so that significant squeeze film effects can be generated by relatively modest squeeze velocities. Nevertheless, because of the computational difficulties involved in predicting these non-steady events, it is only relatively recently that they have been incorporated into the design procedure of hydrodynamic bearings.

In addition to the time variation of the magnitude of the load, journal bearings can be divided into two classes depending on whether or not the load is rotating relative to the bearing. Our examples so far have been restricted to those cases in which the load vector is stationary. The effect of a rotating load (providing that it is of a constant magnitude) is relatively simple to take into account. In Fig. 7.28(a) the journal is rotating with a clockwise angular velocity ω, while the load W rotates at Ω and the bearing is at rest. The situation can be brought back to one of a *steady* load vector by imposing a rotation of Ω in the *counter-clockwise* direction on all the components, as in Fig. 7.28(b). The surface velocity of the journal relative to a stationary external observer is now $-(\omega - \Omega)R$ while that of the bearing surface is ΩR. Thus the magnitude of the entraining velocity $\bar{U}$, which is the value which must be used to calculate the effective Sommerfeld number of the bearing, is given by

$$\bar{U} = \tfrac{1}{2}\left|\Omega R + \left[-(\omega - \Omega)R\right]\right| = \tfrac{1}{2}(\omega - 2\Omega)R. \qquad (7.118)$$

Clearly, if the load were to be rotating at half the speed of the journal, that is, $\omega = 2\Omega$, then $\bar{U}$ will go to zero and the load-carrying capacity of the bearing, which depends on $\bar{U}$, disappears. The situation is rather akin to that of *half-speed whirl* — see Section 8.2.4. Some failures in the main bearings of internal combustion engines have been attributed to this effect. At some stages of the firing cycle the load may be rotating at exactly half the speed of the crankshaft; if this situation persists for more than about 20° of shaft rotation then collapse of the bearings may ensue.

If neither the journal nor the bearing are rotating but a steady load is suddenly applied to the shaft, then the two solid surfaces will move together as the lubricant is squeezed out of the gap between them. The load is now carried by the squeeze film effect, referred to above, and this can be described by setting the tangential components of velocity to zero in the general form of Reynolds equation (7.57), thus

$$\frac{\partial}{\partial x}\left\{h^3\,\frac{\partial p}{\partial x}\right\} + \frac{\partial}{\partial y}\left\{h^3\,\frac{\partial p}{\partial y}\right\} = 12\eta\,(\mathcal{W}_2 - \mathcal{W}_1). \qquad (7.119)$$

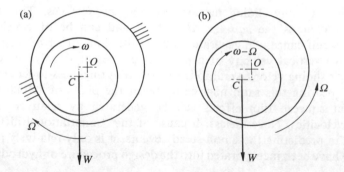

Fig. 7.28 (a) Journal bearing rotates at ω; the journal is stationary but the load rotates at Ω. (b) The load vector can be brought to rest by imposing a counter-rotation of Ω throughout.

To convert this equation into the more convenient cylindrical polar form for a long bearing we can again write $x = R\theta$ and set $\partial p/\partial y = 0$. In addition, if the bearing is stationary while the journal moves radially then we can set $\mathcal{W}_1$ to zero and $\mathcal{W}_2$ equal to $c\dot{\varepsilon}\cos\theta$, so that eqn (7.119) becomes

$$\frac{\partial}{\partial\theta}\left\{(1+\varepsilon\cos\theta)^3\frac{\partial p}{\partial\theta}\right\} = 12\eta\frac{R^2}{c^2}\dot{\varepsilon}\cos\theta. \qquad (7.120)$$

This equation can be integrated directly to give

$$p = 6\eta\frac{R^2}{c^2}\frac{\dot{\varepsilon}}{\varepsilon(1+\varepsilon\cos\theta)^2} + C,$$

where C is a constant. Now although in general the pressure p will be zero in directions perpendicular to the load line, that is, at $\theta = \pi/2$ and $3\pi/2$, little numerical accuracy is lost and the integrations are much simpler if p is set to zero at $\theta = 0$. In this case (Cameron 1966)

$$p = 6\eta\frac{R^2}{c^2}\frac{\dot{\varepsilon}}{\varepsilon}\left\{\frac{1}{(1+\varepsilon\cos\theta)^2} - \frac{1}{(1+\varepsilon)^2}\right\}$$

and W/L, the load carried per unit length, is given by the relation

$$\frac{W/L}{\eta R\dot{\varepsilon}}\left(\frac{c}{R}\right)^2 = \frac{12\pi}{(1-\varepsilon^2)^{3/2}}. \qquad (7.121)$$

In the case of bearings of a finite length, that is, $L/D \neq \infty$, where the effects of the ends cannot be ignored, a corresponding empirical equation is

$$\frac{W/L}{\eta R\dot{\varepsilon}}\left(\frac{c}{R}\right)^2 = \frac{1}{a(1-\varepsilon)^b} \qquad (7.122)$$

in which a and b are constants which can be related to the ratio D/L according to

$$a \approx \{0.6D/L - 0.4\} \qquad \text{and} \qquad b \approx \{1.5 + 0.3\sqrt{D/L}\}$$
$$\text{provided } 0.75 \leq D/L \leq 5.$$

As a rule we will wish to estimate the rate at which the solid surfaces approach each other after the application of the load. Returning to the case of a long bearing, if we assume that at time $t = 0$ the journal is concentric with the bearing, that is, $\varepsilon = 0$, then the eccentricity ratio after some interval Δt can be obtained by the integration of eqn (7.121), thus

$$\int_0^\varepsilon \frac{1}{(1 - \varepsilon^2)^{3/2}} \, d\varepsilon = \frac{W/LD}{6\pi\eta} \left(\frac{c}{R}\right)^2 \int_0^{\Delta t} dt \tag{7.123}$$

that is,

$$\frac{\varepsilon}{(1 - \varepsilon^2)^{1/2}} = \Delta t \left\{ \frac{6\pi\eta}{W/LD} \left(\frac{R}{c}\right)^2 \right\}^{-1}. \tag{7.124}$$

This relationship is shown as curve A in Fig. 7.29 and can be used to estimate the time that the surfaces in a full 360° bearing take to approach one another after the steady load is applied. A similar argument applied to the case of a long half, or 180°, bearing leads to the corresponding curve B.

For the full 360° bearing, a second case of interest is that of an *alternating* load of amplitude W_0 and angular frequency ω', such that

$$W = W_0 \cos\omega' t. \tag{7.125}$$

If the resultant oscillation of the journal is taken to be centred on the geometric centre of the bearing then the corresponding equation to (7.124) is

$$\frac{\varepsilon}{(1 - \varepsilon^2)^{1/2}} = \frac{W_0/LD}{6\pi\eta\omega'} \left(\frac{c}{R}\right)^2 \sin\omega' t. \tag{7.126}$$

Thus, if in our definition of the Sommerfeld number S we replace the angular rotational frequency ω by the frequency of oscillation ω', then the maximum value of the resultant eccentricity ε_{max} is given by

$$\frac{\varepsilon_{max}}{(1 - \varepsilon_{max}^2)^{1/2}} = \frac{1}{6\pi S}. \tag{7.127}$$

This implies that curve A in Fig. 7.29 can also be used to evaluate ε_{max} with an appropriate change of the value of the ordinate.

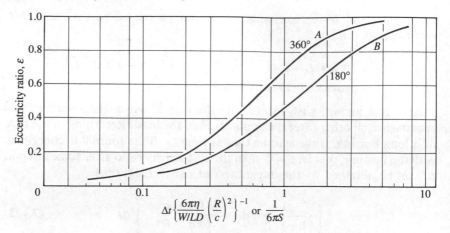

Fig. 7.29 Eccentricity ratio ε versus non-dimensional time $\Delta t \left\{ \dfrac{6\pi\eta}{W/LD} \left(\dfrac{R}{c}\right)^2 \right\}^{-1}$ or oscillatory Sommerfeld number for the squeeze film effect in a non-rotating journal bearing.

Squeeze films between flat plates

If the radius of a long journal bearing becomes very large then the situation becomes effectively that of two long, parallel plates approaching one another. We can then conveniently return to the Cartesian form of eqn (7.119): if the gap between the plates is h, and we assume that the upper plate is falling towards the stationary lower plate then $\mathcal{W}_1 = 0$ and $\mathcal{W}_2 = -\dot{h}$. Reynolds' equation thus becomes

$$\frac{\partial}{\partial x}\left\{ h^3 \frac{\partial p}{\partial x} \right\} = -12\eta\,\dot{h}. \tag{7.128}$$

If the plates remain parallel then h is not a function of position x, so that

$$h^3 \frac{\mathrm{d}^2 p}{\mathrm{d}x^2} = -12\eta\,\dot{h} \tag{7.129}$$

and on integrating,

$$p = -\frac{6\eta\dot{h}x^2}{h^3} + \mathcal{A}x + \mathcal{B}, \tag{7.130}$$

where $\mathcal{A}$ and $\mathcal{B}$ are constants.

Referring to Fig. 7.30 symmetry means that there must be a zero pressure gradient at $x = 0$, and that pressure p is zero when $x = \pm B/2$. Hence, applying these conditions to (7.130),

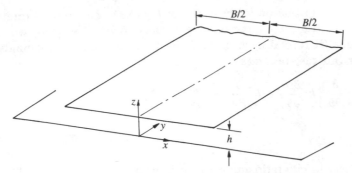

Fig. 7.30 Squeeze films between a pair of long parallel flat plates of width B.

$$p = \frac{6\eta\dot{h}}{h^3}\left\{\frac{B^2}{4} - x^2\right\} \tag{7.131}$$

and the load per unit length W/L is given by

$$\frac{W}{L} = \int_{-B/2}^{B/2} p\,\mathrm{d}x = \frac{\eta\dot{h}B^3}{h^3}. \tag{7.132}$$

In the case of a pair of *circular* plates of radius a we can convert Reynolds' equation (7.128) into polars:

$$\frac{1}{r}\frac{\partial}{\partial r}\left\{rh^3\frac{\partial p}{\partial r}\right\} = -12\eta\,\dot{h}. \tag{7.133}$$

Integration of this equation leads to corresponding expressions for the pressure p and total load W, namely

$$p = -\frac{3\eta\dot{h}}{h^3}(a^2 - r^2) \qquad \text{and} \qquad W = \frac{3\pi\eta\dot{h}a^4}{2h^3}. \tag{7.134}$$

7.9.2 *Dynamic loads: the mobility method*

Many journal bearings, for example those in reciprocating engines and compressors, are subject to loads whose magnitude and direction *both* vary in time. Consequently, the centre of the journal in such a bearing does not arrive at some stable equilibrium position, but rather traces out a locus as the conditions applied to it vary so that the size and position of the minimum film thickness varies in a cyclic manner. With increasing economic pressures to reduce energy losses by using lubricants of lower viscosity the prediction of such minima is an increasingly important aspect of bearing design.

The general situation is illustrated in Fig. 7.31. The journal rotates with angular speed ω, while the bearing rotates at Ω'. The load, which is of magnitude W, is rotating at speed Ω. In the usual notation, Reynolds' equation for this case becomes

$$\frac{\partial}{\partial\theta}\left\{h^3\frac{\partial p}{\partial\theta}\right\} + R^2\frac{\partial}{\partial y}\left\{h^3\frac{\partial p}{\partial y}\right\} = 12\eta\frac{R^2}{c^2}\left\{\dot{\varepsilon}\cos\theta + \varepsilon\left[\dot{\psi} + \Omega\right.\right.$$

$$\left.\left. - \frac{1}{2}(\omega + \Omega')\right]\sin\theta\right\}. \qquad (7.135)$$

ε is the eccentricity ratio and ψ the attitude angle. We can note that setting $\dot{\varepsilon}$, $\dot{\psi}$ and Ω' all to zero will generate the case of the steadily rotating load, as in eqn (7.118), while setting $\dot{\varepsilon}$, ω and Ω' to zero but $\dot{\psi}$ to $\frac{1}{2}\omega$ will produce the conditions for half-speed whirl, see Section 8.2.4. If all the angular velocities, ω, Ω' and Ω, are put to zero then the squeeze film action is isolated to give effectively eqn (7.120).

Once the pressure distribution is established the components of the load W supported can be found in the usual way by eqns (7.85) and (7.86). Of course, usually it is the inverse of this problem that is to be solved. The loads applied to the bearing during its cycle of operation are specified; it is the film geometry that is unknown. The unknowns we wish to evaluate are the bearing eccentricity ratio ε, the attitude angle ψ, and their rates of change with time, that is, the values of $\dot{\varepsilon}$ and $\dot{\psi}$.

Equation (7.135) can be written as

$$\dot{\varepsilon} = \frac{W(c/R)^2}{\eta LD}M^x(\varepsilon,\psi,L/D,\theta_1,\theta_2) \qquad (7.136)$$

and

$$\varepsilon\left[\dot{\psi} + \Omega - \frac{1}{2}(\omega + \Omega')\right] = \frac{W(c/R)^2}{\eta LD}M^z(\varepsilon,\psi,L/D,\theta_1,\theta_2) \qquad (7.137)$$

where M^x and M^z are functions of the variables listed. θ_1 and θ_2 are the values of the angular position coordinate θ corresponding respectively to the start and finish of the lubricant film. In a bearing with a full film $\theta_1 = 0°$ and $\theta_2 = 360°$. Since M^x and M^z are the dimensionless ratios of velocities-to-force they are referred to as *mobilities* and can be thought of as the components of a 'mobility' vector (for further details see Booker (1965, 1971)).

Equations (7.136) and (7.137) can be solved, in some simple cases analytically, but more generally numerically, and the results plotted out as a polar diagram of the journal centre orbit. Figure 7.32 shows some results for a main bearing of a four-cylinder 1.8 litre in-line gasoline engine. In Fig. 7.32(a) the load on this bearing, calculated from the combination of

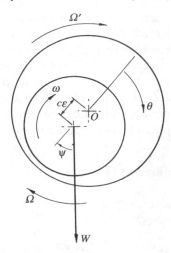

Fig. 7.31 Plain journal bearing in which the journal rotates at ω, the bearing at Ω', and the load at Ω.

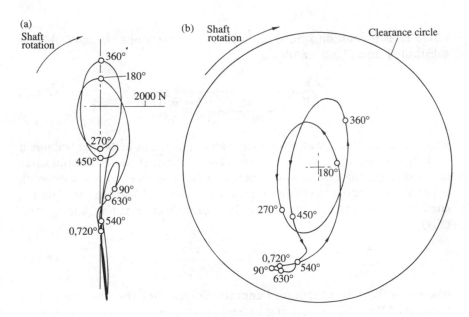

Fig. 7.32 (a) Polar load diagram for an internal combustion engine main bearing; (b) journal centre orbit predicted by the mobility method (from Booker 1965).

combustion and inertia loads, is shown as a polar plot. The angles refer to the rotation of the crank shaft. Notice that most of the load is concentrated an the bottom half of the bearing and that the two major load loops are rotating counter to the direction of shaft rotation. Figure 7.32(b) shows the calculated journal centre orbit for a full-film short bearing; this must of course lie within the clearance circle which corresponds to $\varepsilon = 1$. From such a diagram both the minimum film thickness and the oil flow rate can be estimated; both are important design parameters.

7.10 Lubricated point contacts

In 1955 Kapitza (see Cameron 1966) observed that a solution to Reynolds' equation in two dimensions,

$$\frac{\partial}{\partial x}\left\{h^3\frac{\partial p}{\partial x}\right\} + \frac{\partial}{\partial y}\left\{h^3\frac{\partial p}{\partial y}\right\} = 12\eta\bar{U}\frac{dh}{dx} \qquad (7.58 \text{ bis})$$

is provided by the expression

$$p = -\frac{\mathcal{K}x}{h^2}, \qquad (7.138)$$

where $\mathcal{K}$ is a constant whose value is to be determined. If eqn (7.138) is substituted into (7.58) we obtain

$$\frac{2\mathcal{K}}{\bar{U}\eta} = \cfrac{1}{1 + \cfrac{2x}{\partial h/\partial x}\left\{\cfrac{\partial^2 h}{\partial x^2} + \cfrac{\partial^2 h}{\partial y^2}\right\}} \qquad (7.139)$$

which must be true for all values of x and y. $\bar{U}$ is the surface entrainment velocity between the two solids in the Ox-direction. If we restrict ourselves to considering the case of two spheres loaded together, with a separation between the solids at the point of closest approach of h_c, then the equation corresponding to (7.66), which describes the separation at the general point (x,y), is

$$h = h_c\left\{1 + \frac{x^2 + y^2}{2Rh_c}\right\}, \qquad (7.140)$$

where R is the reduced radius of contact. Substituting this into eqn (7.139) shows that Reynolds' equation is indeed satisfied by a pressure variation of the form of (7.138) provided that

$$\mathcal{K} = \frac{12\bar{U}\eta}{5}, \qquad (7.141)$$

which means that

$$p = -\frac{12\bar{U}\eta x}{5h^2}.$$ (7.142)

Figure 7.33(a) illustrates the form of the positive region of this pressure curve. In the figure, pressure p is plotted non-dimensionally as $p^* = pR/\bar{U}\eta$ and the position coordinate x is normalized by the radius R. The pressure distributions at various values of the parameter y/R are illustrated in Fig. 7.33(a), while Fig. 7.33(b) is a contour plot of the values of p^* in the $(x/R, y/R)$ plane.

The distribution of eqn (7.142) is thus once again antisymmetric about the centre point along the x-axis, producing negative pressures in the outlet region which are of the same magnitude as the positive pressures generated in the convergent inlet zone (these subambient pressures are not shown in the figure). In order to estimate the normal load W we can adopt the half-Sommerfeld conditions and effectively neglect the subambient pressures. In this case it can be shown that

$$W = \frac{12\sqrt{2\pi}}{5}\eta\bar{U}R\sqrt{\frac{R}{h_c}}.$$ (7.143)

Rearranging, the minimum central film thickness h_c is given by

$$\frac{h_c}{R} = \frac{288\pi^2}{25}\left\{\frac{\eta\bar{U}R}{W}\right\}^2.$$ (7.144)

Although most studies on hydrodynamics have concentrated either on nominal line or circular point contacts, solutions to the intervening range of problems, that is, that of ellipsoidal contacts, have also been investigated. If, instead of a sphere on a flat, we consider the case of an ellipsoid with principal radii of curvatures R_x and R_y then the corresponding equation to (7.144) is

$$\frac{h_c}{R_x} = \frac{288\pi^2\alpha}{(3+2\alpha)^2}\left\{\frac{\eta\bar{U}Rx}{W}\right\}^2,$$ (7.145)

where $\alpha = R_x/R_y$. The geometry is illustrated in Fig. 7.34. If $R_y = R_x$ then $\alpha = 1$ and equation (7.144) is recovered.

Naturally these solutions, all based on the half-Sommerfeld idealizations, suffer from the drawback of infringing compatibility at the point at which the pressure p falls to zero (see Section 7.6). It is possible to introduce the Reynolds boundary conditions, that is, to assume that the film finishes at some boundary (which must be established) along which both the pressure and the pressure gradient normal to the boundary are simultaneously zero.

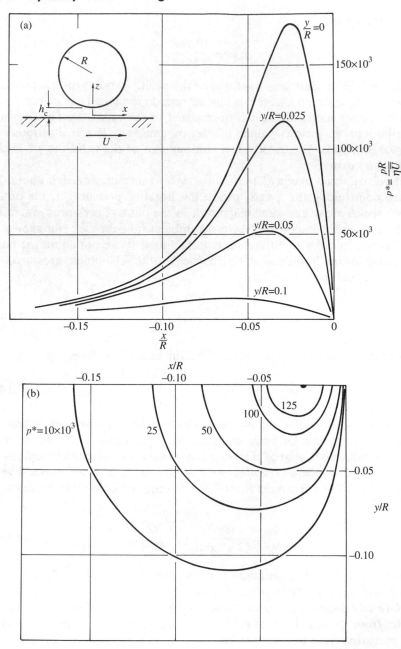

Fig. 7.33 Pressure profiles, plotted non-dimensionally, at a hydrodynamically lubricated point contact between two rigid spheres. The origin of coordinates is taken at the point of closest approach. The inset shows the equivalent geometry of a sphere radius R on a flat. $\bar{U}$ is the entraining velocity.

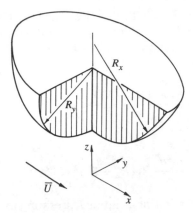

Fig. 7.34   Ellipsoidal contact; R_x and R_y are the principle radii of curvature and $\bar{U}$ is the entraining velocity in the Ox-direction.

An empirical formula relating h_c to the operating conditions on this basis has been found to be

$$\frac{h_c}{R_x} = \frac{1.64\pi^2\alpha}{(3+2\alpha)^2} \left\{\frac{\eta \bar{U} R x}{W}\right\}^2 \{\arctan \tfrac{1}{2}\alpha + 12.8\}^2, \qquad (7.146)$$

and this can be used as a general expression for the minimum film thickness between any rigid ellipsoidal solids, ranging from two spheres in nominal point contact to infinitely long cylinders in nominal line contact.

7.11   Turbulence in hydrodynamic films

When bearings are operated at sufficiently high speeds or large clearances (or perhaps with very-low-viscosity lubricants) flow conditions may cease to be entirely laminar. Transition to *superlaminar*, and eventually *turbulent*, flow occurs when the forces arising from viscosity are insufficient to overcome those associated with fluid inertia. The regular pattern of fluid motion begins to break up into local circulations and vortices; this implies an increase in bearing friction and hence in power loss. The two basic problems are, firstly, to determine under what conditions this transition is likely, and secondly, once laminar flow ceases, to calculate in some average manner such quantities as flow rate, load support, and friction. In fluid mechanics the breakdown of laminar flow is usually associated with a critical value of Reynolds number, Re; typically flow is laminar provided Re is less than approximately 2000. In a slider bearing Re is equal to $\rho U h/\eta$ while in a journal bearing the appropriate group is $\rho \omega R c/\eta$.

The pioneering work of G. I. Taylor (see Cameron 1966) showed that, for the case of a long cylinder of radius R_1 rotating concentrically at speed ω relative to a stationary cylindrical shell of radius R_2, laminar flow will continue provided that

$$\mathrm{Re} = \frac{R_1 \rho \omega (R_2 - R_1)}{\eta} < 41.1 \sqrt{\frac{R_1}{R_2 - R_1}}$$

that is,

$$\frac{\omega^2 \rho^2}{\eta^2} R_1 (R_2 - R_1)^3 < 1690. \tag{7.147}$$

If we think of this as a journal, radius R, rotating in a stationary bearing with radial clearance c, then this condition for continued laminar flow is effectively that

$$\left\{ \frac{\omega \rho}{\eta} \right\}^2 R c^3 < 1690. \tag{7.148}$$

Thus, to keep in this regime fluid viscosity should be high, and the diameter and clearance both low. When this condition is exceeded, orderly features, known as Taylor vortices, are formed initially and the flow regime is known as superlaminar; only when these begin to break down to give an apparently random or chaotic pattern is the fully turbulent region entered. The group on the left-hand side of inequality (7.148) is sometimes known as the *Taylor number*, Ta. Of course, in a real journal bearing the shaft usually rotates about a point which is *not* coincident with the bearing centre and it has been found that as the eccentricity ratio ε increases then so does the critical value of the Taylor number at which flow ceases to be laminar. The transition to superlaminar flow is only likely to be a problem with conventional lubricants in large-diameter bearings rotating at high speeds. At a rotational speed of 10,000 rpm, using an oil of viscosity 0.07 Pa s and specific gravity 0.95 in a bearing of clearance ratio 0.001, turbulence is unlikely unless the bearing diameter is more than 400 mm. However, the onset of turbulence can be a very real problem in much smaller bearings operated at more modest speeds if a low-viscosity process fluid is used as the lubricant. For example, Freon has viscosity $\eta = 0.46 \times 10^{-3}$ Pa s and density $\rho = 1440$ kg m^{-3}; thus with $c/R = 0.001$ and $R = 25$ mm the critical speed from eqn (7.147) is only just over 3000 rpm. From the designer's point of view a most important aspect of the onset of turbulence is the increase in friction or power loss that follows. Figure 7.35, which applies to a 'square' bearing, that is, one with length equal to diameter, illustrates the influences of the bearing design and operating parameters. The upper part of the figure relates the eccentricity ratio ε to Reynolds number Re for two load cases; the load is incorporated in the non-dimensional group $\{ W\rho/\eta^2 \} \{ c/R \}^3$. In

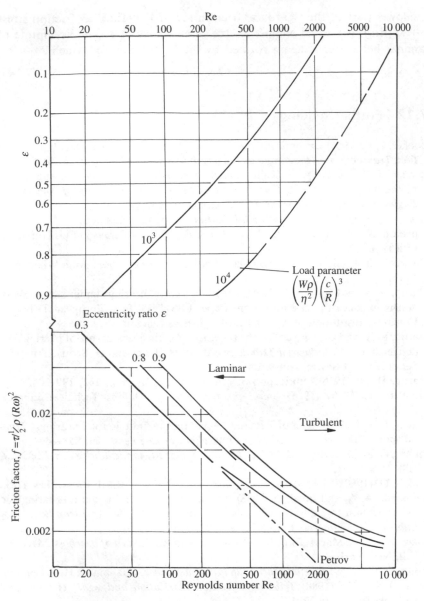

Fig. 7.35 Bearing eccentricity ε and friction factor f as functions of Reynolds number Re for laminar and superlaminar or turbulent flow.

the lower part of the figure the friction factor f, defined as friction stress $\tau \div \frac{1}{2}\rho(R\omega)^2$ is plotted against Re for three values of ε. The asymptote of laminar behaviour is again formed by the Petrov line for which

$$f = 2 \div \text{Re}. \tag{7.149}$$

7.12 Further reading

Booker, J. F. (1965) Dynamically loaded journal bearings–mobility method of solution. *Transactions of the American Society of Mechanical Engineers*, **87**(3), 537–46.

Booker, J. F. (1971) Dynamically loaded journal bearings–numerical application of the mobility method. *Transactions of the American Society of Mechanical Engineers, Journal of Lubrication Technology*, **93**, 168–79.

Cameron, A. (1966) *Principles of lubrication*. Longman, London.

Cameron, A. and Ettles, C. M. Mc. (1981) *Basic lubrication theory*. Ellis Horwood, Chichester.

Hamrock, B. J. and Dowson, D. (1981) *Ball bearing lubrication*. John Wiley, New York.

Martin, F. A. and Garner, D. R. (1974) Plain journal bearings under steady loads: design guidance for safe operation. Paper LB371/73. First European Tribology Congress, Institution of Mechanical Engineers, London.

Martin, F. A. and Garner, D. R. (1983) Design of plain bearings, use of bearing data design charts. In *Industrial Tribology* (ed. M. H. Jones and D. Scott). Tribology Series No. 8, Elsevier, Amsterdam.

Martin, H. M. (1916) Lubrication of gear teeth. *Engineering*, **102**, 199–204.

Muijderman, E. A. (1966) *Spiral groove bearings*. Philips Technical Library, Eindhoven.

Pinkus, O. (1958) Solution of Reynolds' equation for finite journal bearings. *Transactions of the American Society of Mechanical Engineers*, **80**, 858–64.

Pinkus, O. and Sternlicht, B. (1961) *Theory of hydrodynamic lubrication*. McGraw-Hill, New York.

Pinkus, O. (1990) *Thermal aspects of fluid film tribology*. ASME Press, New York.

Raimondi, A. A. and Boyd, J. (1958) A solution for the finite journal bearing and its application to analysis and design. *Transactions of the American Society of Lubrication Engineers*, **1**, 159–74, 175–93, 194–209.

Shaw, M. C. and Macks, E. F. (1949) *Analysis and lubrication of bearings*. McGraw-Hill, New York.

Smith, D. M. (1969) *Journal bearings in turbomachinery*. Chapman & Hall, London.

Szeri, A. Z. (ed.) (1980) *Tribology: friction lubrication and wear*. Hemisphere/McGraw-Hill, New York.

Tipei, N. (1962) *Theory of lubrication*. Stanford University Press, Stanford.

Trumpler, P. R. (1966) *Design of film bearings*. Macmillan, New York.

Walowit, J. A. and Anno, J. N. (1975) *Modern developments in lubrication mechanics*. Applied Science Publishers, London.

Wilcock, D. F. and Booser, E. R. (1957) *Bearing design and application*. McGraw-Hill, New York.

8
Gas bearings, non-Newtonian fluids, and elasto-hydrodynamic lubrication

8.1 Introduction

All real materials possess some microstructure which becomes apparent when a representative specimen is viewed with a sufficient degree of magnification. However, engineers usually work at a comparatively large scale and so use macroscopic models of material behaviour: these are likely to involve a very large number of the individual microstructural units which contribute to the overall material response. Engineering properties of the bulk represent *average* or *integrated* values, and when dealing with either solids or liquids the material is often treated as a homogeneous continuum. In the case of the linear elastic theory of solids what is required is Hooke's law and the observation that most metals undergo only a very small strain before yielding. Similarly, a Newtonian compressible fluid is described by the linear relationships between the applied shear stress and the resultant shear strain rate, and between density and the applied hydrostatic pressure. Many liquid lubricants, including hydrocarbon mineral oils, are effectively incompressible at moderate pressures so that their density remains constant, and this simplifies the analysis still further.

There is no fundamental reason why the lubricant in a hydrodynamic bearing should not be a vapour or a gas (rather than a liquid such as oil or water); indeed considered as a potential lubricant, air has a number of advantages — such as cleanliness and ease of supply — and it has been used in *aerostatic* and *aerodynamic* bearings for several decades. However, it is clear that since the viscosity of air is so low (about 0.01% of typical oils) the speed of a gas-lubricated bearing must be many times that of its oil lubricated counterpart to generate either a similar Sommerfeld number or the same degree of load support. Air bearings are widely used in engineering devices in which there are high peripheral sliding speeds such as machine tool spindles, turbo-machinery, instrument and gyroscope bearings, and in high-speed precision machinery such as dental drills and textile processing machines. In Section 8.2 the ideas already in connection with conventional hydrodynamic bearings are extended to cover a number of these more specialized, but nevertheless technologically important, cases.

While many conventional lubricants can be considered to exhibit Newtonian behaviour under a wide range of service conditions, there are a number of circumstances in which predictions based on this simple model are inadequately accurate. It is well known, for example, that liquids with a complex structure (such as polymer solutions or melts, soap solutions, and solid suspensions or slurries) can respond to applied loads or strain rates in rather surprising ways, and that such fluids are not always at all well modelled by the constitutive equation for Newtonian behaviour. Section 8.3 gives a brief outline of some aspects of non-Newtonian behaviour which are relevant to tribology. Under conditions of extreme pressure the behaviour of conventional mineral lubricating oils can also become distinctly non-Newtonian. The branch of tribology that considers these effects, together with the associated elastic deformation of the bounding solid surfaces, is known as elastohydrodynamics and is treated in Section 8.4.

8.2 Lubrication by gases and vapours; air bearings

Gas-lubricated bearings operate on the same principles as those using liquid lubricants. Examination of any of the equations describing the pressure distribution within hydrostatic or hydrodynamic bearings, shows that, to a first approximation at least, pressure and thus load capacity are both proportional to lubricant viscosity. Since the viscosity of a gas is generally so much less than that of a liquid it follows that pressures and specific loads in gas bearings are very low: frictional or traction stresses will be similarly reduced. However, it is worth noting that *coefficients* of friction, depending on the *ratio* of tractive to normal stresses, are much the same in bearings lubricated by gases as those utilizing liquids. As well as possessing very low viscosities, gases differ from liquids in that they are inherently compressible, and this can lead to gas bearings having significantly different behaviour from those lubricated by liquids. These differences becoming more important as the service conditions become more severe.

Air as an operating fluid has the advantages that it is plentiful and cheap and will not tarnish or corrode solid surfaces. It can also be exhausted safely to the atmosphere without the need for special arrangements for its collection and return. In addition air bearings have a very wide range of temperature operation, as air will not freeze at low temperatures or boil at high ones. On the other hand, since the pressures in air bearings are typically 10^5 Pa, that is, about 1% of those in oil-lubricated bearings, an air bearing must occupy about 100 times the area of an oil-lubricated design in order to support a comparable load. It is also generally the case that the gap between the two opposing solid surfaces is very much smaller in air- than in oil-lubricated bearings. In the case of aerostatic bearings there is clearly a design trade-off between the demands of volumetric flow rate (and therefore the associated pumping power) and the minimum clearance over the bearing land. In aerodynamic bearings the minimum gap is related to the speed of relative motion. Cleanliness, in particular the exclusion of potentially damaging contaminant hard particles from the external environment, becomes even more essential in air bearings than in other tribological devices. Air bearings also suffer from the disadvantage of being inclined to dynamic instabilities. All self-acting fluid bearings are prone to various forms of self-induced vibration but in gas bearings this problem can more easily become acute. The mechanisms that can lead to these instabilities are described more fully in Section 8.2.4.

8.2.1 Reynolds' equation for compressible fluids

A form of Reynolds' equation describing the conditions between two solid moving surfaces separated by a gaseous film can be developed in an entirely analogous fashion to that for incompressible fluids described in Chapter 7. Once again it is usual to assume that there is no slip at the boundaries between the solids and fluid, and that the flow through the bearing gap is laminar. However, in the case of gases, density is not constant and so it is essential to consider continuity of *mass* rather than that of *volume* as we move from section to section along the bearing. Using the notation of Chapter 7 we can write, in place of eqn (7.58), an equation for the general two-dimensional case as

$$\frac{\partial}{\partial x}\left\{\frac{\rho h^3}{\eta}\frac{\partial p}{\partial x}\right\} + \frac{\partial}{\partial y}\left\{\frac{\rho h^3}{\eta}\frac{\partial p}{\partial y}\right\} = 12\bar{U}\frac{\partial(\rho h)}{\partial x}. \tag{8.1}$$

Notice that now, in contrast to the treatment in Chapter 7, the density ρ of the gas cannot be taken outside the differentiation, since it is likely to vary with position (x,y).

8.2.2 Inclined-pad bearing

Initially, we consider the case of an infinitely wide pad bearing considering all the flow to be in the x-direction. If the lower surface is moving with speed U and the upper is stationary then eqn (8.1) can be written as

$$\frac{\partial}{\partial x}\left\{\frac{\rho h^3}{\eta}\frac{\partial p}{\partial x}\right\} = 6U\frac{\partial(\rho h)}{\partial x}. \tag{8.2}$$

In such a case the entraining velocity $\bar{U} = U/2$. Integrating eqn (8.2) directly we have,

$$\frac{\rho h^3}{\eta} \frac{dp}{dx} = 6U\rho h + C,$$ (8.3)

where C is a constant of integration. If now we identify the physical quantities at the section of the bearing where the pressure has reached a maximum and at which the pressure gradient is zero by $\bar{h}$ and $\bar{p}$ then we can evaluate the constant C in terms of these barred quantities. Equation (8.3) becomes

$$\frac{\rho h^3}{\eta} \frac{dp}{dx} = 6U(\rho h - \bar{p}\bar{h}).$$ (8.4)

If $\dot{m}$ is the mass flow rate per unit width then, since this must be constant at all sections of the bearing, its value will be given by

$$\dot{m} = \tfrac{1}{2} U\bar{p}\bar{h},$$ (8.5)

so that eqn (8.4) can also be written as

$$\frac{dp}{dx} = \frac{12\eta}{h^3} \left\{ \frac{U}{2} h - \frac{\dot{m}}{\rho} \right\}.$$ (8.6)

This form of Reynolds' equation can be used to illustrate the difference between liquid and gaseous lubrication. Because gases are compressible, their density is not constant but will increase as the pressure grows. This means that the final term of eqn (8.6) decreases as we move into the convergent bearing gap. The pressure gradient (dp/dx) consequently gets steeper, and so the shape of the pressure distribution will not be the same for gas- and liquid-lubricated bearings of similar geometries. It is a general rule that the centre of pressure of a slider bearing utilizing a gas film will always *trail*, that is, occur further into the bearing, that of a similar bearing using an incompressible lubricant. For similar reasons, the attitude angle of a gas-lubricated journal bearing will be *less* than that of the corresponding liquid-lubricated bearing.

Notice that in making use of eqn (8.6) it is important to distinguish between absolute and gauge pressures (i.e. differential pressures above ambient). The effect of compressibility is especially noticeable when the pressure variation through the bearing is large compared with the external ambient value p_a. If this variation is small compared to p_a (as may be the case if local ambient pressures are very high) then compressibility effects may well be negligible. This effect is illustrated in Fig. 8.1.

Equation (8.4) can be rearranged as

$$\frac{dp}{dx} = 6U\eta \left\{ \frac{h - \bar{p}\bar{h}/\rho}{h^3} \right\},$$ (8.7)

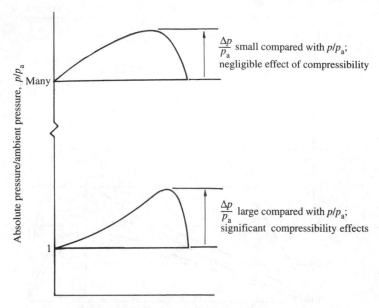

$\frac{\Delta p}{p_a}$ small compared with p/p_a; negligible effect of compressibility

$\frac{\Delta p}{p_a}$ large compared with p/p_a; significant compressibility effects

Fig. 8.1 The influence of ambient pressure on the importance of compressibility effects (from Grassam and Powell 1964).

and this can be conveniently written in non-dimensional terms by defining

$$P = \frac{p}{p_a}, \qquad H = \frac{h}{h_o}, \qquad \text{and} \qquad X = \frac{x}{B}, \tag{8.8}$$

so that

$$\frac{\mathrm{d}P}{\mathrm{d}X} = \frac{6U\eta B}{p_a h_o^2} \left\{ \frac{H - \bar{H}\bar{p}/\rho}{H^3} \right\}. \tag{8.9}$$

The leading group on the right-hand side of this equation is often known as the *bearing* or *compressibility number* Λ, that is

$$\Lambda = \frac{6U\eta B}{p_a h_o^2}. \tag{8.10}$$

Higher values of the bearing number are associated with operating conditions under which the effects of lubricant compressibility become increasingly important. Using these non-dimensional variables, conditions through the bearing are then described by the normalized form of eqn (8.4), thus

$$\frac{\mathrm{d}P}{\mathrm{d}X} = \Lambda \frac{H - \bar{H}\bar{p}/\rho}{H^3}. \tag{8.11}$$

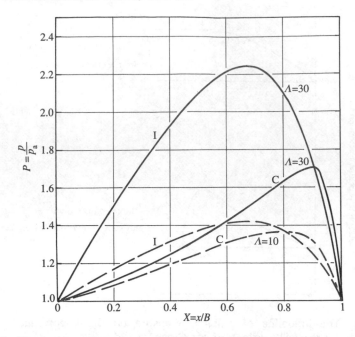

Fig. 8.2 Pressure profiles along an inclined-pad bearing. Curves marked I are for incompressible fluids, and those marked C for compressible. Λ is the bearing or compressibility number; the shape factor $H = 2$ (data from Gross 1962).

The solution of this equation, that is, the variation of P with X, requires some knowledge of, or at least assumptions about, the conditions in the bearing which are likely to influence the variation of density. Since the density of a vapour or gas is greatly influenced by its temperature, eqn (8.11) implies a need to consider thermal conditions through the bearing. Variations in film temperature can arise from both the effects of shearing within the film itself and as a result of heat conduction into the film from external sources. Even for the simplest case of isothermal behaviour, eqn (8.11) is probably best solved numerically; Raimondi (1961) contains some numerical data for a range of cases.

Comparison with the incompressible case can be made by reference to Section 7.3.2. In the current notation eqn (7.32) can be written as

$$P = 1 + \Lambda \frac{X(1-X)(H-1)}{(1+H)[H-(H-1)X]^2} \tag{8.12}$$

where the shape factor $H = h_i/h_o$.

Figure 8.2 shows the pressure profiles for a pad bearing with $H = 2$ for both incompressible (I) and compressible (C) lubricants. The solid curves

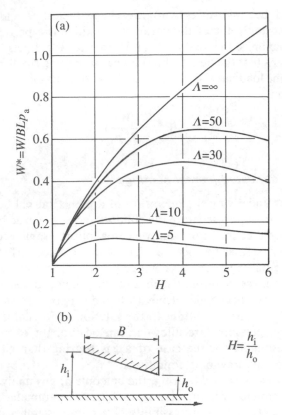

Fig. 8.3 (a) Non-dimensional load W^* carried by an inclined-pad bearing versus parameter H for various bearing numbers Λ. (b) Bearing geometry (data from Gross 1962).

refer to the situation in which the bearing number Λ is equal to 30, and the dashed curves apply when Λ equals 10. It is clear that the peaks in the curves for the compressible fluids occur at rather greater values of the parameter $X = x/B$; as Λ gets very large so the location of the peak pressure moves closer to the trailing edge of the bearing. It is also evident that when the bearing number is small ($\Lambda = 10$) despite the curves I and C being of rather different shapes both the peak pressures and the enclosed areas, which represent the loads carried, are similar. At low bearing numbers the compressibility of the fluid has only a minor effect. However, when Λ is increased to the value 30 the situation is very different: now the peak pressure in the compressible fluid curve C is very much less than that of the corresponding curve I, and the load-carrying capacity is reduced by almost a factor of 3.

Figure 8.3 shows how the load-carrying capacity of an inclined-pad gas-lubricated bearing varies with the shape factor H and the bearing number Λ. At any value of the parameter Λ there is a particular value of H that will maximize the load that can be carried. As Λ gets larger so does H, which means that the outlet film thickness is getting smaller and smaller. In the figure the load has also been plotted non-dimensionally as W^* defined by the relation

$$W^* = \frac{W}{BLp_a}.$$ (8.13)

8.2.3 Gas-lubricated journal bearings

The principles underlying the operation of a plane journal bearing, which uses a gas or air film to separate the solid surfaces, are exactly the same as those of its liquid counterpart, and we can adopt the same geometric parameters to describe its shape; see Fig. 7.13. Gases cannot suffer from cavitation in the region of the negative gauge pressures, and indeed, the region of subambient pressure around such a bearing may provide an additional, and not necessarily negligible, element of load support. To some extent the Sommerfeld boundary conditions to the solution of Reynolds' equation, in which negative pressures are allowed, can be thought of as being more physically reasonable for the case of a gas bearing than either the half-Sommerfeld or the Reynolds conditions.

As was the case for a pad bearing, the outcome of any analysis of journal bearing performance will involve an additional parameter whose value depends on the effect of compressibility. The compressibility number Λ for a journal bearing is conveniently defined as

$$\Lambda = \frac{6\eta\omega R^2}{c^2 p_a}.$$ (8.14)

For an infinitely long bearing (in other words one with a negligibly small amount of end leakage and so large length-to-diameter ratio) and provided compressibility effects are not significant, conditions will be well modelled by the Sommerfeld conditions discussed in Section 7.7.2. The attitude angle, that is, the angle between the load vector and the line of centres, will always be 90°. The total load per unit length on the bearing, W/L, and the eccentricity ratio, ε, will be related by eqn (7.104) which can be written as

$$\frac{W}{2RLp_a} = \Lambda \frac{\pi\varepsilon}{(1 - \varepsilon^2)^{1/2}(2 + \varepsilon^2)}.$$ (8.15)

If the bearing is very lightly loaded so that the eccentricity ratio is small, that is, $\varepsilon^2 \ll 1$, then

$$\frac{W}{2RLp_a} \approx \Lambda \frac{\pi\varepsilon}{2} \tag{8.16}$$

so that S, the Sommerfeld number, and ε are related by

$$S \approx \frac{1}{3\pi\varepsilon}. \tag{8.17}$$

When the bearing is of finite length, so that both axial and circumferential pressure gradients exist, there is no closed-form solution of Reynolds' equation and numerical methods of solution must be employed to cope with intermediate values of Λ. Figures 8.4 and 8.5 illustrate the outcome of such analysis. In Fig. 8.4 the mid-plane circumferential pressure distributions, are plotted for a full journal bearing of length-to-diameter ratio 1 and eccentricity ratio 0.4 for three bearing numbers. The pressure p has been non-dimensionalized by the ambient pressure p_a. As the bearing number increases from zero (which represents incompressible conditions)

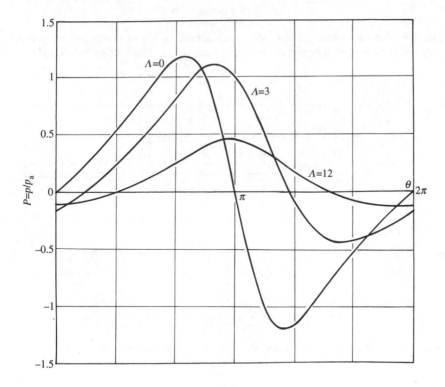

Fig. 8.4 Pressure distribution around the mid-plane of a 360° gas-lubricated journal bearing with eccentricity ratio $\varepsilon = 0.4$ as a function of bearing number Λ (data from Raimondi 1961).

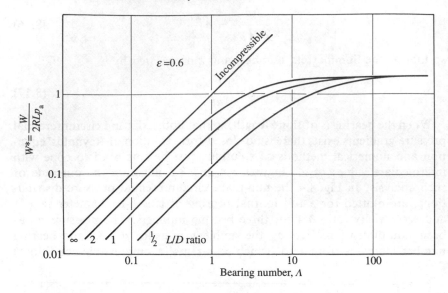

Fig. 8.5 Non-dimensional load W^* versus bearing number for gas-lubricated journal bearings. ε is the eccentricity ratio (data from Raimondi 1961).

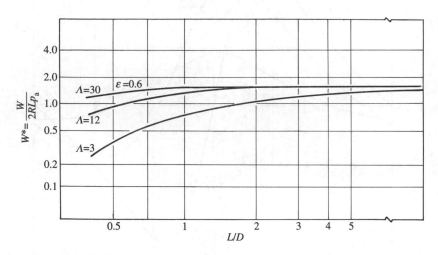

Fig. 8.6 Non-dimensional load W^* versus length-to-diameter ratio L/D for 360° gas-lubricated journal bearings (data from Raimondi 1961).

the pressure distribution gradually changes from one of perfect asymmetry, about the point $\theta = \pi$, to a more symmetrical distribution.

Figures 8.5 and 8.6 serve to illustrate the effects of both compressibility (through the bearing number) and geometry variation (through the length-to-diameter or slenderness ratio L/D). When L/D is very large and the fluid is incompressible then eqn (8.15) applies and so, for a given value of the eccentricity ratio, the curve of non-dimensional load versus bearing number is a straight line on log-log axes; Fig. 8.5 is plotted for a value of $\varepsilon = 0.6$. With increasing values of Λ the discrepancy between incompressible and compressible behaviour becomes greater. As the numerical value of the slenderness ratio reduces so flow from the ends of the bearing (where the pressure must be ambient) becomes more significant and the load-carrying capacity drops; this phenomenon is also illustrated by Fig. 8.5. The same kind of effect can be seen in Fig. 8.6 in which load is plotted against the slenderness ratio of the bearing for various compressibility numbers; again ε is taken as 0.6. Further numerical data on these effects can be found in Raimondi (1961) and Hays (1959).

8.2.4   *Instabilities in self-acting bearings*

Synchronous whirl

In both this and the previous chapter it has been tacitly assumed that the solution of the governing equations represent steady state, stable operating conditions. In practice, whether using a liquid or a vapour lubricant, there are relatively few cases in which the lubricant film is absolutely steady and shows no variation with time—although, fortunately, there are wide ranges of bearing operation over which conditions are stable. For example, the geometric centre of a rotating shaft supported in two journal bearings will not remain at an absolutely steady value of eccentricity but, as a result of the inevitable dynamic out-of-balance of any real shaft, will generally move in an orbit about this position, in the direction of the shaft rotation. This phenomenon is called *synchronous whirl*. For a well-balanced rotor the amplitude of the whirl may be extremely small; however, even when the amplitude of whirl is a comparatively large fraction of the bearing clearance, conditions may still be stable and continuous operation of the device may still be quite safe. As in other vibration problems the prediction of the speeds and mode shapes associated with synchronous whirl requires a knowledge of the inertia of the rotor (which provides the driving out-of-balance force) and the stiffness and damping capacity of the bearings (both of which provide forces in opposition to the motion). The most elementary model of a simply supported rotor is that shown in Fig. 8.7(a).

When the bearing stiffness k is small then the modes of deformation of

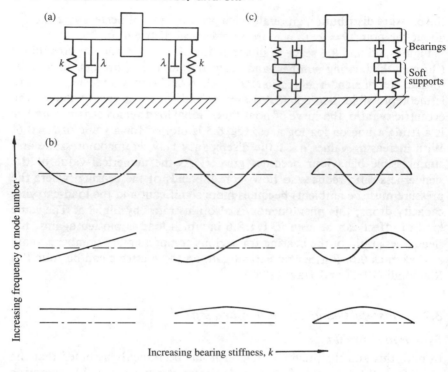

Fig. 8.7 (a) Dynamic model of a rotating shaft supported on two journal bearings; the springs of stiffness k represent the bearings' stiffness and the dampers with constant λ their damping capacity. (b) Whirling modes as functions of stiffness k. (c) Modified model to account for the stiffness of bearing mounts (from Grassam and Powell 1964).

the shaft correspond to the transverse modes of a beam with free end supports, illustrated by the deformed shapes at the left of Fig. 8.7(b). If k is very large then the ends of the shaft cannot move transversely at all, and so the vibration modes become analogous to those of a beam with knife-edge restraints and thus are represented by the modes at the extreme right of Fig. 8.7(b). If the operating speed of the shaft were to be close to one of these *whirling speeds* then the amplitude of the oscillations achieved would be limited only by the damping within the system. If this is small (as will usually be the case in gas bearings) then the scale of vibration may be greater then is acceptable; in extreme cases the bearing surfaces may actually come into mechanical contact and this would invariably constitute bearing failure. Bearing systems are therefore designed so that the critical whirling speeds do not coincide with the anticipated operating rotational speeds. Sometimes this can be achieved by increasing the bearing stiffness so that even the lowest whirling speed is well above the highest likely operating speed. However, this is not often feasible in practice, since the necessary improvement in bearing stiffness may imply unrealistically small bearing clearances. This means that in many applications the lower synchronous resonance speeds are likely to lie within operating speed range of the machine of which the bearing is part. In such cases it is important that the rotor is accelerated (and decelerated) sufficiently rapidly for the critical speeds to be passed through quickly, so that there is insufficient time for the amplitude of oscillations to build up to dangerous levels. The lubricant film itself supplies some damping to the system, the actual damping mechanism depending essentially on squeeze film effects (see Section 8.2.6), but its precise numerical estimation is difficult. Most practical journal designs will have clearance ratios of less than 0.002 and length-to-diameter ratios of around unity. Gas bearings of this sort of geometry are likely to have damping coefficients in the range 0.2–1. The degree of effective damping can be significantly increased by mounting the bearings themselves in suitably soft supports, such as rubber 'O' rings (see Fig. 8.7(c)). When this solution is adopted it is important to appreciate that although this will reduce the amplitude of whirl of the rotor relative to the bearings, the *absolute* amplitude of rotor whirl, relative to the other machine parts, may actually be increased.

Fractional speed whirl

Synchronous whirl is excited by the unbalance of the rotor and would be absent in a rotating system which was perfectly dynamically balanced. However, other instabilities are possible in which the whirling frequency is some simple fraction of the rotating speed of the journal. These are *self-excited*, and are likely to be unstable even at small eccentricities. In contrast to the synchronous case, the system will not pass through such vibration resonances as the speed of rotation is increased; on the contrary, this tends to make the vibration more pronounced and can easily lead to premature bearing failure.

The mechanism of such a fractional (most commonly half-speed) whirl can be understood by remembering firstly that the net restoring force applied by the lubricant film does not act in the same direction as the displacement of the centre of the journal. In general the attitude angle ψ is not zero. This means (see Fig. 8.8(a)) that there is a component of force, $W\sin\psi$, tending to draw the centre of the shaft into the whirl, in other words to give C, the journal centre, a rotation in the same sense as ω. The distribution of velocities within the lubricant film will then be similar to that shown, varying from zero at the bearing surface to ωR at that of the journal. Now consider what happens if the journal centre C has achieved a rotational speed Ω (about the bearing centre O) which is just equal to $\omega/2$. This case is

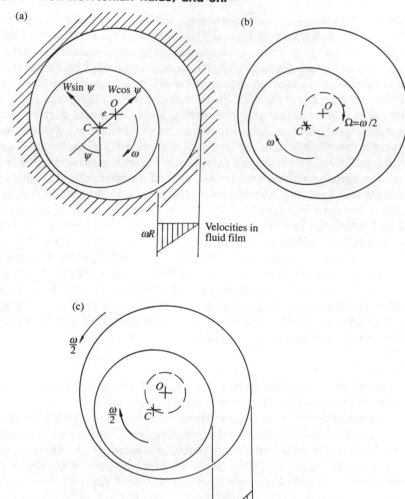

Fig. 8.8 The mechanism of half-speed whirl: (a) forces applied to the journal by the fluid film; (b) the centre of the journal is rotating at a speed Ω which is just equal to $\omega/2$; and (c) imposition of a counter-rotation $\omega/2$ brings the gap to rest.

illustrated in Fig. 8.8(b). The position of the convergent–divergent gap
formed between the two solid surfaces will now also be rotating at speed
$\Omega = \omega/2$. This gap can be thought of as being brought to rest by imposing
a solid body rotation of magnitude Ω to the whole system in the *opposite*
sense to the rotation of the shaft, see Fig. 8.8(c). Thus the system is
equivalent to one in which the journal rotates clockwise at $\omega/2$ while the
bearing also rotates at $\omega/2$ but in the opposite sense. The velocity profile
across the gap is therefore as shown. There is now no net viscous pumping
of fluid into the gap and this means that the basic mechanism of hydrodynamic load support has been lost since the effective entraining velocity
$\bar{U}$ has fallen to zero; the bearing is therefore in grave danger of collapsing
with potentially catastrophic results.

Such half-speed whirl is likely to occur in any hydrodynamic journal bearing whether lubricated by air or oil (or any other fluid). However, its severity
is lowered by any natural damping in the system which can dissipate the
energy associated with the whirl distortion. Air bearings, in which these
damping effects are small, are thus unfortunately especially prone to this
problem. The most widely used methods of suppressing, or controlling, the
extent of fractional speed whirl involve either increasing the damping, or
breaking the circumferential symmetry of the bearing in such a way as to
provide some additional hydrodynamic effects. Multi-lobed (usually three)
bearing designs have been used with some success for some years; see
Fig. 8.9(a). Another successful design is the half-lemon bearing, Fig. 8.9(b),
in which the top half has a difference in radii between the shaft and the bearing of about 2.5 or 3 times that of the bottom but has its centre dropped
so that the actual top and bottom clearances are equal. Yet another alternative is to provide a pocket or step in the top half of the bearing, Fig. 8.9(c).
Further details of these effects can be found in Gross (1962) and Grassam
and Powell (1964).

8.2.5   *Tilting-pad and thrust bearing designs*

A full 360° journal bearing may not be the most appropriate form of bearing
design for the case of a steady load applied to a shaft rotating at a constant
speed even when precautions against whirling are taken. The *tilting-pad
journal bearing*, shown schematically in Fig. 8.10, has a number of advantages over its simple cylindrical competitor.

Firstly, and perhaps most importantly, there is its virtual freedom
form whirling instabilities, particularly if the pivot on one of the pads
is made adjustable or spring loaded. Higher safer running speeds can be
achieved than with plain bearings and the performance is relatively insensitive to the changes in clearance arising from fluctuations in load or
temperature. Secondly, the pivoting arrangement provides greater freedom

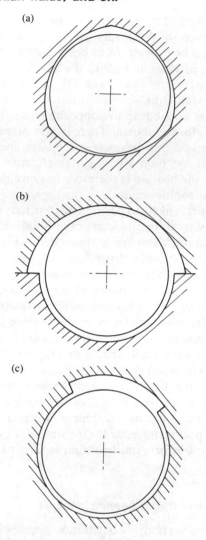

Fig. 8.9 Bearing designs to minimize fractional speed whirl: (a) three-lobed bearing; (b) the half-lemon bearing; and (c) axial groove or pocket bearing.

to accommodate misalignments, either in mounting or loading, than is the case with full cylindrical journals. Finally, but not necessarily of least significance, is the fact that manufacture and installation are likely to be simplified. Only the pad faces require high accuracy whereas in a plain journal both shaft and bearing must be accurately circular and manufactured to very precise tolerances.

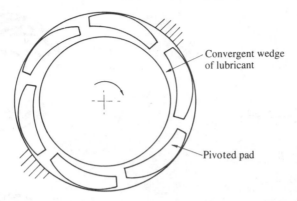

Fig. 8.10 The tilting-pad journal bearing.

For hydrodynamic, or aerodynamic, thrust bearings an analogous choice has to be made between fixed or tilting pad designs. The rotating member in each case is a flat 360° disc, often known as a *runner*. In a fixed-pad design the annular stationary member carries a series of steps, pockets, or grooves, each of which acts rather like an independent miniature Rayleigh step bearing contributing to the overall axial load-carrying capacity. Such a stepped-sector thrust bearing is shown in Fig. 8.11(a); Table 8.1 indicates N, the optimum number of pads, and the location of the step, defined by the angle ratio $\Delta\beta/\beta$ for the particular (but not unreasonable) condition that $(h + \Delta h)/h = 2$, where h is the film thickness over the thinner of the two parallel sections. In the case of a rotating bearing the bearing or compressibility number Λ is defined by

$$\Lambda = \frac{6\eta\omega r_o^2}{h^2 p_a},$$

(8.18)

and a convenient non-dimensional load W^* by

$$W^* = \frac{W}{\pi(r_o^2 - r_i^2)p_a}.$$

(8.19)

The design of this type of bearing can be refined to reduce the leakage of lubricant by providing narrow lands on either side of the pocket, in much the same way as on a conventional Rayleigh step bearing (see Fig. 7.8); this is illustrated in Fig. 8.11(b). The only disadvantage to this design lies in the increased difficulty of manufacture of the pockets which are likely to be only of the order of 10 microns in depth.

If the geometry of the pockets is modified so that they have the form of a number of spiral grooves with closed ends, see Fig. 8.11(c), then the load-carrying capability is built up as a result of the pumping of fluid that occurs

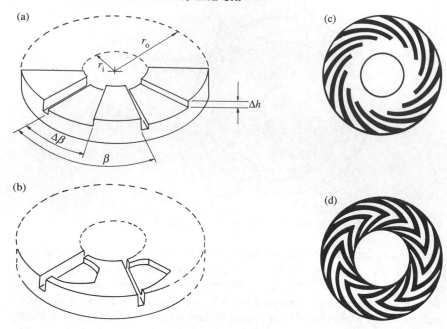

Fig. 8.11   (a) Stepped-sector thrust bearing; (b) modified design with enclosed pockets. Faceviews of (c) spiral groove bearing, or 'viscosity plate', and (d) 'herring-bone' thrust bearing.

Table 8.1   Stepped-sector thrust bearing: optimum design parameters for the condition that $(h + \Delta h)/h = 2$ (data from Grassam and Powell (1964)).

r_o/r_i	Bearing parameters (see Fig. 8.11 and text)	Compressibility number 10	40
2	N	8	7
	$\Delta\beta/\beta$	0.48	0.33
	W^*	0.046	0.22
5	N	4	3
	$\Delta\beta/\beta$	0.45	0.26
	W^*	0.064	0.286

along each path. The spiral grooves may have their open ends on the inner diameter, so that fluid is pumped outwards, or on the outer diameter giving inward flow. In general, inward-pumping bearings possess the greater load capacity but outward-pumping designs have a greater resistance to tilt arising from any off-centre loading. Inward- and outward-pumping features can

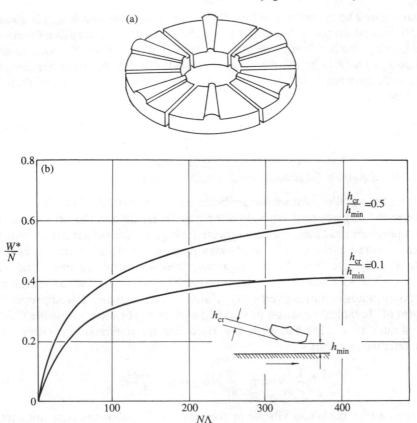

Fig. 8.12 (a) Tilting-pad thrust bearing, and (b) non-dimensional load versus bearing number. N is the number of pads around the circumference, Λ is the bearing number, and W^* is the normalized load defined by eqns (8.18) and (8.19) (data from Grassam and Powell 1964).

be combined in a herring-bone pattern, Fig. 8.11(d). Thrust bearings of this sort can be used with both incompressible and compressible fluids and further details on their design, performance, and construction is available in Muijderman (1966). If the pumping grooves are cut on a cone then the bearing can sustain both axial and radial loads and this form of pumping design has considerable potential for tribological applications in both bearings and seals.

Tilting pads can also be used in aerodynamic thrust bearings and such a design is illustrated diagrammatically in Fig. 8.12(a). The individual sector-shaped pads require a slight convex crown to ensure stability and the dimensions of the pads are best chosen so that they are approximately square in plan view. The optimum position for the pivot (see Section 7.4.2) is about 60% around the pad in the direction of motion. Load capacity as a function of bearing number for this design for two different extents of crowning are shown in Fig. 8.12(b); N, the number of pads around the circumference, is typically between 6 and 12. If the pads are to be close to 'square' then it follows that

$$\frac{r_0}{r_i} \approx \frac{N + \pi}{N - \pi}. \tag{8.20}$$

8.2.6   Squeeze films and compressible fluids

If two flat parallel solid surfaces vibrate relative to one another with a component of displacement normal to their surfaces, then in the presence of a compressible fluid a superambient pressure is generated between them which can provide a useful positive load-carrying capacity. This is a further example of a squeeze film phenomenon described for incompressible fluids in Section 7.9. As a simple example consider the case of two flat, circular discs each of radius a, moving vertically relative to one another. The appropriate form of Reynolds' equation in polar coordinates, for a compressible fluid, is similar to (7.133) but with the fluid density ρ remaining within the differentiation; thus

$$\frac{1}{r}\frac{\partial}{\partial r}\left\{r\rho h^3 \frac{\partial p}{\partial r}\right\} = -12\eta\frac{\partial(h\rho)}{\partial t}. \tag{8.21}$$

Suppose that the plates vibrate at frequency ω' so that the film thickness h is given by the equation

$$h = h_0(1 + \varepsilon\cos\omega't). \tag{8.22}$$

h_0 is the mean film thickness and ε is now known as the *excursion ratio*, and represents the size of the vibration amplitude relative to h_0. Equation (8.21) can be expressed in non-dimensional terms by making the substitutions

$$h/h_0 = H; \qquad r/a = r^* \qquad \text{and} \qquad p/p_a = P \tag{8.23}$$

so that it becomes

$$\frac{H^3}{r^*}\frac{\partial}{\partial r^*}\left\{r^*P\frac{\partial P}{\partial r^*}\right\} = -\Sigma\frac{\partial(HP)}{\partial(\omega't)} \tag{8.24}$$

where Σ is the *squeeze number* defined by

$$\Sigma = \frac{12a^2\eta\omega'}{h_0^2 p_a}. \tag{8.25}$$

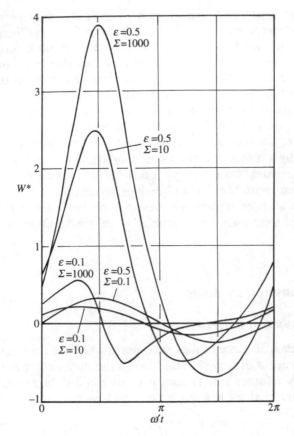

Fig. 8.13 Squeeze film effects between a pair of circular flat plates vibrating at ω'. Plot of normalized load W^* against $\omega't$ (from Salbu 1964).

The squeeze number is a measure of the importance of compressibility effects in squeeze bearings in much the same way as the bearing or compressibility number Λ is in conventional journal or slider bearings. The normal load W carried by the plates can be non-dimensionalised to W^* by defining

$$W^* = \frac{W}{\pi a^2 p_a}. \tag{8.26}$$

The variation of this quantity over one cycle of vibration is shown in Fig. 8.13. The closer the curve is to being symmetrical then the smaller the net capacity for load support. Lack of symmetry in the curve, and therefore increased load capacity, is encouraged by greater values of the squeeze number Σ and excursion ratio ε, as can be seen from the figure.

At low squeeze numbers, and small excursions, the gas is forced radially out of the gap between the plates as the clearance decreases and is drawn back in again as the plates move apart. The pressure and hence forces acting on them arise largely as a result of the viscous resistance to this flow. As Σ and ε increase then these viscous forces grow; this restricts the flow of gas at the periphery of the discs so that an increasing proportion is trapped in the central region where it behaves much like a fixed mass of gas enclosed in a cylinder by a moving piston. On the approach stroke it is compressed and resists movement very strongly; on the return stroke it is rarefied and offers much less resistance. Thrust bearings, spherical bearings, and journal bearings have all been successfully designed on the basis of this principle using both electromagnetic and piezoelectric devices to generate the squeeze motion.

8.3   Non-Newtonian fluids

8.3.1   *Constitutive models*

The simplest class of viscous fluids are those in which the rheological properties are independent of time; in other words the fluid possesses no memory of its previous deformation. The shear stress τ in the fluid is then a simple function of the local shear strain rate $\dot{\gamma}$, and we may write

$$\tau = f(\dot{\gamma}). \tag{8.27}$$

If this function is linear, so that the shear stress is proportional to the shear strain rate, then the fluid is said to be Newtonian and the coefficient of proportionality is the viscosity. The term non-Newtonian is used to describe those fluids for which there is no simple linear relationship between the shear stress and the shear rate, in other words, the viscosity of a non-Newtonian fluid is not constant at given values of temperature and pressure. It may depend on such factors as the local rate of shear or the kinematic history of deformation of the fluid.

When investigated experimentally many fluids show behaviour of one of the types illustrated in Fig. 8.14. Some industrial fluids, such as slurries, drilling muds, and oil-based paints and greases, display the characteristics of a *Bingham material*, Fig. 8.14(a); τ_i is the magnitude of the stress which must be applied in order to *initiate* flow. The existence of this threshold stress (below which the material is undeformed) means the material is, initially at least, really rather more like a solid than a liquid; only if the applied shear stress exceeds τ_i will the material flow, which it then does

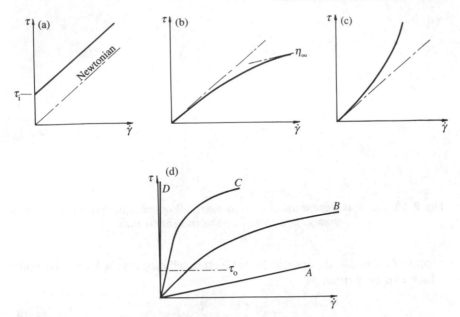

Fig. 8.14 Shear stress versus shear rate: (a) Bingham solid; (b) shear-thinning fluid; and (c) shear-thickening fluid. The Eyring visco-elastic model of material behaviour (d) can generate a range of characteristics. In curves (a)–(c) the chain dotted line represents Newtonian behaviour.

with a constant Newtonian viscosity. *Shear-thinning* fluids are those in which the apparent viscosity falls with increasing rates of shear, as illustrated in Fig. 8.14(b). The characteristic curve may become linear at very high rates of shear and this limiting slope is sometimes designated by, η_∞. When plotted on log-log axes (as opposed to the linear scales of Fig. 8.14) some fluids of industrial importance, such as suspensions of fine particles and polymer melts, often exhibit linear characteristics over several decades of shear rate. In such a case we can write $\tau = k\dot{\gamma}^s$, and the apparent viscosity is given by

$$\eta = \frac{\tau}{\dot{\gamma}} = k\dot{\gamma}^{s-1}, \tag{8.28}$$

where k and $s(<1)$ are constants. k is a measure of the *consistency* of the fluid and s a measure of its departure from Newtonian behaviour (for which $s = 1$). Fluids of the type of Fig. 8.14(c), which show an *increase* in viscosity with shear rate, occur less frequently than types (a) or (b). They can usually be adequately modelled by eqn (8.28) but with the exponents s taking values greater than unity.

Several other empirical fits to curves of the forms of Fig. 8.14 have been

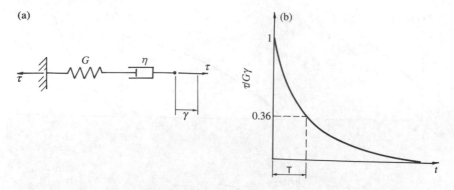

Fig. 8.15 (a) Maxwell model of a visco-elastic fluid, and (b) response to a step change in strain. T is the relaxation time.

proposed. One that is of particular interest in tribology is the Eyring formula which can be written as

$$\tau_0 \sinh\left(\frac{\tau}{\tau_0}\right) + \eta\frac{\dot{\tau}}{G} = \eta\dot{\gamma}, \tag{8.29}$$

and can produce the characteristics illustrated by curves $A\text{--}D$ in Fig. 8.14(d). This model incorporates two non-Newtonian effects which have been observed to occur in heavily loaded lubricated contacts (see Section 8.4.4). The initial term on the left-hand side of eqn (8.29) in which τ_0 is a material constant (the *Eyring stress*) represents the non-linearity in the relationship between τ and $\dot{\gamma}$.

In Newtonian fluids shear stress is proportional to rate of shear, while in a classical elastic solid there is a linear relation between the applied stress and the resultant shear strain. In some materials the effects of both viscosity and elasticity can be apparent simultaneously, and such materials are then said to be visco-elastic; their behaviour can be modelled by that of an elastic spring and viscous dashpot in series as illustrated in Fig. 8.15. This type of simulation is an example of a *Maxwell* model.

The spring of stiffness G represents material elasticity, while the dashpot, with constant η, represents the viscous resistance to motion. Equating the force in each of the components of this system shows that at all times the stress τ must satisfy the equation

$$\tau + T\dot{\tau} = \eta\dot{\gamma}, \tag{8.30}$$

where $T = \eta/G$ and has the dimensions of time. If, at time $t = 0$, a step change in deformation γ is applied (so that at all $t > 0$, $\dot{\gamma} = 0$) then it follows that the variation of τ is given by

$$\tau = G\gamma \exp(-t/\mathsf{T}). \tag{8.31}$$

This variation is illustrated in Fig. 8.15(b). T is known as the *relaxation time* and represents the interval during which the shear stress falls to the fraction $1/e$, that is, about 37%, of its initial value.

To some extent the distinction between a solid and a liquid is made on the subjective assessment of the value of T. For example, at room temperature, glass is actually a saturated and exceedingly viscous liquid but, because its relaxation time is of the order of centuries, it can be for all practical purposes regarded as a solid with an elastic shear modulus G, so that $\tau = G\gamma$. A material such as silicone putty has a relaxation time of a few seconds. It will bounce on a hard surface like an elastic solid because the impact process that this involves is complete in a small fraction of T so that there is insufficient time for it to respond in a viscous fashion. On the other hand, under much lower stresses it will flow like a viscous fluid and deform in relatively short, quite observable, times. In a stress relaxation experiment, a solid can be safely treated as ideally elastic only if the duration Δt of the event of interest is very much shorter than the relaxation time of the material. The ratio of these two times is known as the *Deborah number*, De, so that

$$\mathrm{De} = \frac{\text{relaxation time}}{\text{duration of event}} = \frac{\mathsf{T}}{\Delta t}. \tag{8.32}$$

If De is very much greater than unity the material can be treated as being simply elastic, as in the case of the bouncing silicone putty or almost any feasible experiment involving glass. If De is of the order of 1 (or less) then visco-elastic effects are likely to be of significance.

A material described by the Eyring equation (8.29) can be thought of as a non-linear Maxwell material. In circumstances in which τ/τ_0 is significantly less than unity $\sinh(\tau/\tau_0)$ becomes approximately equal to τ/τ_0, and so eqn (8.29) reduces to (8.30). If, in addition, η/G is small then this equation simplifies to $\tau = \eta\dot{\gamma}$, that is, that of a straightforward Newtonian fluid; this behaviour is shown as Curve A in Fig. 8.14(d). If conditions are such that while η/G is still small the ratio τ/τ_0 is greater than unity then, although the elastic term in eqn (8.29) remains negligible, the characteristic will be dominated by the hyperbolic sine, giving the behaviour illustrated by Curve B. As the value η/G grows then the small strain rate portion of the curve becomes increasingly influenced by this elastic term, as in Curve C. Eventually this influence dominates so that

$$\frac{\dot{\tau}}{G} = \dot{\gamma} \tag{8.33}$$

that is, we have a purely elastic material for which $\tau \approx G\gamma$, shown as Curve D in Fig. 8.14(d).

Finally, we can note that some real fluids are very imperfectly represented by relationships between shear stress and shear strain rate which are independent of time. Their viscosity depends not only on the rate of shear but also on the time for which that shear rate has been applied. In other words, they possess a memory of their previous deformation. Such rheological behaviour is known as *thixotropy* and if such a thixotropic material is sheared at a constant rate after a period of rest then, as the structure breaks down, the viscosity decreases; however, the rate of decrease, that is, the shape of the τ–$\dot{\gamma}$ characteristic, depends on the length of the preceding rest or recovery period. Synovial fluid, the naturally occuring lubricant in human and animal joints, is an example of such a thixotropic fluid.

8.4   Elasto-hydrodynamic lubrication

8.4.1   Line contact

Variable viscosity

It is common experience that the viscosity of a lubricating oil decreases as its temperature is raised and this is in no way in conflict with its behaviour as a Newtonian fluid. What is not so immediately obvious, but can be verified experimentally, is the effect of local high pressures on the numerical value of its Newtonian viscosity. For many mineral oils, and indeed for some other fluids used in tribological situations, this dependence on pressure can be modelled by the exponential or *Barus* equation, namely

$$\eta = \eta_0 \exp \alpha p, \tag{1.10 bis}$$

where η_0 is the base viscosity at very low gauge pressures, and the constant α, which in SI units is measured in Pa^{-1}, is known as the *pressure–viscosity index*.

Many other, principally empirical, relations have been suggested for describing the variation of viscosity with pressure (and, in some cases, temperature), see for example Briant *et al.* (1991), Gobar (1988), and Jacobson (1991). In some cases the observed increase in viscosity is slightly less than the exponential would indicate, and a simple power law can be used:

$$\eta = \eta_0 (1 + Cp)^n, \tag{8.34}$$

where C is a constant and the index n is of the order of 16. This equation is rather less 'powerful' than the exponential Barus relation. In the case of mineral lubricating oils, the most successful expression — albeit at the cost of some algebraic complexity — is probably the Roelands equation (Roelands *et al.*, 1963) which, considering only pressure sensitivity, can be written as

$$\eta = \eta_0 \exp \left\{ [9.67 + \ln \eta_0] \left[\left(1 + \frac{p}{p_0^*} \right)^\beta - 1 \right] \right\} \tag{8.35}$$

where β is a constant characteristic of the lubricant such that $0.5 < \beta < 0.7$ and the constant $p_0^* \approx 2 \times 10^8$ Pa. At lower ranges of pressure eqn (8.35) reduces to one of the form of (8.34).

As far as load support is concerned in a typical bearing situation, the local increases in viscosity that equations of this sort attempt to describe are clearly working in our favour. We might thus reasonably suspect that in practice it may be possible to successfully lubricate heavily loaded sliding contacts at conditions of greater severity than would be predicted by the simpler, constant-viscosity, hydrodynamic analysis. As an example, consider the following conditions which are characteristic of those between a heavily loaded pair of steel gear teeth:

tooth load per unit length, W/L 2.5×10^6 Nm^{-1}
reduced radius of curvature, R 0.02 m
entraining velocity $\bar{U}$ 5 ms^{-1}
base viscosity of mineral oil 0.07 Pa s
   lubricant at room
   temperature and pressure

We can estimate the minimum film thickness h_c at the centre of the contact on the basis of constant viscosity from eqn (7.79), thus

$$h_c = 4.9 \frac{\bar{U} \eta R L}{W} = 0.014 \text{ micron.} \tag{8.36}$$

This dimension is at least an order of magnitude less than the roughness, or R_a value, of the best of the regular range of surface finishes that can be achieved on engineering components (see for example the data of Table 2.3). The predicted oil film thickness is therefore very much less than the amplitude of the likely roughnesses on the surfaces, and it would be sensible to expect significant interference and interaction between opposing solid asperities which could well result in unacceptable damage and wear during use. However, gear teeth, and other devices with this sort of effective geometry, run quite satisfactorily under these sort of loads for very long periods with no evidence of excessive surface distress. The complete explanation for this is not obvious, but one potentially important effect is the increase in oil film thickness which might arise from the thickening of the oil actually *within* the contact due to the locally very high pressures. (The maximum Hertzian pressure between *dry* steel surfaces of this geometry is, from eqn (3.49), equal to just over 2.1×10^9 Pa.) In fact, it is relatively easy to make an estimate of the contribution that this local increase in viscosity might make to the separation of the surfaces.

Reynolds' equation in one dimension, which is based on continuity of volume flow, still holds, so that

$$\frac{dp}{dx} = 12\eta\,\bar{U}\,\frac{h-\bar{h}}{h^3}, \tag{8.37}$$

but now this must be combined with a second equation describing the variation of lubricant viscosity with pressure. Using, for simplicity, the Barus equation (1.13), we have

$$\eta = \eta_0\,\mathrm{exp}\alpha p$$

so that, substituting,

$$\frac{dp}{dx} = 12\eta_0\,\bar{U}\,\mathrm{exp}\alpha p\,\frac{h-\bar{h}}{h^3}. \tag{8.38}$$

This equation can be solved by making the substitution

$$q = \frac{1-\exp(-\alpha p)}{\alpha} \tag{8.39}$$

so that

$$\frac{dq}{dx} = 12\,\eta_0\,\bar{U}\,\frac{h-\bar{h}}{h^3}. \tag{8.40}$$

Equation (8.40) has exactly the form of Reynolds' equation for a constant-viscosity or *isoviscous* fluid but with q replacing pressure p; the variable q is sometimes known as the *reduced pressure* and eqn (8.40) as the reduced Reynolds' equation for a piezoviscous (or variable-viscosity) fluid. The equation of this form has been solved in Section 7.6 for the geometry of two discs running together, so that we can say that the maximum value of q is effectively given by eqn (7.78) as

$$q_{max} = 2.15\,\bar{U}\eta\sqrt{\frac{R}{h_c^3}}. \tag{8.41}$$

The pressure p between the surfaces must always be greater than zero; it follows from (8.39) that if p is large then the *maximum* value that the parameter q can have approaches $1/\alpha$. Equation (8.41) can thus be rearranged to give an estimate of the *minimum* film thickness corresponding to this maximum pressure; setting $q_{max} = 1/\alpha$ we have

$$h_{min} = 1.66\,(\alpha\eta_0\,\bar{U})^{2/3}R^{1/3}. \tag{8.42}$$

The load per unit length on the cylindrical line contact for these circumstances can be evaluated by integrating the area under the pressure curve. When this load is compared to that of the isoviscous case, eqn (7.79), it is found that the load-carrying capacity of the contact has increased because of this change of viscosity by a factor of about 2.3, that is,

$$W \approx 2.3 \times 4.89 \frac{\bar{U}\eta RL}{h_{\min}}. \tag{8.43}$$

In practice, it is usually the load that is the independent variable so that rearranging eqn (8.43) shows that, for a given load, the phenomenon of oil thickening or piezoviscosity has led to an increase in the oil film thickness of 2.3 times the isoviscous case. Although this is a significant increase, it is still not sufficiently large to explain the satisfactory long-term performance of many heavily loaded contacts. In our example, even increasing the oil film thickness by a factor of 2.3, from 0.014 to approximately 0.03 microns, would still lead us to expect significant amounts of asperity interaction and so associated surface damage. The full explanation must involve effects over and above local increases in lubricant viscosity.

Elastic deformations: entry conditions and the Ertel–Grubin analysis
So far we have imagined the surfaces of the solid components to remain rigid, and so to retain their original profiles at all times. This idealization must now be called into question. At even the very lightest loads the surfaces must deform, initially elastically, and although these deformations may be very small they can play a vital role in the generation of the all-important protective hydrodynamic film. Adopting our usual simplification that around the contact region the undeformed cylindrical profile can be satisfactorily modelled as a parabola, we can write, following the notation of Fig. 3.1(b), that the magnitude of the gap h between the solid surfaces can be written as

$$h = \frac{x^2}{2R} + w_z - \delta \tag{8.44}$$

where w_z represents the deformation of the cylindrical surface and δ is the normal relative displacement of distant points in the two surfaces. We have seen (in Section 3.3.2) that in turn w_z must satisfy the relation

$$\frac{\partial w_z}{\partial x} = \frac{2}{\pi E^*} \int_{-\infty}^{+\infty} \frac{p(s)}{x-s} \, ds \tag{8.45}$$

where $1/E^*$ replaces $2(1-v^2)/E$ for like-on-like surfaces.

Equations (8.44) and (8.45), together with the reduced form of Reynolds' equation, (8.40), form a set of equations which must be solved simultaneously to yield expressions for the film shape h and either the reduced pressure variation q or the true pressure variation p with position through the bearing measured by the x-coordinnate. In recent years an increasing number of numerical investigations of these equations have been carried out for various surface geometries, since no complete simple analytical solution is available. However, a great deal of physical insight into the mechanics of the generation of the film, and thus the load-carrying capacity of the contact, together with a reasonable degree of numerical accuracy, can be achieved by following a similar line of argument to that first put forward by Grubin (based on the work of Ertel) as long ago as 1949 (Grubin (1949); see also Cameron (1966) and Gohar (1988)). These workers realized that in a high-pressure contact supplied with a pressure-sensitive lubricant and in which conditions are achieved where αp exceeds unity (say $\alpha p > 5$) eqn (8.39) indicates that q will be effectively constant: the reduced Reynolds' equation (8.40) requires that if $dq/dx = 0$ then the film must be essentially *parallel* in the high-pressure region with a constant thickness equal to $\bar{h}$. They made the imaginative leap of supposing that even in the presence of the lubricant the elastic deformations of the solids are exactly the same as they would have been in a dry Hertzian contact extending over the same contact width and that the pressure distribution over the lubricated contact remains very close to the Hertzian semi-ellipse.

These simplifications of what is rather a complex problem linking hydrodynamics and solid mechanics imply that the numerical value of the film thickness $\bar{h}$ will be governed by the shape of the convergent wedge at the *entry* to the parallel section of the contact and that $\bar{h}$ will not depend on conditions at the exit. On the basis that the elastic deformations of the solids are, to a good approximation, the same as would occur in dry conditions, we can write, following eqn (3.51), that within the convergent wedge region at the entry to the contact,

$$h = \bar{h} + \frac{4\sqrt{2}}{3}\frac{a^2}{2R}(\xi - 1)^{3/2} \tag{8.46}$$

where $\xi = |x/a|$ and so $\xi - 1$ measures the distance from the start of the parallel region; see Fig. 8.16.

Substituting this expression into that for the reduced pressure, eqn (8.40), and considering for simplicity only the lower surface to be in motion with velocity U, leads to the equation

$$\frac{B}{a}\frac{dq}{d\zeta} = \frac{6\eta_0 U}{\bar{h}^2}\frac{\zeta^{3/2}}{(1 + \zeta^{3/2})^3} \tag{8.47}$$

where

$$B = \left\{\frac{4\sqrt{2}}{3}\frac{a^2}{2R\bar{h}}\right\}^{2/3} \quad \text{and} \quad \zeta = B(\xi - 1). \tag{8.48}$$

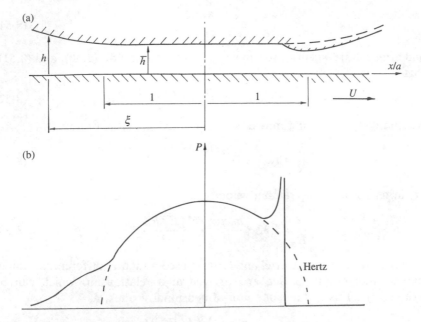

Fig. 8.16 Diagrammatic representation of the conditions within an elasto-hydrodynamically lubricated line contact. (a) Film profile: the film thickness is $\bar{h}$ across the central parallel region which extends from $-1 < x/a < +1$. The presence of the constriction or 'nip' in the exit region of the bearing is also shown. (b) Corresponding pressure distribution: the Hertzian semi-ellipse is indicated by the dotted curve, and a pressure 'spike' is shown near the exit of the bearing.

This equation can be integrated directly for the build-up of pressure in the entry region by taking $\zeta = 0$ at some large distance from the contact, that is, at $\zeta \Rightarrow -\infty$. At the start of the parallel section, where $\zeta = 0$, the reduced pressure q approaches the value $1/\alpha$ and so here

$$q \approx \frac{1}{\alpha} = \frac{6\eta_0 Ua}{\bar{h}^2 B} \int\limits_{-\infty}^{0} \frac{\zeta^{3/2}}{(1 + \zeta^{3/2})^3} \, d\zeta. \tag{8.49}$$

The integral on the right-hand side of this equation can be evaluated analytically leading to

$$\frac{1}{\alpha} \approx \frac{6\eta_0 Ua}{\bar{h}^2 B} \times 0.255. \tag{8.50}$$

But we have shown in Section 3.3.4 that for a Hertzian line contact the dimension a is given by the relation

$$a^2 = \frac{4WR}{\pi E^* L},$$ (8.51)

and after a little algebra, we can write from, eqns (8.48), (8.50), and (8.51), that

$$\bar{h} = 1.37 (\eta_0 \alpha U)^{0.75} (W/L)^{-0.125} E^{*0.125} R^{0.375}.$$ (8.52)

Alternatively, collecting into non-dimensional groups,

$$\frac{\bar{h}}{R} = 1.37 \left\{ \frac{\eta_0 \alpha U}{R} \right\}^{3/4} \left\{ \frac{E^* R L}{W} \right\}^{1/8}.$$ (8.53)

or, in terms of the entraining velocity $\bar{U}$,

$$\frac{\bar{h}}{R} = 2.32 \left\{ \frac{\eta_0 \alpha \bar{U}}{R} \right\}^{3/4} \left\{ \frac{E^* R L}{W} \right\}^{1/8}.$$ (8.53)

In their original work Ertel and Grubin used a slightly different, although similar, integration routine and arrived at a relationship which can be written, in terms of the same non-dimensional groups, as

$$\frac{\bar{h}}{R} = 2.08 \left\{ \frac{\eta_0 \alpha \bar{U}}{R} \right\}^{8/11} \left\{ \frac{E^* R L}{W} \right\}^{1/11}.$$ (8.54)

Returning to our numerical example (see eqn (8.36)) we can now make a second estimate of the film thickness $\bar{h}$. For mineral oils a typical value of the pressure–viscosity index α is found experimentally to be close to the value $2 \times 10^{-8} \text{Pa}^{-1}$ and for steel-on-steel surfaces $E^* = 115 \times 10^9$ Pa; hence from eqn (8.53), the predicted value of $\bar{h}$ is now about 1.6 microns. This is of the order of one hundred times as great as the estimate arrived at on the basis of *rigid* surfaces lubricated by an *isoviscous* lubricant and would be sufficient to prevent significant asperity contacts and hence lead to an acceptable component life.

Exit conditions: constrictions and pressure spikes

The essential step in the Ertel–Grubin analysis is the assumption that the surfaces have the deformed shape of an unlubricated contact, given by the Hertz theory, but are moved apart by a fixed displacement $\bar{h}$. On this basis, estimates of the film thickness are very much greater than on simple hydrodynamic theory alone and are in general agreement with the order found experimentally. However, it is clear from an examination of Reynolds' equation (8.37) that the *actual* film profile be rather more complex than a simple parallel strip. This is because, if the film were everywhere at least of thickness $\bar{h}$, then the pressure gradient would have to be everywhere positive. The pressure would, as we have seen, build up to very high values

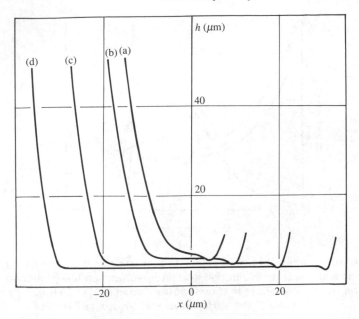

Fig. 8.17 Film profiles in an elasto-hydrodynamically lubricated contact (from Dowson and Higginson 1959).

in the inlet, remain at very high values through the Hertzian region, but then increase still further in the outlet, where again $h > \bar{h}$. This is clearly physically not the case as the pressure must fall to zero again in the exit region of the bearing to the local external ambient value. The explanation to this apparent paradox relies on the fact that when the pressures are so high that the value of $\exp \alpha p$ is very much greater than unity then *very small* changes in the film thickness permit *large* variation in pressure; consequently, a semi-elliptical pressure distribution is not in conflict with an essentially parallel film. However, the pressure can only fall to very low values if the film thickness falls below $\bar{h}$: in order to re-establish ambient pressure at the exit of the contact there must be some local constriction of the film. The existence of this elasto-hydrodynamic 'nip' has been verified both in numerical solutions to the elasto-hydrodynamic lubrication problem and in experimental investigations of the film shape. Figure 8.17 illustrates some numerical results; they refer to a particular arrangement in which $U = 1 \, \text{m s}^{-1}$, $R = 2 \, \text{m}$, $\eta_0 = 0.137 \, \text{Pa s}$, $E^* = 5.93 \times 10^{10} \, \text{Pa}$, and $\alpha \approx 1.9 \times 10^{-8} \text{Pa}^{-1}$. Curves (a), (b), (c), and (d) correspond to total loads on the bearing varying in the ratio 1:2:4:6.

There are two notable features demonstrated by these results. Firstly, the

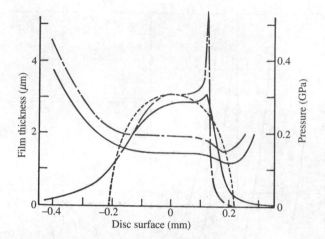

Fig. 8.18 Comparison between calculated (chained line) and measured (solid line) shape and pressure distribution in elasto-hydrodynamically lubricated contacts. The Hertz semi-ellipse is shown dotted. Load is 100.3 kN m^{-1}; speed is 1.3 m s^{-1} (from Hamilton and Moore 1971).

way in which with increasing normal load the elastic flattening of the surface produces a proportionately much longer section of substantially parallel film; and secondly, the fact that quite *large* changes in load bring about comparatively *small* variations in the minimum film thickness. In each case, the elasto-hydrodynamic nip near the exit is apparent; within this local constriction the film thickness falls to about 75% of that of the parallel section.

Numerical solutions to the problem have also verified that the pressure variation across the contact is close to the semi-elliptical form of the Hertzian distribution. They show, however, that in some circumstances, there is an additional and distinctive feature, namely a pressure spike, with a maximum value that can be considerable greater than the peak Hertzian pressure. This spike occurs towards the outlet, at or near the film constriction, and its occurence has been verified experimentally, although the observed peak values tend to be rather lower than theory would predict. Some characteristic experimental data are shown in Fig. 8.18. The measured film shapes and pressure distributions are shown as solid lines while the chained curves are from elasto-hydrodynamic calculations.

The closer that the overall pressure distribution approximates to Hertzian then the more apparent is the pressure spike: its presence is critically dependent on the lubricant exhibiting some dependence of viscosity on local pressure (i.e. on having a positive value of the pressure–viscosity index α) and on the contact being suiticiently heavily loaded for this variation to be significant, in other words, on the value of αp exceeding some critical value. Empirically it has been found that, if p_0 is the value of the peak Hertzian pressure, then pressure spikes only become of significance if $\alpha p_0 > 2.2$.

8.4.2   Regimes of lubrication

Non-dimensional elasto-hydrodynamic lubrication groups

Equations (8.53) and (8.54) express the film thickness in an elasto-hydrodynamically lubricated contact in a compact non-dimensional form. This particular formulation has the disadvantage that it is not really possible to give any immediate physical significance to the non-dimensional groups that have been chosen. Viewed from an operational standpoint we could form four suitable groups, each associated with an important physical aspect of the problem that can be chosen by the designer. In their pioneering work in this area Dowson and Higginson (1959, 1968) adopted an approach of this sort which has also been followed by later investigators in the field. In a particular situation we can regard the material properties and initial geometry of the surfaces as constant, and are principally concerned with how the film thickness depends on the independently chosen variables of load and speed. It is thus sensible to relate the size of the fluid film, normalized by the reduced radius R, to a load parameter equal to $W/2E^*R$, a speed parameter $\bar{U}\eta_0/2E^*R$, and a material parameter $2\alpha E^*$. Dowson and Higginson's semi-empirical power law equation is then of the form

$$\frac{h_{min}}{R} = 2.65\{2\alpha E^*\}^{0.54}\left\{\frac{\bar{U}\eta_0}{2E^*R}\right\}^{0.7}\left\{\frac{W}{2E^*RL}\right\}^{-0.13} \tag{8.55}$$

where h_{min} is the minimum thickness of the fluid film which occurs at the exit nip or constriction. The film thickness over the larger part of the contact is greater than this — usually by about 30%: Dowson and Toyoda (1979) have proposed that for situations involving a wide range of materials, loads and speeds, the film thickness $\bar{h}$ on the centre line of the contact can be obtained from the formula

$$\frac{\bar{h}}{R} = 3.11\{2\alpha E^*\}^{0.56}\left\{\frac{\bar{U}\eta_0}{2E^*R}\right\}^{0.69}\left\{\frac{W}{2E^*RL}\right\}^{-0.1}. \tag{8.56}$$

These equations indicate two significant features of elasto-hydrodynamic contacts. Firstly, the comparatively small influence of the load on the film thickness (already noted with reference to Fig. 8.17); the film thickness parameters are proportional only to $W^{-0.13}$ and $W^{-0.1}$ respectively. This lack of dependence of the film thickness on load contrasts to the situation in hydrodynamic lubrication when film thickness is proportional to $W^{-0.5}$. Secondly, the virtual independence of the film thickness on the contact modulus E^*; this is easily demonstrated by rearranging eqns (8.55) or (8.56) to show that film thickness depends only on E^* raised to a power of no more than ± 0.03. This weak dependence on the elastic modulus has been verified in experiments in which the film thickness between glass and steel surfaces was found to be much the same as that between two hard steel surfaces of the same radii operating under the same conditions. Of course, the *width* of the contact zone and the variation of pressure across it are very different: they both remain very dependent on the elastic moduli of the solids.

A large number of other non-dimensional groups associated with load and speed can be formed (see Johnson (1970)); for example, several authors have suggested that the non-dimensional load group g_2, defined by the relation

$$g_2 = \left(\frac{\alpha^2 W E^*}{\pi R L}\right)^{1/2}, \tag{8.57}$$

is physically meaningful as it is equal to αp_0, the ratio of the peak Hertzian pressure to a characteristic fluid pressure. This pressure is equal to that required to increase the viscosity by the exponential factor e. Speed can be incorporated into the non-dimensional group:

$$g_4 = \left(\frac{8\eta\alpha^4 \bar{U} E^{*3}}{R}\right)^{1/4}, \tag{8.58}$$

and this group is sometimes known as the *character* of the system. We should expect the effects of variable viscosity to be more important as the numerical values of both g_2 and g_4 increase in any particular contact.

From a physical perspective it makes sense to choose the *independent* parameters in such a way that it is possible to separate out the effects of elasticity of the solid surfaces (as measured by E^*) from the variation of fluid viscosity with pressure (as measured by α) and to choose the *dependent* thickness parameter so that E^* and α are excluded from it. In this case, in which $h_{\min}$ is the minimum film thickness, appropriate groups are;

as a viscosity parameter, $g_1 \equiv \left(\dfrac{\alpha^2 W^3}{\eta_0 R^2 L^3 \bar{U}}\right)^{1/2}$

as an elasticity parameter, $g_3 \equiv \left(\dfrac{W^2}{2\eta_0 R \bar{U} E^* L^2}\right)^{1/2}$, and (8.59)

as a film thickness parameter, $h^* \equiv \dfrac{W h_{\min}}{\eta_0 R L \bar{U}}$.

The parameter g_3 is a measure of the ratio of the pressure generated hydrodynamically in the film to the pressure in the underlying substrate. The value of g_3 increases as E^* falls and the elastic deformation becomes more

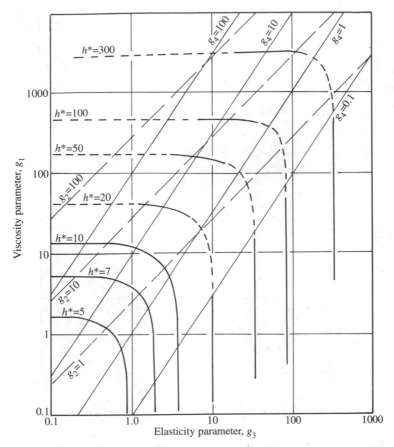

Fig. 8.19 Johnson chart, that is, a plot of g_1 versus g_3, showing contours of the parameters h^*, g_2, and g_4 (defined in eqn (8.58)) (from Johnson 1970).

significant. The parameter g_1, on the other hand, is a measure of the significance of the change of the viscosity of the fluid within the contact; as the susceptibility of the lubricant to thicken with pressure increases so the numerical value of the parameter g_1 increases.

The different regimes of lubrication can be displayed by plotting the viscosity parameter g_1 against the elasticity parameter g_3 on rectangular logarithmic axes so producing what is sometimes called a *Johnson chart* or map; see Figs. 8.19 and 8.20 (Johnson 1970). The position of a point on this map defines uniquely the conditions at an elasto-hydrodynamic contact. Since only two of the parameters are independent it follows that lines, or contours, of any other non-dimensional group (such as h^* or g_2 or g_4) can

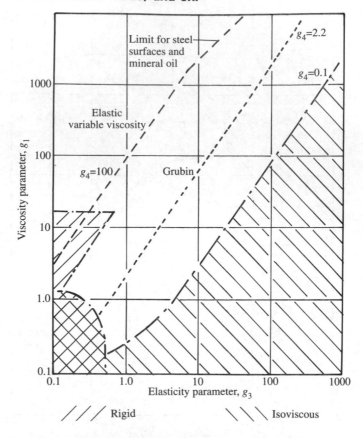

Fig. 8.20 Johnson chart, showing regimes of bearing behaviour.

be added to the chart. Figure 8.19 shows such a map on which these contours have been drawn.

When g_1 and g_3 are small then the influences of variations in both viscosity and elasticity can be neglected. This region of rigid-isoviscous behaviour thus lies at the bottom left-hand corner of the map; see Fig. 8.20.

Maintaining a low value of g_3 while increasing g_1 moves the operating point into the region of rigid-piezoviscous behaviour shown hatched in Fig. 8.20. There is an effective limit to the value of the parameter g_1 beyond which the effects of elasticity can no longer be overlooked and this corresponds to the horizontal boundary of this region shown on the figure.

Conversely, if g_1 is kept small and g_3 increased, then the operating point moves into a region of elastic-isoviscous behaviour. The upper boundary to this region (shown shaded in Fig. 8.20) has been found to correspond approximately to the contour $g_4 = 0.1$. If the parameter g_4 is less than this value then variations in viscosity can be neglected. Above this boundary we are in a regime where the conditions of lubrication are influenced by both variable viscosity and material elasticity. Increasing the speed of the contact whilst keeping everything else constant causes a shift towards the rigid-isoviscous region parallel to the g_4 contours, while increasing the load moves the operating point away from the bottom left-hand corner of the map parallel to a g_2 line.

It is natural to ask what practical considerations limit the possible regions within which real devices might lie. Taking first steel lubricated by a mineral oil, limitations of strength will restrict g_2 to a value of about 100. At lower loads the limitation is one of speed; taking a maximum value of the entraining velocity $\bar{U}$ as 150 ms^{-1} and using an oil of base viscosity 0.01 Pa s limits g_4 to less than 100. The contour corresponding to this limitation has been added to Fig. 8.20. Operating points to the left of this line are not allowable. A much more compliant material (such as rubber) lubricated by any fluid (or conversely, almost any solid lubricated by water) would give $g_4 < 0.1$ and so would operate well within the elastic-isoviscous regime. Conditions of elasto-hydrodynamic lubrication in which both variable viscosity and elasticity play a part lie in the band $0.1 < g_4 < 100$.

Within the rigid-isoviscous region the analysis of Section 7.6 can be used, so that from eqn (7.79)

$$h^* = 4.89, \tag{8.60}$$

and this condition can be used as an effective boundary to this region on the map.

In the elastic-isoviscous region the results of Herrebrugh (1968) show that, to a reasonable approximation,

$$h^* = 3.1g_3^{0.8} \tag{8.61}$$

while in the variable-viscosity region (and, in fact for rather higher values og g_2)

$$h^* = g_1^{2/3}.$$

Within the central band of the plot, where both elasticity and variable viscosity are important, the Ertel–Grubin analysis becomes appropriate. Dowson and Higginson's semi-empirical formula (eqn (8.55)) can be written as

$$h^* = 2.65g_1^{0.54}g_3^{0.06}. \tag{8.62}$$

Within the range $5 < g_2 < 30$ the data can be fitted almost as well by the simpler equation

$$h^* = 2.1g_1^{0.60}. \tag{8.63}$$

It is clear from Fig. 8.19 that the non-dimensional film thickness h^* changes from being principally dependent on the group g_1 to being dependent on g_3 as the character g_4 changes from about 0.1 to something over 2. The value of g_4 thus indicates the relative importance of the groups g_1 and g_3. Finally, in this consideration of elasto-hydrodynamic line contact, we can observe that the incidence of a pressure spike in the exit region depends on the value of αp_0 exceeding approximately 2 and so corresponds to the group g_2 being greater than 2, which, as we have seen, indicates that we are approaching conditions in which the film thickness is not appreciably influenced by the elasticity of the solid surfaces.

8.4.3 *Point contacts*

It is equally well established that in lubricated point contacts, as occur between two spheres (or an equivalent sphere and plane), the high pressures generated can lead to both corresponding enormous magnifications in the lubricant viscosity and significant elastic deformations of the solid surfaces. In the same way as occurs in line contacts, the deformed surfaces take on contours similar to those of the equivalent dry Hertzian contact but with an intervening oil film whose thickness is very much greater than would be expected on the basis of classical hydrodynamic analysis. The most important practical aspect of any theoretical model of elasto-hydrodynamic point contact is the determination of the minimum film thickness in terms of the physical properties of the surfaces and the lubricant and operating parameters of the system. Extensive use can again be made of non-dimensional operating groups. The problem of simultaneously solving Reynolds' equation in two dimensions and the equations of equilibrium and compatibility in the underlying solids is not amenable to a full analytical treatment and consequently approximate numerical methods have been employed. A typical equation, based on a number of specific solutions, and corresponding to (8.56), for h_c, the film thickness on the centre line of such a point contact between a sphere and a flat, is of the form

$$\frac{h_c}{R} = 1.90\{2\alpha E^*\}^{0.53} \left\{\frac{\bar{U}\eta_0}{2E^*R}\right\}^{0.67} \left\{\frac{W}{2E^*R^2}\right\}^{-0.067}. \tag{8.64}$$

There is not yet unanimity over either the precise value of the initial constant or that of the indices to which the non-dimensional groups are raised; this is perhaps not surprising as equations of this sort represent power law curve fits to a limited number of numerically computed cases. Such data points can once again be plotted on a set of axes, one of which represents a viscous parameter g_3' and the other an elastic parameter g_1'. The appropriate definitions are now

$$g'_1 = \frac{\alpha W^3}{\eta^2 \bar{U}^2 R^4} \quad \text{and} \quad g'_3 = \left\{ \frac{W^4}{2\eta^3 \bar{U}^3 E^* R^5} \right\}^{2/3}. \tag{8.65}$$

However, this sort of plot highlights one drawback of the Johnson map, namely that for a given material–lubricant combination (e.g. steel surfaces and mineral oils) all possible observations fall within a very narrow triangular region: the directions of varying load and varying speed being only 3.2° apart for point loading and 11.3° for line contact. Consequently other independent parametric plotting groups, which do not compress the observations to the same extent, bave been proposed.

Just as in elasto-hydrodynamic line contacts, there is in elasto-hydrodynamic point contacts a local constriction in the oil film, together with an associated pressure spike, towards the downstream end of the contact; this constriction actually extends around the side of the contact forming a 'horse-shoe' shaped constraint (Gohar 1988, Jacobson 1991). The thinnest parts of the lubricant film occur actually at either side of the central high-pressure region. In plan view this region, which is close to being of constant thickness, has a more or less rectangular footprint and is contained within the nominal Hertzian contact circle; see Fig. 8.21. Formulae are also available for $h_{\min}$, the minimum film thickness at this constriction; for example, those due to Dowson and Hamrock (1981):

$$\frac{h_{\min}}{R} = 1.79 \{2\alpha E^*\}^{0.49} \left\{ \frac{\bar{U}\eta_0}{2E^* R} \right\}^{0.68} \left\{ \frac{W}{2E^* R^2} \right\}^{-0.073}. \tag{8.66}$$

If the effective shape of the asperity is elliposoidal with principle radii of curvature R_x and R_y (see Fig. 8.22) then, because the situation is no longer axisymmetric about the Oz-axis, we also need to specify the direction of lubricant entrainment, for example, by specifying the angle θ in the figure. The simplest procedure for estimating the film thickness $\bar{h}$ is that suggested by Greenwood (1990), namely that

$$h(\theta) = h(0)\cos^2\theta + h(90)\sin^2\theta, \tag{8.67}$$

where the film thicknesses $h(0)$ and $h(90)$ are evaluated for the same load and entraining speed but the appropriate surface radii. Further details of this form of analysis can also be found in Evans and Snidle (1982) and Chittenden, Dowson et al. (1985).

8.4.4 Rheology in concentrated contacts

We have seen that in both line and point contacts operating under conditions of elasto-hydrodynamic lubrication the fluid film is very thin, perhaps less than a micron, and that the pressures developed are very large, comparable

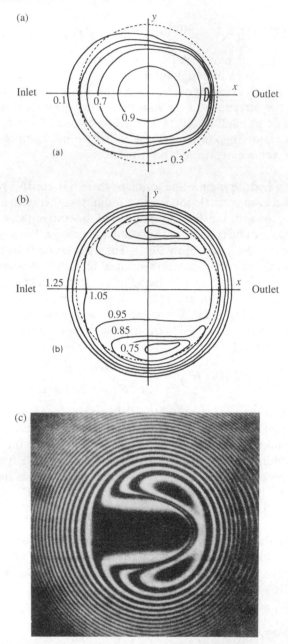

Fig. 8.21 (a) Elasto-hydrodynamically lubricated circular point contact: (a) contours of constant pressure; and (b) film thickness contours in which the dotted line is the Hertzian dry contact area. Figures are fractions of the central value (from Evans and Snidle 1982). (c) Optical interference pattern in a rolling elasto-hydrodynamically lubricated contact (from Foord *et al.* (1969–70)).

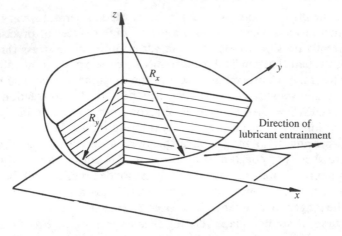

Fig. 8.22 Geometry of elasto-hydrodynamic lubrication of ellipsoidal contacts.

with the Hertzian stresses in a dry contact of the same overall geometry. In the case of steel surfaces the numerical example of Section 8.4.1 indicates a (not untypical) peak pressure of just over 2 GPa. While it has been established that the thickness of the elasto-hydrodynamic film is determined in the comparatively low-pressure entry zone, where the lubricant is drawn into the contact by the rolling velocity and behaves as a Newtonian fluid (albeit with a viscosity sensitive to the local pressure), there is a good deal of evidence that mineral oils can show deviations from Newtonian behaviour under the extreme conditions present further into the contact. Such variations are very significant if we wish to estimate the friction or traction force required to maintain the sliding motion of the two components separated by an elasto-hydrodynamic film. The traction force per unit width F/L of a slider bearing of length B will be given by

$$F/L = \int_0^B \tau(x)\,\mathrm{d}x \tag{8.68}$$

and so its evaluation requires some knowledge of how the traction stress τ depends on the applied shear rate at the conditions of pressure and temperature existing in the contact. The search for such a realistic constitutive relation has been the subject of considerable research in recent years. Although the simple Barus model of rheological behaviour will generate reasonable estimates of film thickness in elasto-hydrodynamic contacts it is much less successful in modelling the associated surface tractions.

The principal, although certainly not the only, method of investigating the rheological properties of lubricant films has been by the use of the disc machine (see Section 9.5). The main drawback to this technique lies in the fact that the pressure and, to a lesser extent the temperature, vary through the contact in ways which are not known to high degrees of precision. Nor is it generally possible to measure the variation in shear stress throughout the film; the best that can be done is to measure the resultant overall traction force. The standard procedure is to run the machine at a constant load W and entraining velocity $\bar{U}$ and to measure the variation in traction F as the sliding speed $\Delta U = U_1 - U_2$ is increased from zero under nominally isothermal conditions. The mean pressure in the film (equal to W/A where A is the load-bearing area) and the film thickness h (either measured or calculated) are therefore constant during a test. The shear rate $\dot{\gamma}$ in the film, which is approximately parallel through the loaded section, will be given by $\Delta U/h$ and the mean shear stress τ by F/A. Thus a traction curve which relates the tangential F to the sliding speed ΔU can immediately be converted into a curve of similar shape relating mean shear stress τ to shear rate $\dot{\gamma}$.

On intuitive grounds we might expect any material to possess an ultimate strength which cannot be exceeded whatever the imposed conditions. Evidence from experimental work on polymer films suggests that when the shear stress reaches a value of the order of one-thirtieth of the shear modulus, the mechanism of flow changes from one involving thermally activated diffusion to a situation in which deformation concentrates in distinct shear bands. The fact that in traction tests the mean shear stress rarely exceeds one-tenth of the mean normal pressure has led to the suggestion that at sufficiently high hydrostatic pressures even lubricant films can behave rather as idealized solids exhibiting such a characteristic 'plastic' flow stress. Thus on a traction plot of τ versus $\dot{\gamma}$ we might expect a cut-off value τ_{max} above which we would find no further experimental values.

In Section 8.3 mention was made of the Eyring formulae for visc-oelastic behaviour and this has been found to be a satisfactory model for a number of tribologically significant fluids in heavily loaded concentrated contacts. Here shear stress τ and strain rate $\dot{\gamma}$ are related by an equation of the form

$$\tau_0 \sinh\left(\frac{\tau}{\tau_0}\right) + \eta\,\frac{\dot{\tau}}{G} = \eta\dot{\gamma}. \qquad (8.29 \text{ bis})$$

τ_0, η and G are materials constants which may vary with temperature and pressure. The relative importance of visco-elastic effects can be gauged by introducing the Deborah number, De, into this equation: De is the ratio of the relaxation time τ $(= \eta/G)$ of the fluid to the duration of the event of interest Δt, which in this case will be the time of passage of a typical fluid element through the contact and therefore of the order of $B/\bar{U}$. Equation (8.29) can then be written as

$$\tau_0 \sinh\left(\frac{\tau}{\tau_0}\right) + \text{De}\,\frac{d\tau}{d(x/B)} = \eta\dot{\gamma}. \qquad (8.69)$$

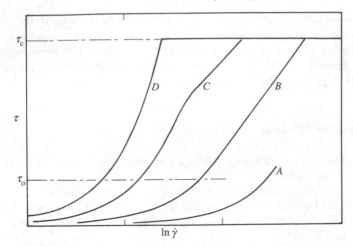

Fig. 8.23 Traction curves for a fluid film governed by Eyring behaviour together with a limiting stress τ_c plotted to a logarithmic base.

Curve A $De \ll 1$ $\tau/\tau_0 \ll 1$ Newtonian
Curve B $De \ll 1$ $\tau/\tau_0 > 1$
Curve C $De > 1$ $\tau/\tau_0 > 1$
Curve D $De \gg 1$ elastic
(after Evans and Johnson 1986).

Visco-elastic effects are likely to be significant when $De > 1$. For mineral oils G is of the order of 10^9 Pa and in a typical contact $B/\bar{U}$ will be less than 10^{-3} s, so for visco-elasticity to be important η must be of the order of 10^6 Pa s. Although this is many orders of magnitude greater than typical lubricant viscosities under ambient conditions it is not at all uncommon under conditions of elsto-hydrodynamic lubrication. For example, following our numerical example, with a base viscosity of 0.07×10^9 Pa s and a pressure–viscosity index of 2×10^{-8} Pa^{-1} the Deborah number will equal 1 when the local pressure is close to 300×10^6 Pa. Since the peak Hertzian pressure in this particular case is more than 2×10^9 Pa we are certainly in a situation where visco-elastic effects are significant.

The relative siginilficance of the various terms of eqn (8.69) are illustrated in Fig. 8.23 in which shear stress is plotted against the logarithm of strain rate (compared to Fig. 8.14 which uses linear scales). If the Deborah number is much less than unity and the shear stress does not exceed τ_0 then the fluid is Newtonian giving curve A. If, while maintaining $De < 1$, the shear stress exceeds τ_0, then the elastic term in eqn (8.69) remains negligible and so the traction curve follows that of the hyperbolic sine function to give curve B. As De is increased through unity the linear, small-strain-rate portion of the curve is increasingly influenced by the elastic term so that, if $De > 1$ and $\tau > \tau_0$, then the expected form will be similar to curve C. In case D the viscous region has become vanishingly small and the rheological behaviour is that of an elastic solid. In each case, the existence of a limiting shear stress τ_{max}, puts a limit on the value of τ as shown.

8.5   Further reading

Briant, J., Denis, J., and Parc, G. (1991) *Rheological properties of lubricants.* Editions Technip, Paris.

Cameron, A. (1966) *Principles of lubrication.* Longmans, London.

Chittenden, R. J., Dowson, D., Dunn, J. F. and Taylor, C. M. (1985) Ehl film thickness in concentrated contacts. *Proceedings of the Royal Society*, **A387**, 245–69, and 271–94.

Dowson, D. (1968) Elasto-hydrodynamics. *Proceedings of the Institution of Mechanical Engineers*, **182** (3A), 151–67.

Dowson, D. and Higginson, G. R. (1959) A numerical solution to the elasto-hydrodynamic problem. *Journal of Mechanical Engineering Science*, **1**, 6–20.

Dowson, D. and Higginson, G. R. (1966) *Elasto-hydrodynamic lubrication.* Pergamon Press, Oxford.

Dowson, D. and Toyoda, A. (1979) A central film thickness formula for ehd line contacts. *Proceedings of 5th Leeds-Lyon Symposium, 1978.* Mechanical Engineering Publications, London.

Evans, C. R. and Johnson, K. L. (1986) The rheological properties of ehd lubricants. *Proceedings of the Institution of Mechanical Engineers*, **200** (3A), 303–12.

Evans, H. P. and Snidle, R. W. (1982) The ehl of point contacts at heavy loads. *Proceeding of the Royal Society*, **A382**, 183–9.

Foord, C. A., Wedeven, L. D., Westlake, F. J. and Cameron, A. (1969–70) Optical elasto-hydrodynamics. *Proceedings of the Institution of Mechanical Engineers*, **184**, 487–506.

Gohar, R. (1988) *Elasto-hydrodynamics.* Ellis-Horwood, Chichester.

Grassam, N. S. and Powell, J. W. (1964) *Gas lubricated bearings.* Butterworth, London.

Greenwood, J. A. (1988) Film thicknesses in circular elasto-hydrodynamic contacts. *Proceedings of the Institution of Mechanical Engineers*, **202** C1, 11–17.

Greenwood, J. A. (1990) Elasto-hydrodynamic film thickness in point contacts for arbitrary entraining angle. *Journal of Mechanical Engineering Science*, **204**, 417–20.

Gross, W. (1962) *Gas film lubrication.* Wiley, New York.

Grubin, A. N. (1949) *Investigation of the contact of machine components.* Central Scientific Research Institute for Technology and Mechanical Engineering, Moscow. 30, 1893. DSIR Translation No. 337.

Hamilton, G. M. and Moore, S. L. (1971) Deformation and pressure in an elasto-hydrodynamic contact. *Proceedings of the Royal Society*, **A322**, 313–30.

Hammock, B. J. and Dowson, D. (1981) *Ball bearing lubrication.* Wiley, London.

Hays, D. F. (1959) A variational approach to lubrication problems and the solution for the finite journal bearing. *Transactions of the American Society of Mechanical Engineers, Journal of Basic Engineering*, **86**, 13–23.

Herrebrugh, K. (1968) Solving the incompressible and isothermal problem in elasto-hydrodynamic lubrication through an integral equation. *Transactions of the American Society of Mechanical Engineers, Journal of Lubrication Technology*, **90**, 262–78.

Jacobson, B. (1991) *Rheology and elasto-hydrodynamic lubrication*. Elsevier, Amsterdam.

Johnson, K. L. (1970) Regimes of elasto-hydrodynamic lubrication. *Journal of Mechanical Engineering Science*, **12** (1), 9–16.

Kapitza, P. L. (1955) Hydrodynamic theory of lubrication during rolling. *Zhurnal Tekhnicheskoi Fiziki*, **25**, 759–75.

Muijderman, E. A. (1966) *Spiral groove bearings*. Philips Technical Library, Eindhoven.

Powell, J. W. (1970) *Design of aerostatic bearings*. Machinery Publishing Company, Brighton.

Raimondi, A. A. (1961) A numerical solution for the gas lubricated full journal bearing of finite length. *Transactions of the American Society Lubrication Engineers*, **4**, 131–55.

Roelands, C. J. A., Vlugter, J. C., and Waterman, H. I. (1963) The viscosity-pressure-temperature relationship of lubricating oils. *Transactions of the American Society of Mechanical Engineers, Journal of Basic Engineering*, **85**, 601–7.

Salbu, E. O. J. (1964) Compressible squeeze films and squeeze bearings. *Transactions of the American Society of Mechanical Engineers, Journal of Basic Engineering*, **86**, 355–66.

Tanner, R. I. (1985) *Engineering rheology*. Oxford University Press.

9
Boundary lubrication and friction

9.1 Introduction

The satisfactory operation of both hydrostatic and hydrodynamic bearings requires that the solid surfaces which constitute the bearing faces are completely separated by the intervening fluid film. Since the bearing surfaces are then not physically touching, the resistance to their tangential motion, that is, the force of friction, is directly attributable to viscous losses in the lubricant. If the lubricant (whether liquid or gas) exhibits Newtonian rheological behaviour with constant viscosity then the value of this frictional force, and the associated coefficient of friction, will increase with the value of the tangential sliding velocity. We have seen (eqn (7.27)) that the coefficient of friction within a hydrodynamically lubricated bearing is generally dependent on the square root of the group $UL\eta/W$ where U is the relative sliding speed of the surfaces, W/L the normal load supported per unit length, and η the Newtonian viscosity. A reduction in speed, or an increase in the specific load on the bearing, leads to a fall in the friction coefficient. However, there is a limit to this process: when the specific load is very high, or the relative sliding speed small, it is difficult to build up a sufficiently thick film to entirely separate the bearing faces, and so there will be some mechanical interaction between opposing surface asperities. This is inevitable, even allowing for the large increase in effective lubricant viscosity and the elastic flattening of the surface profiles that can occur in the elasto-hydrodynamic regime. Eventually the highest regions of the surfaces may be protected by lubricant films literally only a few molecules thick and it is this state of affairs that is known as *boundary lubrication*. Within this range of operation the bulk properties of the fluid, such as its density and viscosity, are of relatively little importance, while its chemical composition, as well as that of the underlying metals or substrates, become increasingly significant. The coefficient of friction μ can increase substantially with the transition to boundary conditions, perhaps by as much as two orders of magnitude over the minimum value which occurs within the fully hydrodynamic regime. For the important case of a journal bearing these changes can be displayed on a plot of μ versus the non-dimensional group $\eta\omega/p$ where ω is the rotational speed and p the nominal bearing pressure, that is, the radial load

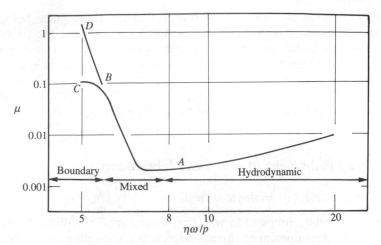

Fig. 9.1 The Stribeck diagram for a journal bearing. η is the lubricant viscosity, ω the rate of rotation, and p the nominal bearing pressure.

divided by the projected bearing area. This curve, which has the characteristic shape shown in Fig. 9.1, is known as the *Stribeck curve* (or sometimes as the $z\mathfrak{N}-p$ curve following the general American notation in which z is the lubricant viscosity and $\mathfrak{N}$ the rotational speed; the group $z\mathfrak{N}/p$ is sometimes known as the Hersey number). The service parameter $\eta\omega/p$ is in some ways comparable with the bearing Sommerfeld number. Above the point A in Fig. 9.1, bearing operation is quite adequately predicted by the various forms of hydrodynamic analysis described in Chapter 7, while below point B considerations of surface physics and chemistry predominate. If the group $\eta\omega/p$ has a small numerical value, then considerable amounts of solid-to-solid contact are implied. Under these circumstances the value of the coefficient of friction depends much on the particular materials involved; it may vary from less than unity for some material pairs (curve BC), to the very large values characteristic of seizure and gross adhesion for others (curve BD). Surfaces operating near points C or D generally show a good deal of surface damage and distress and the breakdown of effective lubrication within this range is sometimes known as *scuffing*. The intermediate region, from A to B, has been given a wide variety of names, but is most usually known as the region of *mixed lubrication*. It encompasses elasto-hydrodynamic conditions (described more fully in Chapter 8) as well as the general transition into boundary lubrication.

The minimum point in the Stribeck curve is associated with that point at which both the viscous shear losses in the lubricant film bave fallen to a low value (usually by virtue of the low sliding speed) and yet the forces of solid friction have not grown large. At first sight it might seem that tribological pairs should be designed to operate close to this point, since the frictional power losses in the bearing will also be minimized. However, this would be a short-sighted design policy as the minimum is in practice unpredictable and somewhat unstable; relatively small changes in operating conditions can shift the bearing operating point along the path *ABC* or *ABD* with serious potentially detrimental consequences to machine components.

9.2 The mechanism of boundary lubrication

9.2.1 *Importance of molecular films*

The term boundary lubrication was first coined by Sir William Hardy, a pioneer in the development of surface science. It distinguishes those circumstances in which it is the cleanliness or *greasiness* of the surfaces and the inherent *oiliness* or *lubricity* of an exceedingly thin film of the lubricant that are the crucial factors in controlling the frictional force rather than the bulk properties of the lubricant. The term oiliness is difficult to define with any precision, the American Society of Automotive Engineers consider it to signify *differences in friction greater than can be accounted for on the basis of viscosity when comparing different lubricants under identical test conditions*. It is important to appreciate the very small scale at which these considerations, which depend on the molecular structure of the lubricant, become important. Bearing surfaces are generally 'smooth' by engineering standards, perhaps with a peak-to-valley roughness dimension of a few tenths of a micron. By contrast the molecular lengths of the paraffinic hydrocarbon chains, characteristic of lubricating oils (see Section 1.3.1) are of the order of 0.0025 microns, that is, only about 1 per cent of typical asperity heights. Although very small by engineering standards such a dimension is very long in molecular terms. Commercially available lubricating oils virtually always contain some small additions of chemical compounds specifically designed to enhance their boundary lubricating performance by forming protective surface films of this sort of thickness. At first sight it may seem unlikely that such thin surface layers can possibly provide any real degree of protection to real heavily loaded sliding surfaces, but it is a comparatively easy matter to demonstrate that this is indeed the case. If a drop of stearic acid, one of the 'classical' boundary lubricants, is floated on the surface of a bath of clear water it will spread until it is everywhere one molecule thick. In addition all the molecules will be oriented with their longest dimension perpendicular to the surface, like the pile on a carpet. Such a film may then be floated off on to a metal plate for subsequent friction testing. Figure 9.2 shows some results from experiments in which a

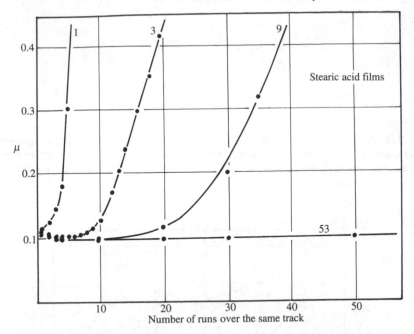

Fig. 9.2   The friction of stearic acid films deposited on a stainless steel surface (from Bowden and Tabor 1950).

stainless steel slider was run repeatedly over a flat plate of the same material on which one or more molecular layers of stearic acid had been deposited. Although a single monolayer of the lubricant was soon worn away in only a few traversals, low-friction behaviour was maintained for much longer periods if a number of such layers were laid down on top of one another.

The best boundary additives for use in lubricated contacts are long chain molecules with an active end group, typically organic alcohols, amines, or fatty acids. Such molecules readily absorb on to metal surfaces to form oriented layers with the polar end groups attached to the surface and, in addition, there may also be strong forces of lateral attraction between the chains. If two such coated surfaces are loaded together, they tend to slide over each other at their outermost faces. Although some penetration of the film, allowing true metallic contact, may occur it is generally much less than would happen if only the base mineral oil were present on the surface with no active additive. Figure 9.3 illustrates this mechanism of boundary lubrication with the aligned surface film providing a protective coating over much of the interface.

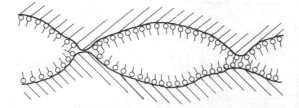

Fig. 9.3 Solid surfaces under conditions of boundary lubrication, greatly magnified and highly diagrammatic. The extent of true metallic contact and the associated effects of adhesion on friction and surface damage are much reduced by the presence of the boundary lubricant.

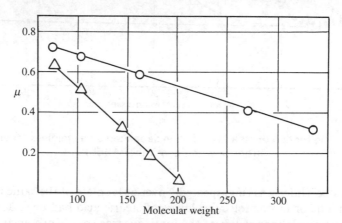

Fig. 9.4 Relationship between coefficient of static friction μ and molecular weight for paraffin hydrocarbons $\bigcirc$ and fatty acids $\triangle$ (from Bowden and Tabor 1950).

Oiliness additives

Generally speaking, the longer the chain length of the attached molecule the greater the degree of separation of the rubbing surfaces and the lower the resulting coefficient of friction. This effect can be seen in Fig. 9.4 which illustrates the behaviour of two classes of organic compounds. Paraffinic hydrocarbons are typical of base, refined mineral oils; fatty acids occur naturally in much smaller quantities but are characteristic of the additives, present by design, in commercial lubricating oils.

The effectiveness of a boundary lubricant depends greatly on the tenaciousness of the bond between the active end of the protecting molecule and the metal surface to which it adheres. This bond can be derived physically, from electrostatic or dipole forces, or can be formed as a result

Table 9.1. Coefficients of sliding friction at room temperature; data from Bowden and Tabor (1950).

Surfaces	Lubricant		
	paraffin oil	paraffin oil + 1% lauric acid	
Nickel	0.3	0.28	
Chromium	0.3	0.3	unreactive
Platinum	0.28	0.25	
Silver	0.8	0.7	
Copper	0.3	0.08	
Cadmium	0.45	0.05	reactive
Zinc	0.2	0.04	
Magnesium	0.5	0.08	

of chemical reaction between surface and lubricant. This is an important practical distinction, and not just an academic one: *physical* mechanisms are completely reversible so that a physically adsorbed boundary lubricant can be detached from the solid leaving its surface unchanged. On the other hand, *chemical* reactions are irreversible and involve some permanent modification in the state of the surface. The differences in the behaviour of physically and chemically formed films can be investigated by varying either the reactivity of the solid surface or the temperature at which the sliding experiments are carried out.

Table 9.1 shows some results on the slow-speed friction of a steel slider loaded against lower specimens of various metals lubricated either by a straight paraffin oil (which will be effectively chemically unreactive in all circumstances) or by the same oil used as a base or carrier to which a small addition of the highly polar compound, lauric acid ($C_{12}H_{23}COOH$), has been added. The upper four metals in the table would not be expected to react with lauric acid; by contrast, the lower four are more reactive metals and would be expected to interact chemically with the acid. It is clear that when lubricated by the simple paraffinic oil all the experiments give similar, comparatively high, values of the friction coefficient μ. However, in the presence of the base oil *and* the additive the observations fall into two distinct groups. The unreactive metals are not well lubricated by the fatty acid solution; on the other hand, there is a considerable fall in the value of μ for the reactive group. These results suggest that, at least under these conditions of sliding where bydrodynamic effects are minimized, lubrication is not affected by the fatty acid itself but rather by the product of reaction between the metal substrate and the acid which form what is technically known as a *metallic soap*.

This view is corroborated by experiments looking at the effect of temperature on the lubrication of metal surfaces by both paraffins and fatty acids. On unreactive surfaces (for example, those typified by the noble metal platinum) both compounds cease to have any significant lubricating effect at temperatures above the melting-point of the solid paraffin. However, on a reactive surface, such as copper, although the behaviour of paraffins is similar, that of fatty acids is different in that they continue to lubricate to temperatures considerable higher than their own melting-points. For example, in the particular (but still representative) case of lauric acid on copper, the breakdown of effective lubrication occurs at just about the melting-point of the metallic soap (copper laurate) which is formed as a result of chemical activity between the active fatty acid and the metal surface. This temperature, 110°C, is very much higher than the melting-point of pure lauric acid, 44°C. The idea that the effective lubricant is the solid reaction product is confirmed by the fact that a film of copper laurate will also lubricate unreactive platinum surfaces again up to a temperature of approximately 110°C.

We can thus summarize the requirements that any additive to the base oil must satisfy if it is to provide a reasonable degree of boundary lubrication. Firstly, the molecule of the additive should contain an active group to provide a strong bond to the metal substrate; it will usually be of the long straight chain type with the active group at one end. Secondly, the additive concentration must be sufficient to provide adequate surface coverage, but without being so great as to adversely affect the bulk properties of the carrier lubricant. Finally, the temperature at which either the additive (or its reaction product with the metallic surfaces) undergoes significant softening should be greater than the anticipated operating temperature of the loaded surfaces.

9.2.2   *Effect of temperature on surface films*

At rubbing speeds at anything more than the smallest values, we have seen that the surface temperatures of the solids can be significantly greater than those in the bulk (see Section 3.8). When the surfaces are relying for protection on physically adsorbed molecules, as will be the case for a simple paraffinic oil, this temperature increase will result in the molecules becoming more mobile and so will tend to gradually *reduce* the extent of the surface coverage and thus the effectiveness of the lubricant; the coefficient of friction correspondingly increases. The degree of surface coverage, the density of the 'pile on the carpet' formed by the adsorbed molecules, is also influenced by the concentration of the surface active molecules (the solute) in the base oil (the solvent). The lower this concentration then the greater will be the tendency for the solvent to redissolve the solute and so leave the surface unprotected. There is always a dynamic equilibrium between the

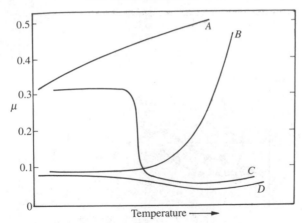

Fig. 9.5 Idealized frictional behaviour of various lubricants as a function of temperature: *A* paraffin oil; *B* fatty acid; *C* extreme-pressure (e.p.) lubricant; and *D* fatty acid + e.p. lubricant.

mechanisms of attachment and detachment from the substrate. It is the essentially reversible nature of these physical effects that account for the gradual increase in μ with temperature of surfaces lubricated by the paraffin oil. This variation is illustrated by curve A in Fig. 9.5.

On the other hand, if the lubricant and the metal are capable of reacting together *chemically* then an increase in temperature will tend to speed up the rate of reaction. We have seen that such a reaction process can provide protection up to temperatures at which the reaction product ceases to be a solid film and melts. Behaviour of this sort, characterized by that of fatty acids, is shown diagrammatically in Fig. 9.5 as curve *B*; since this effect depends on a chemical reaction involving the surface of the component it can to some extent be thought of as a mild form of surface corrosion. Once the surface temperatures reach values much over 150°C the degree of physical adsorption is likely to have fallen to a negligible value and, in addition, most metallic soaps have started to undergo thermal decomposition. This means that at service temperatures much above this sort of level there is likely to be a significant increase in frictional forces with an associated escalation in surface damage; the surfaces start to *scuff*; see Section 9.4.

To provide surface protection under these more difficult conditions some other form of stable low-shear-strength film is required, and it is within this range of operation that commercial lubricants rely on the so-called *extreme-pressure* or e.p. additives. The real functions of these additions is to lubricate at comparatively high temperatures (rather than pressures) perhaps up to as much as 300–400°C. Satisfactory operation in this higher temperature range

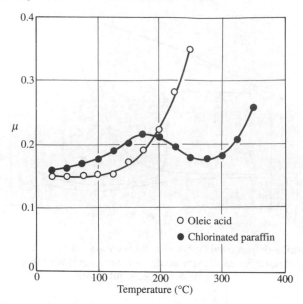

Fig. 9.6 Comparison of the effects of temperature on lubrication with a paraffin oil containing a boundary lubricant (5% oleic acid), open symbols, and as e.p. oil closed symbols (from Fuller 1984).

depends on the addition of small quantities of organic compounds usually containing chlorine, sulphur, or phosphorus. At lower temperatures these additives remain substantially inert, but they react with hot metallic surfaces to form protective films of solid metal chloride, sulphide, or phosphide. The behaviour of a lubricant containing such an e.p. additive is illustrated as curve *C* in Fig. 9.5. In principle it is possible to combine suitable boundary (i.e. low-temperature) and extreme-pressure (high-temperature) additives to give good lubrication over a wide range of temperatures; indeed, this is the basis of a large area of lubricant technology. Figure 9.5 curve *D* shows the ideal characteristic of an oil loaded with both a fatty acid and an e.p. additive. Particular care has to be taken in practice to ensure that the valuable bearing surface does not become exposed and vulnerable around the transition temperature from boundary to e.p. activity. There is always the danger that there will be some delay between losing the physically adsorbed layer and the formation of the the protecting chemical layer. This so-called *temperature distress gap* can be seen as the 'hump' in the friction trace of the chlorinated. paraffin in Fig. 9.6.

Lubricant technologists have responded to the demands for increased working temperatures by developing a range of additives capable of an even greater degree of surface protection; these are known as *antiwear* additives. Almost every commercial motor oil, for example, contains a small addition (typically 1 per cent) of the organic compound zinc dialkyl dithiophosphate, generally known as ZDDP. Experimental work has suggested that such phosphorus-containing antiwear additives act in rather a different way from conventional e.p. additions. They form comparatively thick (i.e. in the range 0.05–0.5 micron in thickness) polymeric, viscous, or even varnish-like layers. As these are of the same order of thickness as the surface rough-nesses, they are able to protect the surfaces from seizure and excessive wear by actual bodily separation which can augment any conventional hydro-dynamic or elasto-hydrodynamic load support. Evidence for such films has come both from measurements of electrical contact resistance and capaci-tance between rubbing solids as well as from microscopic examination and chemical analysis of the rubbed surfaces.

9.3 Elasto-hydrodynamic lubrication on the microscale

There is now a body of work concerned with exploring how the mechanisms of hydrodynamics and elasto-hydrodynamics might operate not only at the scale of the contact as a whole, but also at the much smaller scale of the many individual asperity interactions contained within the comparatively macroscopic contact area. Conventional theories of elasto-hydrodynamic lubrication consider the lubricated surfaces to be perfectly smooth, and consequently there is no feature of the theory that leads to a criterion of film failure; as the load is increased so the film gets thinner and thinner – in principle, without limit. Yet in practice, there is a critical load above which a given pair of surfaces, operating at a particular sliding speed, are liable to show evidence of distress, and eventually failure. Initially, it would seem reasonable to imagine that this is in some way dependent on the ratio of the minimum film thickness (calculated from smooth elasto-hydrodynamic theory) to the roughness of the surfaces (usually measured by the root mean square roughness value R_q). This ratio is often designated Λ, the 'lamda ratio', so that

$$\Lambda = \frac{h_{min}}{\sqrt{R_{q1}^2 + R_{q2}^2}} \tag{9.1}$$

where h_{min} is the minimum lubricant film thickness and R_{q1} and R_{q2} are the respective RMS surface roughnesses of the two solid surfaces. If this ratio is very much greater than unity, so that the film is thick, then the traction or friction force will be a function of the bulk rheological properties of the lubricant at the appropriate operating conditions of load, temperature, shear rate, and so on, and the influence of the surface roughness will be negligible. When the film thickness and the surface roughness are of comparable dimensions, that is, when $1 < \Lambda < 5$, then the traction will still be determined by the bulk properties of the lubricant but now we need to consider the *local* contact conditions of asperity interaction. This is the regime now referred to as 'micro-elasto-hydrodynamic lubrication'. At still lower values of Λ (<1) it is conventionally supposed that asperity interaction becomes more severe and that the shear properties of the films on the solid surfaces, whether formed by adsorption or reaction, become significant and the situation is one of boundary lubrication.

Recent experience, however, suggests that the transition from fluid film conditions to boundary lubrication is much less clear-cut than this. Within the range of micro-elasto-hydrodynamic lubrication any analysis of the contact must take into account the statistical nature of the surface roughness as well as the physical properties of the lubricant. The very high hydrodynamic pressures generated in the thin lubricating film will simultaneously affect both the surface topography of the solids and the rheology of the fluid so leading to a coupled analytical problem which is not straighforward to solve. Assuming that the behaviour of the lubricant remains Newtonian (albeit with a viscosity dependent on the local ambient pressure) and that the surface displays a statistical distribution of asperity heights, analysis suggests that although increasing the surface roughness may reduce the mean separation of the surfaces it has relatively little effect on the value of the *minimum* film thickness. As the surfaces are pushed into closer proximity small volumes of the lubricant, which is becoming increasingly viscous, are trapped in the valleys and pockets between the asperities, which are themselves becoming flattened by the local stress fields. There is some evidence that these effects conspire to produce an absolute minimum film thickness which, under the sorts of conditions that are likely to occur between even heavily loaded gear teeth or cams and followers, is often about 0.5 micron. The analysis is further complicated by evidence from experimental work, largely on disc machines (see below), which suggests that under these very high pressures (which can be comparable with the yield stress of the metal surfaces) the behaviour of tribological fluids can become distinctly non-Newtonian, the relation between the applied shear rate and the resulting shear stress generated in the film passing through a number of regimes of behaviour until, at the very highest ranges of pressures, the fluid is actually behaving like a very thin film of a perfectly plastic material 'flowing' at a constant shear stress whose value is independent of the shear strain rate. These transitions are considered in a little more detail in Section 8.3.1.

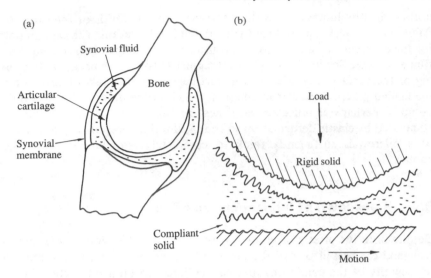

Fig. 9.7 (a) Representation of an animal joint; (b) equivalent engineering bearing. Under load the roughness of the cartilage layer is much reduced (from Dowson 1977).

9.3.1 *Lubrication of soft surfaces*

The interaction between the compliance and the topography of the bearing surface is particularly striking in those situations in which one of the lubricated solid surfaces is coated with a relatively thin layer of a comparatively soft elastic material. Natural examples of this sort of bearing are those occurring in animal joints. Figure 9.7 shows schematically a representation of such a joint together with an 'engineering' model of the equivalent bearing.

The load-transmitting bones have ends which are enlarged to form bearing surfaces. Some, such as the hip or shoulder, are best represented by spherical or ellipsoidal shapes; others, such as the knee and finger joints, can be modelled as cylindrical contacts. Within the joint the bone surfaces are covered by a layer of relatively soft and porous articular cartilage a few millemetres thick. These surfaces are in turn separated by a film of synovial fluid, which is only marginally more viscous than water, and the whole bearing is enclosed by the synovial membrane. The surface of the cartilage is comparatively rough by engineering standards, with a peak-to-valley height of the order of a micron, and the relative sliding speeds of the two surfaces are low. Conventional hydrodynamics would predict film thicknesses very much less than the height of the surface asperities, that is, lamda ratios of less than 1, and so comparatively high friction and wear. However, in normal healthy human joints the coefficient of friction is of the order of 0.001 which is very low even for hydrodynamic lubrication. The explanation for the efficiency of animal joints would seem to involve both the squeeze film effect (see Section 7.9) and, most significantly, the remarkable flattening of the surface asperities on the cartilage layer that takes place within the loaded patch. Surface roughnesses, which in the unloaded state would be up to perhaps as much as ten times the film thickness, are effectively eliminated by elastic deformation so that a film thickness of only a fraction of a micron is sufficiently thick to completely eliminate solid-to-solid contact.

9.4.   Breakdown of lubrication: scuffing

Despite both the effects of elasto-hydrodynamic lubrication on the micro-scale and the best efforts of the lubricant technologists, there is a limit on the severity of the conditions that can be imposed on a lubricated sliding contact without the initiation of catastrophic failure. The first obvious signs of severe distress are usually a sudden increase in noise and vibration followed by a progressive increase in the operating temperature. This increase may even be sufficiently severe to cause smoke and fumes to be given off from the lubricant. Examination of the component surfaces at this stage will generally show evidence of gross adhesion and material transfer from one to the other and an obliteration of the original surface topography. This sequence of events is known as *scuffing*, which can be formally defined as 'localized damage caused by the occurrence of solid-phase welding between sliding surfaces, without local surface melting'.

9.4.1   *The Blok temperature hypothesis*

The onset of scuffing between pairs of heavily loaded, usually ferrous, surfaces imposes a very real limit on the operation of a number of important machine elements where there is an element of sliding as well as rolling; these include gears, cams, and followers. Despite intensive research into the topic over recent decades there is still no fully reliable way of predicting whether scuffing will be encountered in a particular situation.

Over fifty years ago the suggestion was made that scuffing is initiated, at least in the case of a chemically inactive lubricant, when the *total* contact temperature reaches some critical value (Blok 1937, 1970). The total contact temperature is envisaged to be made up of two contributions. One is the bulk temperature of the machine components (this can usually be either measured, calculated, or at least estimated without undue difficulty). The other component is the instantaneous increase in the temperature of the surface, as it passes through the contact zone. This is known as the *flash temperature* and can be estimated from a knowledge of the interfacial mechanical and thermal conditions (see, for example, Section 3.8.2). In its simplest form this total temperature hypothesis supposes that the value of the critical total temperature at which scuffing is initiated depends only on the nature of the surfaces and the lubricant between them. It is supposed not to be dependent on the particular values of individual parameters such as load, surface roughness, slide:roll ratio, and so on, but only on their combined effect. Despite the beguiling simplicity of this idea there is now considerable experimental and practical evidence that, although a simple temperature criterion can be quite successful over a limited range of conditions, total temperature alone is not a sufficiently accurate predictor of surface distress, and that the onset of scuffing is rarely entirely independent of the specific values of the individual factors. It has, for example, been found that the availability of oxygen in the vicinity of contacts lubricated with a plain mineral oil is important in delaying the onset of gross failure. This suggests that the formation of protective surface films is important even when using a lubricant which does not contain specific e.p. additives. Recent work on the prediction of scuffing failure attempts to marry the effects of film temperature on lubricant properties (principally its base viscosity) with the development of local elasto-hydrodynamic effects that is, those on the scale of the surface roughness.

An alternative method to that of the Stribeck diagram of representing the transition from one regime of lubrication to another is the transition diagram or map suggested by the OECD International Research Group on Wear of Materials. A typical example is shown in Fig. 9.8 relevant to the sliding of two lubricated steel surfaces sliding under isothermal conditions. In region I the coefficient of friction is low (<0.05) as is the wear rate ($<10^{-9}\,\text{mm}^3\,\text{N}^{-1}\,\text{m}^{-1}$) and the regime is one of mild wear. Conditions are such that the lamda ratio Λ is of the order of 2 or more so that the load is carried essentially by elasto-hydrodynamic lubrication. In region II boundary lubrication dominates. After a period of running-in the coefficient of friction settles at about 0.1 and the wear rate, although higher than that in region I (perhaps in the range 10^{-6} to $10^{-8}\,\text{mm}^3\,\text{N}^{-1}\,\text{m}^{-1}$) may well still be tolerable for many engineering applications. As wear progresses, the surfaces actually become smoother so that Λ increases. Under these steady state conditions wear is largely oxidative in nature rather than adhesive or abrasive. The transition from region I to II is principally dependent on mechanical and geometric effects, for example, the roughness of the surfaces, the generation of the elasto-hydrodynamic lubrication film and so on, with only a subsidiary dependence on the chemical nature of the surfaces and the lubricant.

In region III conditions are so severe that no lubricant film (boundary

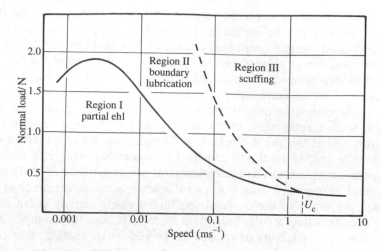

Fig. 9.8 An OECD International Research Group transition diagram for the non-conformal lubricated contact of steel surfaces (from de Gee *et al.* 1985).

or micro-elasto-hydrodynamic lubrication) can prevent intimate metallic contact. Rapid wear (more than $10^{-5} \, mm^3 \, N^{-1} \, m^{-1}$) and high friction prevail—the surfaces have scuffed. The transition from region II to III depends mainly on chemical effects and is much less dependent on surface roughness or sliding speed than the boundary between regions I and II (hence the steeper gradient of the curve delineating regions II and III than that between regions I and II). Scuffing of the contact can be avoided by ensuring a low rate of frictional energy dissipation, through low friction, low sliding speed, or low load. The diagram illustrates the fact that at low sliding speeds there may be some warning of incipient failure as the load is increased, since the system must pass through a regime of boundary lubrication before full scuffing is initiated. However, above the speed U_c, increasing the load on the contact will move conditions from the desirable state of region I to the catastrophic conditions of region III with no such warning period.

9.5 Test methods in boundary lubrication

As a result of the active involvement of many researchers into the mechanics of boundary lubrication a wide range of test machines, using various component geometries, has been developed. The qualitative assessment of the extreme-pressure properties of lubricating fluids is often carried out on one of three commercially developed instruments, namely, the Shell four-ball machine, the Falex pin-and-vee-block machine, or the Timken extreme-pressure tester. However, the conditions of contact in each of these are not really suitable for more fundamental studies into the active tribological mechanisms in heavily loaded contacts. Such investigations are more usually made on twin-disc (or cylinder) machines where, by controlling the slide:roll ratio, there is more precise control over the hydrodynamic or elasto-hydrodynamic conditions. When dealing with solid or dry boundary lubricants, conventional reciprocating or pin-on-disc machines can be used (see Section 4.3).

9.5.1   *Four-ball tests*

This technique is widely used in the evaluation of lubricating oils designed to operate under severe conditions. The test machine consists of three hardened steel balls (usual 12.7 mm or 0.5 inch in diameter) sitting in a stationary cup which is supplied with the test lubricant. An upper fourth ball, mounted in a chuck, is loaded symmetrically against the lower three. The arrangement is illustrated in Fig. 9.9(a). Loads of up to several kN can be applied at rotational speeds in excess of 10,000 rpm. The lower trio of balls can either be clamped, so that contact conditions are those of pure sliding, or be free to rotate and so generate rolling contact. When investigating boundary lubrication the lower balls are fixed so that contact occurs in three small, circular patches. The normal load is usually applied by dead weights, and the frictional or traction force at each contact evaluated from measurements of the driving torque. The oil temperature is monitored by a thermocouple. In a standard test the speed of rotation is 1760 rpm and load is increased incrementally until the lubricant ceases to be effective and the balls either scuff or, in an extreme case, actually weld together. The test has the advantage of being rapid (the standard test time is ten seconds), inexpensive, and uses only small quantities of oil. It is an extremely effective test for making comparisons between different lubricants but is less satisfactory, because of the comparatively ill-defined conditions within the contact spots, at evaluating absolute lubricant properties. Alternative set-ups using the contact of a ball on three flats or a cone and cylinders are also sometimes used, as is also illustrated in Fig. 9.9(a).

9.5.2   *Falex and Timken machines*

Both these devices produce nominal line contacts, rather than the point contact of the four-ball machine. The essential features of a Falex pin-and-vee-block machine are illustrated in Fig. 9.9(b). The test consists of running a rotating steel journal (standard speed 290 rpm) against two stationary vee blocks immersed in the test lubricant. In test method *A* (the 'run-up' test) an increasing load between the vee blocks is applied continuously (through

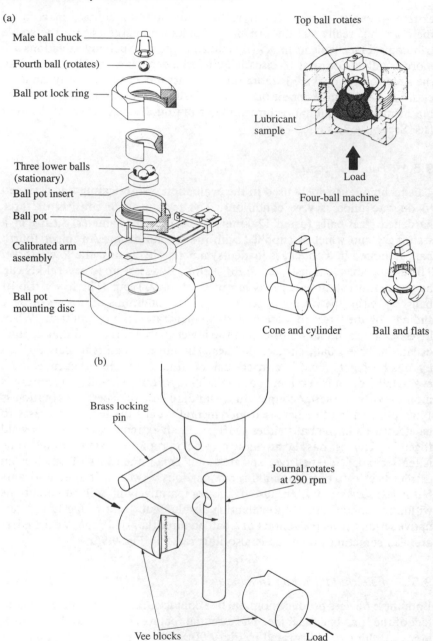

(a)

Male ball chuck

Fourth ball (rotates)

Ball pot lock ring

Top ball rotates

Lubricant sample

Load

Four-ball machine

Three lower balls (stationary)

Ball pot insert

Ball pot

Calibrated arm assembly

Ball pot mounting disc

Cone and cylinder

Ball and flats

(b)

Brass locking pin

Journal rotates at 290 rpm

Vee blocks

Load

Fig. 9.9 (a) Shell four-ball test machine. (b) Falex pin-and-vee-block test (from Barwell and Jones 1983).

a ratchet arrangement) while in method B (the 'one-minute step' test) the load is applied in 1112 N (250 lbf) increments at intervals of 1 min. In both cases the lubricant is deemed to have failed when the brass pin which retains the rotating journal shears. In the Timken machine a rotating steel cup is loaded against a stationary rectangular block at a sliding speed of 124 mn min^{-1}; the standard test time is 10 min. The load is increased in intervals of 44.5 N (10 lbf) until the surfaces scuff. These tests are described more fully in ASTM standards D2782, D2783, and D3233.

9.5.3 *Disc machine tests*

In essence a disc machine consists of two discs loaded against one another in nominal line contact. The discs are driven independently so that their peripheral speeds are not necessarily the same: in this way the contact kinematics can be varied from pure sliding through a range of combinations of sliding and rolling to pure rolling. The slide:roll ratio S, (eqn (7.65)) can be varied from 0 to 1. The discs are usually supported in self-aligning bearings and loaded by dead weights. Instrumentation is provided to measure the traction load, and temperatures can be monitored by thermocouples sometimes fitted both within the discs and in the lubricating oil supply. If the discs are electrically insulated from one another it is possible to measure the contact resistance or capacitance between them, and from such measurements the film thickness can be estimated. The disc surfaces can be ground or polished to a range of surface finishes and by grinding one disc to a 'crowned' profile, contact conditions can be changed to those of a nominal point rather than a line contact. If the axes are slightly skewed then a component of spin can also be introduced into the contact.

A disc machine is a versatile device which can be used to investigate conditions in a variety of heavily loaded tribological contacts. For example, by choosing appropriate disc radii and slide:roll ratios it is possible to simulate conditions at various stages in the contact between individual gear teeth in a gear drive or between the faces of the two components in a cam and follower mechanism. The effects of load and temperature on the onset of scuffing can be carefully examined for a range of potential lubricants with or without e.p. or antiwear additives. The influences of changes in material composition, heat treatment, or surface topography can similarly be investigated using the comparatively inexpensive and easily prepared disc specimens rather than those of the much more complex geometry of the real engineering components being simulated. A disc machine can also be used to investigate the rheological behaviour of lubricants and other fluids under conditions of very high hydrostatic stress and high rates of shear.

9.6 Solid lubricants

When a thin, coherent surface film of a material with a low shear strength (say τ) which has a thickness comparable with the size of the surface asperities, is interposed between two metal surfaces, the true area of the contact A will be little modified; it will continue to be primarily controlled by the mechanical properties (principally the indentation hardness $\mathcal{JC}$) of the substrate. On the other hand, the resistance to tangential motion, the friction force, will depend on the product of the area A and the stress required to shear the bond between the film and the opposing solid surface; this stress can be no more than τ. The coefficient of friction μ can thus be no greater than the ratio of $A\tau$ to $A\mathcal{JC}$, that is,

$$\mu \leqslant \frac{\tau}{\mathcal{JC}}. \tag{9.2}$$

Minimum coefficients of friction will be provided by relatively soft films deposited on hard substrates and it is really this idea that underlies the operation of all boundary lubricants. In a lubricated bearing, when the non-dimensional service group ($\eta\omega/p$) becomes small (i.e. we move to the left of the Stribeck diagram of Fig. 9.1) successful protection of the bearing surfaces relies increasingly on the existence of a thin surface layer which exhibits behaviour which becomes more and more like that of a solid. It is hardly surprising, therefore, that a range of true solids has been investigated and found, in some circumstance, to be successful as boundary lubricants. Such solid lubricants may be classified under the general headings of organic films, polymeric materials, lamellar solids, and thin metallic films.

9.6.1 *Organic films; soaps and greases*

We have seen that the formation of a thin layer of a metallic soap has been established as the mode of operation of a range of active boundary lubricants which are often incorporated, as additives, in hydrocarbon mineral oils. Rather than relying on *in situ* soap film formation, the solid soap can, in appropriate circumstances, be used as the active lubricant in its own right. A *grease* consists of a base hydrocarbon oil which has been thickened to almost a solid by the addition of a combination of metallic soap (or soaps) together with additional inorganic fillers. The soaps are present as microscopic fibres, known as *micelles*, which form a matrix holding the base oil by either a sorption or a swelling process. Inorganic thickening agents, usually finely divided solids such as bentonite (derived from clay) or silica, are used to control the consistency, and hence the useful operating temperature range, of the resulting grease. In very-high- or low-temperature applications the base fluid may be a synthetic fluid rather than a natural hydrocarbon (see Section 1.3). In those applications where loads are especially high (or speeds low) small quantities of finely divided molybdenum disulphide, graphite, or PTFE may also be incorporated (see below).

Greases are most often used instead of fluids in situations in which the lubricant is required to maintain its original position in the device, perhaps because opportunities for frequent replenishment may be either limited or economically unjustified. This might be due to the physical configuration of the mechanism, the type of motion, or perhaps because the lubricant has to perform some or all of the sealing function in preventing either lubricant loss or contaminant entry to the bearing. As a result of their complex structure the rheology of greases is not always easily modelled, though their behaviour can often be adequately approximated by that of a Bingham solid (see Section 8.3.1), that is, a material which flows with a constant Newtonian viscosity η once the applied shear stress has exceeded some critical value τ_i, but which is effectively rigid below this threshold. Thus,

$$\tau - \tau_i, = \eta \dot{\gamma} \qquad \text{provided} \qquad \tau > \tau_i, \qquad (9.3)$$

and if $\tau \leqslant \tau_i$, then $\dot{\gamma} = 0$

9.6.2 *Polymeric materials*

Many long chain polymers such as the polythenes, polypropylenes, nylons, and so on, have sufficiently low shear strengths to make them attractive candidates as thin-film solid lubricants. However, the friction properties of many of these resemble metals in that adhesion and relatively gross transfer and wear can occur when they are slid either against themselves or against some metallic counterfaces (see Section 4.4).

Probably the most important exception to this general tendency is PTFE (polytetrafluoroethylene) or Teflon which exhibits a comparatively low tendency to interfacial adhesion, hence its widespread use as a dry boundary lubricant. When rubbed against a smooth metal counterface a very thin film of PTFE is transferred to the opposing surface, and it appears that the molecules immediately in contact with one another become highly orientatated with their major axes aligned in the direction of sliding; this leads to the resultant coefficients of friction being very low (typically less than 0.05). Because it exhibits low forces of adhesion, it is of course not easy to form such a coherent thin film in the first place. The most successful techniques for forming mechanically robust PTFE films involve dispersing very fine, submicron particles of the polymer in a phenolic resin which is then sprayed on to the component surface. Subsequent curing at about 300°C produces a coherently bonded layer, typically about 5 microns in thickness, which is capable of withstanding temperatures up to about 250°C; above this PTFE starts to thermally decompose. As well as providing low-friction pairs on

(a)   (b)

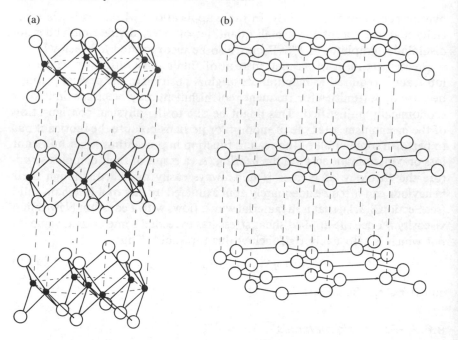

Fig. 9.10 Structure of two commonly used solid boundary lubricants: (a) molybdenum disulphide; and (b) graphite, showing the arrangement of parallel basal planes.

all manner of mechanical assemblies which cannot be conventionally lubricated, boundary films of this sort can now be made sufficiently tough and abrasion-resistant to be successfully applied to the loaded surfaces of such things as hand and power saw blades, and industrial shears and guillotines.

9.6.3 *Lamellar solids*

Solids having a lamellar or plate-like crystal structure exhibit pronounced directional or anisotropic mechanical properties, and sometimes these can be exploited by using them as thin-film lubricants. The two most widely used examples are graphite and molybdenum disulphide (MoS_2) and the crystal structure of these two materials is illustrated in Fig. 9.10. A characteristic of this type of solid is that the strength of the interatomic bonding *between* successive layers is very much weaker than that *within* the planes themselves. However, this property, which is shared by quite a large number of both inorganic crystals and condensed organic solids is insufficient on its own to guarantee successful performance as a boundary lubricant. It is additionally important that a strongly bonded adherent coating of the lubricant is formed on the opposing surfaces. This usually means that, if the junction is going to rely on the presence of a solid lubricant, the components must be 'run-in', under comparatively light loads, to pressure-bond the film to the metal substrates before the full working loads are applied. Running-in is often achieved by rolling or burnishing the bearing surfaces in the presence of the lubricant which is applied directly as a finely divided powder.

Despite the similarity of their layered structure the actions of graphite and molybdenum disulphide as boundary lubricants applied in this way are actually rather different; the mode of action of MoS_2 is rather more straightforward than that of graphite. Figure 9.10(a) illustrates the arrangement of individual sheets within the layered structure of MoS_2; the layers are either all pure molybdenum or pure sulphur. Bonding between the adjacent layers of sulphur atoms is comparatively weak and this leads to low values of friction which are maintained over a wide range of operating conditions including surface temperatures varying from $-150°C$ to more than $800°C$. Of particular value is the fact that lubrication by a burnished film of MoS_2 is relatively insensitive to the nature of the surrounding environment. The solid performs as well in inert atmospheres, or *in vacuo*, as in ambient air and this has made it a particularly suitable lubricant for use in space and satellite applications. As well as being formed from the burnishing technique, MoS_2 can also be applied by spraying in combination with an organic resin in the same way as PTFE. The main limitation to films formed in this way, when compared to those from the application of a dry powder, is that thermal decomposition of the resin binder imposes a more modest limitation on the maximum tolerable surface temperature and thus to the maximum possible loads and speeds.

At room temperatures in air the coefficient of friction associated with a thin film of mechanically transferred graphite on to a metallic surface is very similar to that for an MoS_2 or PTFE film. However, if neither oxygen nor water vapour are available in the surrounding atmosphere, the coefficient of friction is liable to rise to several times this value with an even more dramatic increase in the amount of surface damage and wear. This phenomenon is known as 'dusting'. The introduction of even a minute trace of oxygen or water vapour (or one of a number of other organic vapours or gases) re-establishes the lower values of friction and wear. It is thought that this behaviour of graphite is brought about by the considerable difference in surface energies of the edge and basal planes of its structure, Fig. 9.10(b). The surface energy of the basal planes is low $(40-80\,MJ\,m^{-2})$, about one tenth of that of the edge planes. *In vacuo*, friction on the edge planes will be high as a result of the strong forces of adhesion associated with the high value of the surface energy. Even though in a transferred graphitic surface the edge sites may only constitute a small proportion of the whole, the high friction forces which they can generate can give the total surface a relatively high mean value of μ. If the surface becomes damaged, as a result of the applied tangential motion, further edge sites are exposed and so the frictional force continues to rise. If now a trace of oxygen or water vapour is introduced this is immediately preferentially adsorbed on the high-energy edges so lowering their surface energies and reducing dramatically the potential for adhesion and friction at these positions. The amount of water vapour required to sustain low friction has been found to be the amount needed for about 10 per cent of the surface to be covered by a monolayer, reflecting the fact that most of the surface has an inherent low-friction basal plane structure. Despite these environmental and temperature limitations, graphite is still widely used as a solid lubricant, both as an additive to hydrocarbon oils and greases, and (like PTFE) as a constituent of monolithic solid bearing materials designed to operate under conditions of less than full film fluid lubrication (see Section 10.2).

Another lamellar solid with potential as a solid lubricant is graphite fluoride which has the chemical formula $(CF_x)_n$ where $0.3 < x < 1.1$. The fluorine atoms are inserted between the basal planes of the graphite structure and are covalently bonded to the carbon atoms. The material is chemically stable, has good resistance to oxidation, and attractive frictional properties.

9.6.4 *Thin metallic films*

Even journal bearings that are designed to operate well within the range of comparatively thick hydrodynamic films may undergo short periods when the numerical value of the group $\eta\omega/p$ pushes their operating point into the mixed lubrication regime. For example, this might occur during start-up or shut-down of the machine of which the bearing is a part. Some amelioration of the potential component damage that might arise from these periods can be provided by coating, or plating, one of the bearing surfaces with a thin layer of a low-shear-strength metal or alloy. The plating alloy must obviously be chosen not only to have a low shear strength but also to have the combination of an inherently weak interaction with the opposing sliding surface and yet good adhesion with the underlying substrate material. Copper–lead or lead–bronze automotive bearing shell surfaces are coated with an overlay of lead–tin or some similar alloy to a depth of a few tens of microns. Not only does this provide some degree of boundary lubrication but also has the advantage of protecting the copper-based alloy from possible corrosion from the acidic constituents that can be formed in the oil (see also Section 1.3.2).

Lead, tin, indium, silver, and gold have all been used as solid lubricant films usually deposited by electroplating, physical vapour deposition, or sputtering. Such films, often as thin as 0.1 micron, can successfully be used to lubricate rolling-element bearings for use in space or ultra-high-vacuum

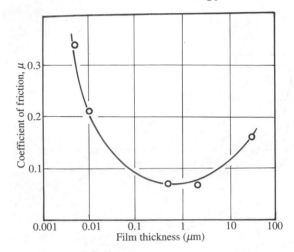

Fig. 9.11 The coefficient of friction for a steel slider sliding against a counter-face of tool steel carrying thin films of indium (from Bowden and Tabor 1950).

applications. Coefficients of sliding friction between a steel slider and a steel counterface plated with a thin soft metallic film can be very low (e.g. 0.06 for indium, 0.09 for tin) whereas the sliding friction for bulk specimens would be very much greater. Figure 9.11 shows the behaviour of indium films on a hard tool steel. There is an optimum film thickness for minimum friction: films thinner than this cannot prevent asperity interaction while for thicker films the behaviour begins to approach that of a bulk specimen of the softer material.

9.7 Tribology in metal-working

9.7.1 *Friction in metal-working*

A metal-working process involves, by definition, sufficient plastic flow of the workpiece to bring about some permanent change in its form. The tooling which imposes this change of shape must also carry the forces involved but be of such a geometry and material that these loads can be carried entirely elastically; unlike the workpiece the tooling must not deform permanently. With only a very few exceptions, there is always some relative motion between the tool (or die) and the workpiece, and these interfaces are pressed together by high normal stresses. It therefore follows that there are important tribological considerations to be taken into account. Tool or die surfaces are invariably lubricated, although the thickness of the film may

(a)

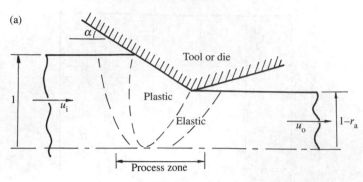

(b)

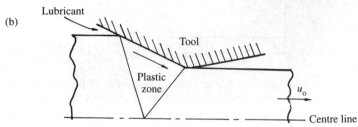

(c)

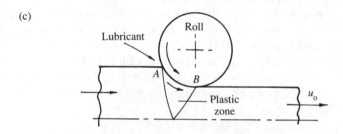

(d)

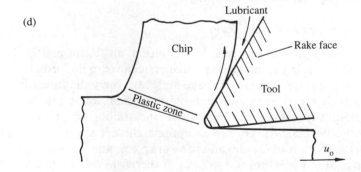

often mean that the dominant mechanism is more akin to boundary lubrication than one depending on hydrodynamic effects.

Despite the fact that many practical processes are geometrically complex, and involve three-dimensional material flow, they can for the purposes of illustration be simplified to the two-dimensional scheme of Fig. 9.12(a) which shows the upper half of an assumed symmetrical deformation process. Material enters the process zone (shown with a unit half-width) at mean speed u_i, and leaves with mean speed u_o having undergone a fractional reduction of cross-sectional area r_a. The geometry of the rigid tooling is simply described by the die semi-angle α.

Within the process zone, there is a region of intense plastic deformation which is surrounded by a hinterland in which the stresses are purely elastic and consequently the strains comparatively small. These elastic strains are often neglected in comparison with the overall change of shape so that the material can be treated as rigid–perfectly–plastic. This greatly simplifies any analysis of the process (see Section 3.5), the aim of which is to relate the final state of the workpiece (its dimensions, hardness, and so on) to its initial state, the geometry of the tooling, and to the loads applied. Since the loads are transferred from the tool to the workpiece across the interface between them, any analysis requires some form of tribological model of this boundary.

Metal-working processes can be classified by various criteria, one of the most significant being the temperature at which the deformation occurs. *Hot* metal-working processes are those carried out with the workpiece heated to a mean temperature Θ in excess of about 70 per cent of its melting-point Θ_{mp}, that is, those in which the *homologous* temperature $\Theta/\Theta_{mp} > 0.7$. Typical hot working temperature ranges for a number of materials are shown in Table 9.2.

Cold working is carried out with no intentional preheating, although the temperature of the process zone will generally rise above ambient because of the energy dissipated during deformation. As a general rule, hot working has the advantage of a lower material flow stress, permits heavier reductions or changes in shape, and offers numerous ways (through combinations of thermal and mechanical processing) to control the properties of the issuing product. Cold working yields products of closer dimensional tolerance and

---

Fig. 9.12 Typical two-dimensional metal-deforming processes: (a) the region of intense plastic deformation is surrounded by an elastically stressed region; (b) drawing (or extrusion) through a fixed die; (c) in rolling, the tool or roll rotates in the direction shown; and (d) in cutting or machining, the deformation is restricted to a thin surface layer. In processes (b) and (c) lubricant is drawn into the process zone; it is swept out of this region in (d).

Table 9.2 Typical hot-working temperature ranges for various metals and alloys.

Metal	Alloy	Temperature range for hot working (°C)
Carbon steels		900–1200
Copper		750–950
	Leaded brasses	625–800
Aluminium		250–550
	Aluminium alloys	300–550
Titanium alloys		900–1100
Nickel alloys		900–1200

better surface finish and the (sometimes useful) strength acquired by strain hardening can be retained in the part.

Deformation and machining

The purpose of any metal-forming process is a permanent change in shape. In some cases this is achieved by causing the plastic zone to extend across the total width of the workpiece. These processes of *plastic deformation* are exemplified by drawing or extrusion (Fig. 9.12(a) and (b)) and rolling (Fig. 9.12(c)). When the new shape is obtained by removing the surface layer in the form of a chip, the process is one of *machining* (Fig. 9.12(d)). All these operations nearly always require the provision of a lubricant to prevent severe pick-up and galling between the tool and the workpiece, although it is clear that there is a distinction between the lubricating conditions in drawing and rolling and those in metal cutting. In the first two of these operations the entraining velocity (shown arrowed in Fig. 9.12) is such as to draw lubricant into the interface, whereas in machining the relative motion of chip and tool is such that any hydrodynamic activity will tend to sweep the lubricant away from the contact zone so that effective lubrication is much more difficult. Successful cutting lubricants possess a distinctly different combination of physical and chemical properties from successful drawing or rolling lubricants (see below). It is also of note that, although usually carried out without any form of preheating, the specific energy density in machining is extremely high so that very high local temperatures can be generated (e.g. with steels often in excess of 1000°C).

For convenience in calculation, the interfacial tangential or friction stress τ in metal-working processes can be related to the die or interface pressure p in one of two ways. In cold-forming operations the assumption of a constant coefficient of friction μ (equal to the ratio τ/p) may be adequately accurate, especially under reasonably well-lubricated conditions; a typical value might be 0.2. However, there must be an upper limit to the value of μ which can operate under poor lubrication since, when the product of μp

reaches k, the shear flow stress of the adjacent material, deformation will proceed by shear within the bulk of the material rather than at the interface. This condition is known as *sticking* friction. The alternative description of the interface supposes that it is characterized by the ratio of the surface shear stress τ to the yield stress k. This ratio m is known as the *friction factor* and clearly for the same physical limitation must be such that

$$m \leqslant 1. \tag{9.4}$$

The effects of friction in metal forming are often intertwined with those of other factors. In the most general way the deformation load can be considered to be a function of three effects, so that

$$\text{deformation load} = f_1(k) \times f_2(\tau) \times f_3(\text{geometry, etc}). \tag{9.5}$$

The function $f_1(k)$ represents the work of pure deformation: it can be thought of as the work that would have to be done to bring about the required change of shape in the most efficient or homogeneous manner, that is, with every element of the workpiece being subject the same strain history. The second term in eqn (9.5) represents the effect of friction. Work done against friction does not contribute directly to the desired deformation but nevertheless absorbs some of the applied load. In a truly frictionless process $\tau \to 0$ and $f_2(\tau) \to 1$. The third term in eqn (9.5) reflects the difficulty in practice of achieving homogeneous deformation. The additional effort involved as a result of the introduction of inhomogeneity in the deformation is often referred to as 'redundant work'. The value of the multiplying factor f_3 can be quite large (i.e. in excess of 2); its value depends principally on the geometry of the tooling but is also influenced by the frictional conditions, which therefore play a part in *both* factors f_2 and f_3.

Although the work done against friction does not make a direct contribution to the useful reduction of area it may not always be a totally negative effect. For example, in rolling it is only the friction or traction stresses acting at the interface between the roll and the strip which maintain continuous motion of the workpiece. In the entry region, A in Fig. 9.12(c), the surface of the roll is moving more rapidly than that of the stock being deformed so that friction is dragging material into the roll gap. Conversely, at the exit B the speed of the rolled strip exceeds that of the surface of the roll so that the direction of the friction force is reversed. For the process to continue there must be a sufficient forward force on the strip. If μ (or m) falls below some critical value then the material slips in the gap and rolls fail to bite.

9.7.2 *Metal-working lubricants*

Under some very specific conditions, particularly those carried out at high speed, it is possible to maintain a full fluid film between the die and the plastically deforming workpiece. The lubrication conditions could then be properly referred to as *plasto-hydrodynamic*. However, in most metal-forming situations, because of both the more modest surface speeds and the fact that the surface of the workpiece in contact with the die is often growing in area, conditions are much more likely to be in the mixed regime. The surface topography of the workpiece (and also in some circumstances that of the die) can be important because, as the workpiece surface grows, small quantities of the fluid film may get trapped in surface pockets formed between its surface asperities. These pockets act as local reservoirs providing a source of lubricant protection to the surrounding expanding surface. However, these reservoirs may fail to keep up with the demands of surface extension and as a result the lubricant film thickness may fall to almost molecular dimensions; conditions become essentially those of boundary lubrication. Consequently, there are industrial examples of metal-forming processes lubricated both by fluids heavily loaded with e.p. additives and those in which the separation of die and workpiece depends on a film of effectively solid lubricant of the sorts described in Section 9.6.

Because of the nature of the geometry and the relative surface velocities, rolling imposes relatively mild duties on the lubricant. On the other hand, the smooth continuation of the process and, in particular, the surface topography of the product are extremely sensitive to small variations in lubrication. For this reason, and because of the great economic importance of sheet rolling, considerable efforts have been made to understand rolling lubrication. The hot rolling of most industrially important metals is carried out at temperatures beyond the operating range of organic lubricants and most undergo this operation without the application of a specific rolling lubricant, the oxide layer on the surface of the billet being sufficient to prevent gross adhesion between the material being rolled and the steel rollers of the mill. The exception to this is aluminium, whose hard brittle oxide offers insufficient protection, and the hot rolling of aluminium alloys is generally carried out in the presence of an oil-in-water emulsion (at a typical concentration in the range 2–25 per cent). If hot rolling of ferrous or nickel-based alloys is to be lubricated, then lubricants based on synthetic esters fortified by e.p. additives containing phosphorus are used.

The rusting of steel, especially in storage, presents special problem for its subsequent processing, and after picking (to remove the oxide scale) a coat of protective oil is usually applied to the stock. This oil is still present on the surface when the material is subsequently cold rolled. Emulsions of mineral oils or, more commonly, fatty oils in water are often used as specific steel cold-rolling lubricants. Thin gauge strip is often required to have a bright reflective surface and this burnishing is indicative of frictional conditions well within the boundary regime.

When processing aluminium, cold-rolling speeds are high so there is likely to be a significant amount of hydrodynamic load support. Nevertheless, because of the potentially high adhesive forces between the stock and the steel rolls, most cold rolling of aluminium is carried out with mineral-oil-based lubricants containing boundary additives such as fatty acids or alcohols. Staining of the product by an over-active chemical species is always a danger and must be avoided. A substantial portion of aluminium is rolled into very thin foil. Two strips are rolled together; only the outer surfaces are bright. The inner surfaces in contact show the texture characteristic of the free deformation of a polycrystalline material and consequently have a matt finish. Entrapment of more viscous lubricants can lead to the perforation of the foil so that very thin oils, perhaps with viscosities as low as $6 \, cS$ ($\approx 5 \times 10^{-4} \, Pa \, s$) are used.

The geometry of the deformation in drawing and extrusion, Fig. 9.12(b), is similar to that occurring in rolling in that the reduction of area takes place in a convergent or wedge-shaped region. In drawing, the reduction of area that can be achieved is limited by the fact that the deforming metal is being pulled through the die by a tensile stress which can never exceed the material yield stress; maximum reductions of area of 30–50 per cent are typical. In extrusion the starting material, usually in the form of a cylindrical billet, is placed into a container and forced through the reduction die by a punch attached to the ram of a hydraulic or mechanical press. The billet is first upset to fill out the container and then the ram pressure increased still further until flow begins. The extrusion ratio, that is, the ratio of billet cross-sectional area to that of the product, can be as high as 400:1. The use of glass as a lubricant in the hot extrusion of steels at large area reductions and in long lengths has become established in the last few decades. To protect the extrusion with a uniform, heat-insulating, low-friction coating over its entire length, a pad of woven or compressed glass fibres is placed in front of the billet which is usually preheated and rolled in glass powder to provide additional protection. As extrusion begins, a thin film of glass melts off the pad forming a film 10–30 microns in thickness with an effective viscosity of 20–200 Pa s. The speed of extrusion (0.2–0.5 ms^{-1}) must be high enough to generate hydrodynamic conditions and coefficients of friction at the interface as low as 0.002–0.05 can be achieved. By far the greatest quantity of aluminium alloys are hot-extruded without any lubrication at all although shearing along the container wall may account for as much as 50 per cent of the extrusion load. Graphite can be used as a die lubricant over the die faces.

The traditional low-temperature lubricants for ferrous wire or tube drawing are metallic soaps (although rarely of a single fatty acid) often applied dry just prior to entry to the die. There is no doubt that they behave in a non-Newtonian fashion (see Section 8.3). The high-speed drawing of copper wire is done almost exclusively in aqueous lubricants consisting of high-soap, low-fat emulsions; low-speed drawing of copper bar and tube is traditionally lubricated by water-soluble soap–fat pastes. The low cold-flow stress of pure aluminium and its low alloys permits their extrusion at room temperature using natural fats and waxes, such as lanolin, as lubricants.

9.7.3 *Lubrication in machining*

The physical conditions under which cutting or machining lubricants must be effective are somewhat unusual, because fresh and therefore highly reactive surfaces are created at the tip of the cutting tool (see Fig. 9.12(c)). Pressures on the rake face of the tool are high—comparable with the yield stress of the material being cut—and the interface between them is generally relatively inaccessible to any applied cutting fluid. Thus it is hardly surprising that many conventional boundary lubricants are comparatively ineffective as cutting fluids; in particular fatty acids (and the corresponding soaps) are generally considered to be poor machining fluids. Because previously undeformed workpiece material is continuously moving into the zone of intense plastic deformation, there is little heating ahead of the tool and, at production speeds, the greater part of the heat liberated is carried away by the chip; comparatively little is retained in the machined surface of the component. However, the tool itself is in continuous contact with deforming material and consequently very high temperatures can be generated on its surface, often well in excess of 1000°C. Demand for increased production rates and better machine utilization has lead to the use of higher and higher cutting speeds and these have only been possible because of the development of cutting tool materials with improved values of hardness at the high temperatures at which they must operate. Conventional cutting tools use brazed or mechanically clamped inserts of cemented carbides (typically with between 4 and 12 per cent WC in a matrix of cobalt) nearly always coated with one or more thin layers of such ceramic compounds as TiN, TiC, or alumina, Al_2O_3, to provide both lower frictional resistance to chip flow and reduced material loss from the tool faces by diffusional wear mechanisms.

In high-speed machining (e.g. turning on a lathe, high-speed milling or grinding) the principal role of the cutting fluid is one of cooling. The effects of the provision of a coolant on tool wear, and thus tool life, are complex: although the peak temperature, usually reached some way along the rake face of the tool, may not be greatly affected by the nature of the cutting fluid the temperature distribution and temperature gradients are certainly changed, and these can affect the rate of damage or wear to the tool surface. Of course, a coolant that reduces the temperature of the material being cut will also increase its resistance to shear and so tend to increase the cutting forces. However, this effect is usually rather small. In general the physical property of a coolant which has the greatest influence on its ability to remove thermal energy from the cutting zone is its specific heat c. Consequently, water with a value of c rather more than twice that of most hydrocarbon oils forms the basis of many high-speed cutting fluids. These are either classified as *emulsions* or *synthetics*.

Both mineral and natural oils can be broken down into small droplets and dispersed in water by mechanical agitation alone. However, to form stable dispersions or emulsions and prevent the phases separating on standing, emulsifying agents must be added. Most cutting fluids of this sort contain between 2 per cent and 10 per cent oil as well as additional components such as corrosion inhibitors (to prevent rusting of the machine tool) and biological agents (to both prevent or reduce algal souring of the stored fluid and give protection to machine operators who may come into contact with the fluid). True synthetic metal-working fluids have no oil in them. The earliest versions contained simply soda ash but these tended to leave a deposit on machine tool slideways and workpieces. Later versions of this form of cutting fluid have been based on organic amines together with sodium nitrite, $NaNO_2$, as an oxidation inhibitor. Among synthetics are those special fluids based on water-soluble polymers which form stable complexes with metal ions; they have low surface tensions and thus good wettability and so penetrate more easily between the tool and the workpiece surfaces.

At lower cutting speeds, or under more severe conditions, such as in broaching or thread cutting, particularly when the machining process involves some secondary rubbing faces on the tooling, cutting fluids are often based on mineral oils. In principle, the viscosity of the base oil should be low to assist access to the cutting zone but in practice, at least in these very-low-speed operations, quite viscous compounds can be effective. Almost all cutting oils contain boundary or e.p. additives. Chlorine-containing compounds are common, while sulphur is frequently used in severe operations, though it cannot be used with copper-based alloys because of unacceptable staining of the surface. A number of experimental studies have made use of tetrachloromethane (CCl_4) as a model lubricant; this is a particularly effective cutting fluid at low speeds for many metals (although for safety and environmental reasons cannot be used in production). This effectiveness is despite the fact that CCl_4 is a poor boundary lubricant in conventional circumstances: it appears that in machining it has just the optimum combination of high reactivity with the fresh metal surfaces and good mobility at the interface between the chip and the tool as a result both of its compact molecular dimensions and high vapour pressure. Conventional boundary lubricants have much larger molecules and lower vapour pressures and so are less able to penetrate the interface between the chip and the rake face of the cutting tool.

The importance of the nature of the gases or vapours surrounding the cutting zone has been made apparent by experiments in which materials have been cut *in vacuo* or in controlled atmospheres. This kind of experiment can throw light on the fundamental lubricating mechanism of many cutting fluids and also help in the choice of cutting conditions for use in the assembly or fabrication of structures in deep space. The removal of oxygen from the cutting environment generally leads to much higher cutting forces and severe degradation in the quality of the machined surface though there appears to be a complex interaction between surface layers on the cutting tool and the workpiece and the vapour immediately adjacent to them.

9.8   Further reading

ASTM: D2782 Measurement of extreme-pressure properties of lubricating fluids (Timken method).

ASTM: D2783 Measurement of extreme-pressure properties of lubricating fluids (Four-ball method).

ASTM: D3233 Measurement of extreme pressure properties of fluid lubricants (Falex pin and vee-block methods).

Barwell, F. T. and Jones, M. H. (1983) in *Industrial tribology* (ed. M. H. Jones and D. Scott) Elsevier, Amsterdam.

Blok, H. (1937) theoretical study of temperature rise at surfaces of actual contact under oiliness lubrication conditions. General discussion on lubrication. *Proceedings of the Institution Mechanical Engineers*. Vol. 2, pp. 222–35.

Blok, H. (1970) The postulate about the constancy of scoring temperature in Interdisciplinary approach to lubrication of concentrated contacts (ed. P. M. Ku) NASA., 153–248.

Bowden, F. P. and Tabor, D. (1950) *The friction and lubrication of solids.* Clarendon Press, Oxford.

Campbell, W. E. (1969) Boundary lubrication. In *Boundary lubrication: an appraisal of world litertiture*, ASME.

Dorinson, A. and Ludema, K. C. (1985) *Mechanics and chemistry in lubrication.* Elsevier, Amsterdam.

Dowson, D. (1977) in *Tribology of natural and artificial joints* (ed. Dumbleton) Elsevier, Amsterdam.

de Gee, A. W. J., Begelinger, A., and Salomon, G. (1985) *Proceedings of the 11th Leeds-Lyon Symposium on Tribology* (ed. D. Dowson, C. M. Taylor, M. Godet, and D. Berthe) Butterworth, London, pp. 105–16.

Jones, W. R. (1973) *Boundary lubrication revisited.* NASA Tech. Memo 82858.

Kapsa, P. H. and Martin, J. M. (1982) Boundary lubricant films: a review. *Tribology International* **15** 37–42.

Schey, J. A. (1983) *Tribology in metalworking.* ASME, Ohio.

Schey, J. A. (ed.) (1970) *Metal deformation processes: friction and lubrication.* Dekker, New York.

10
Dry and marginally
lubricated contacts

10.1 Introduction

Not all sliding tribological pairs are designed to operate in the presence
of a generous supply of lubricant. The nature of the working environment
may make it impossible, or impracticable, to arrange for the contact to be
lubricated by a full hydrostatic or hydrodynamic fluid film, for example,
in deep space or satellite applications where any liquid lubricant would be
lost or degraded by evaporation, or in food processing or chemical plant
where contamination of either the product or the environment by any escape
of a lubricating fluid would be unacceptable. In other, simpler applications
the design constraint may simply be the cost of the lubricants supply and
handling equipment. Generally, the aim in the design of machinery is to
minimize friction; however, in some devices (clutches, brakes, friction
drives, and so on) friction is beneficial, indeed essential, and often components of this sort are operated unlubricated. In this chapter, as well as
the tribology of dry sliding (taking as examples both brakes and clutches
as well as dry rubbing bearings) we shall also consider the tribological aspects
of some 'marginally' lubricated contacts. These are only intermittently
lubricated by a less than complete fluid film and so rely for their success
on a combination of hydrodynamic, elasto-hydrodynamic, and boundary
lubrication. Bearings and bushes designed to run completely dry are often
manufactured as monolithic solid components and involve at least one non-
metallic material, while marginally lubricated bearings often make use of
porous, sintered metals (usually bronze) impregnated by an appropriate
mineral oil, grease, or solid boundary lubricant.

   Conditions of marginal lubrication can also occur on the surfaces of
rotating and reciprocating seals. Such components, which are usually
lubricated by the process fluid itself, are required wherever a moving shaft
passes through the wall of a pump or pressure vessel. An important design
requirement is the prevention or minimization of the loss of the fluid across
the interface between the moving and stationary components. These may
be separated by a film of fluid generated in a similar way to those in

hydrodynamic bearings. However, in seals this film is much thinner than is the case in hydrodynamic bearings — perhaps only a fraction of a micron in thickness — so that on occasion there will be true dry contact between the moving seal faces. Consequently, successful material combinations for use in seals often show a strong similarity to those chosen for dry bearing applications.

10.2   Friction in dry sliding

10.2.1   *Slideways and bearings*

We have seen in Chapter 4 that dry sliding contacts are characterized by the ratio of F, the friction or tangential force, to W, the force component acting normal to the common interface. This coefficient of friction μ is generally typical of the physical and chemical states of the two surfaces. The angle that the *total* reaction force R makes with the common normal at the point of contact is known as the *friction angle f*; see Fig. 10.1. It follows immediately from the relations between the magnitudes of F, W, μ, and f, shown in Fig. 10.1, that the maximum value of f is given by the relation

$$\tan f = \frac{F}{W} = \mu. \tag{10.1}$$

On occasion it is useful to introduce f into the solution to a design problem. For example, consider the situation illustrated in Fig. 10.2 which might model conditions on a machine tool slideway with the block B representing the moving carriage of the machine. It is drawn between a pair of parallel guides, separated by the distance $2a$, by the force indicated which is applied centrally. Points A_1 and A_2 representing 'high spots' on the guides are separated in the direction of motion by the distance L. The problem is to decide, when the force is applied, whether the block will 'stick' in the position shown or continue to be drawn smoothly through the guides. R_1 and R_2 are the two total reaction forces at points A_1 and A_2; they can make angles of no more than f_1 and f_2 (their respective friction angles) with the local normals, and their lines of action intersect at the point C. If C is to the *right* of the line of action of the applied force, then this provides a counter-clockwise moment on B (consider taking moments about the point C). It follows that there will be no tendency for it to jam between the guides. On the other hand, if C is to the *left* of the line of the force then the effect of increasing the central force is to push the block more firmly against the guides since it now provides a clockwise moment about C; the consequence is that B jams. The critical situation for the block to just slide smoothly,

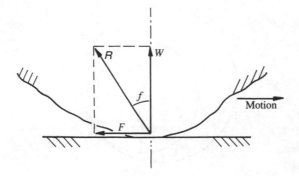

Fig. 10.1 A dry sliding contact. The total reaction force R, which can be resolved into the normal component W and the friction component F makes an angle f with the normal at the point of contact. f is known as the friction angle.

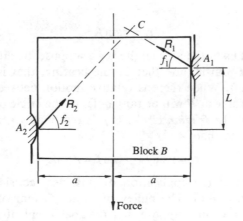

Fig. 10.2 Block B is drawn between two parallel guides on which A_1 and A_2 are 'high spots'.

and not to jam in its guides, is for C to lie on the mid-line of the block; in other words, we can write that for smooth motion

$$\tan f_2 < \tan f_1 + \frac{L}{a}. \tag{10.2}$$

The concept of the friction angle can also be useful in considering the action of dry pivots or unlubricated journal bearings. A typical case is shown in Fig. 10.3. The diagram shows the forces acting on the journal: the applied load W must generate an equal but opposite reaction R at the

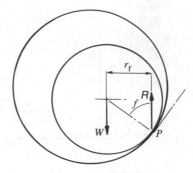

Fig. 10.3 The reaction R must be displaced from the centre of the journal by a distance r_f which is the radius of the friction circle.

point of contact P. This point is, however, displaced from the centre of the journal by a distance r_f. If f is the friction angle at P then, from simple geometry,

$$r_f = r\sin f, \tag{10.3}$$

where r is the radius of the journal (here we suppose the difference between the radius of the journal and that of the bearing, that is, the clearance, is negligibly small). When there is relative motion between the shaft and the bearing, the force at P will be tangential to the circle of radius r_f. This circle is known as the *friction circle*. If μ is reasonably small (say $\mu < 0.4$) then $\sin f \approx \tan f$, so that

$$r_f = r\sin f \approx r\tan f = r \times \mu. \tag{10.4}$$

As an example of the application of this idea, consider the situation illustrated in Fig. 10.4(a). The pulley of radius 50 mm runs on the shaft which is of radius 5 mm. Suppose that the coefficient of friction between the two surfaces is 0.4. In Fig. 10.4(b) the belt tension T is just sufficient to *raise* the 10 kg weight, while in Fig. 10.4(c) T' is the tension value which just allows the weight to descend. We wish to estimate the values of T and T'. We assume that the belt does not slip on the pulley, but rather that the pulley slips on the shaft.

When the system is in static equilibrium the pulley will make contact with the shaft at a point vertically above the centre of the shaft, and the belt tension will be equal to the weight of 10 kg, that is, 98.1 N. If the tension is marginally *increased* then the pulley effectively rolls around the shaft so that the point of contact is slightly displaced to the *right*, that is, to the point P_2 in Fig. 10.4(b). The force R will be tangential to the friction circle which is of radius $r_f = r \times \mu = 5 \times 0.4 = 2$ mm. Thus, taking moments about the point P_2 we can say that $T \times 48 = 98.1 \times 52$, so that

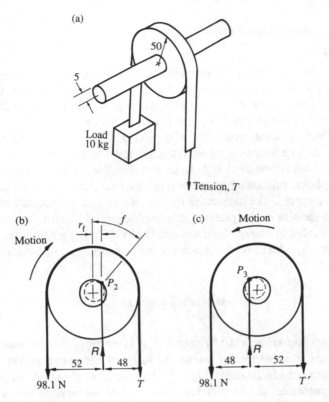

Fig. 10.4   (a) The pulley runs freely on the shaft and supports the load of 10 kg:
(b) raising the load; (c) lowering the load.

$$T = 106.3 \, \text{N} \tag{10.5}$$

and the force R, which is equal to the sum of T and the load, is equal to
204.4 N. On the other hand, if the value of the tension is reduced, to slightly
below that required to maintain the weight in equilibrium then W will start
slowly to descend. The point of contact will move this time to the left of
the shaft centre, say to P_3, as in Fig. 10.4(c). The tension T' will now be
given by

$$T' = 98.1 \times \frac{48}{52} = 90.6 \, \text{N}, \tag{10.6}$$

so that, in this case the total reaction force R' exerted at the shaft is equal
to 188.7 N.

10.2.2 Clutches and brakes

Clutches

When rotary motion is to be transmitted from one shaft to another for inter-mittent periods only, some form of clutch is required between them. The function of the clutch is both to produce a gradual increase in the angular velocity of the driven shaft (so that it can be accelerated to the speed of the driving shaft smoothly and without shock) and, once this has been achieved, to act as a permanent coupling transmitting the full torque of the prime mover without subsequent slip. In its simplest form a clutch consists of two annular plates, one attached to each of the two shafts, pressed together by means of a spring. The maximum torque M that can be transmitted is simply related to the size of the plates, their coefficient of friction, and the normal force generated by the spring pressure. If p is the pressure between the plates of the clutch, which extend from an inner radius r_i to an outer radius r_o then

$$M = \mu \times 2\pi \times \int_{r_i}^{r_o} pr^2 \, dr \qquad (10.7)$$

The integration of eqn (10.7) requires a knowledge of the way in which the pressure p is distributed across the faces of the clutch plates. The most straightforward idealization might be to take p as constant, in which case simple integration shows that for each pair of surfaces in contact,

$$M = \frac{2\mu r_i}{3} \frac{(r_o/r_i)^3 - 1}{(r_o/r_i)^2 - 1} \times P \qquad (10.8)$$

where P is the total force pushing the plates together. However, we know that the friction material covering the driving surfaces will wear during use, and this phenomenon is more likely to lead to the local values of pressure and sliding velocity being related by an equation with the form (see Section 10.3);

$$\text{pressure} \times \text{sliding velocity} = \text{constant.} \qquad (10.9)$$

If this is the case, so that $p \times r\omega = $ constant, then integration of equation (10.7) leads to the result that

$$M = \frac{\mu r_i}{2}\left(\frac{r_o}{r_i} + 1\right) P.$$

The 'effective radius' r_{eff} at which the force μP acts, given by the relation

$$M = \mu P \times r_{\text{eff}}, \qquad (10.10)$$

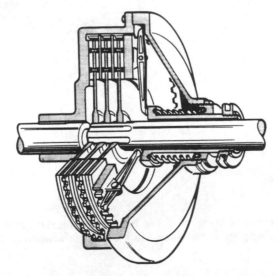

Fig. 10.5 Part section of a small multi-plate clutch showing how the intermediate plates are splined to the outer housing and the friction plates to the driving shaft (from Baker 1992).

is thus equal to $r_i\{1 + (r_o/r_i)\}/2$, and the constant in eqn (10.9) is equal to $P\omega/2\pi(r_o - r_i)$. In practice, the capacity of the clutch can be increased by having more than one pair of driving and driven plates; each of the driving plates must be keyed to the input shaft, while the driven plates are similarly keyed to the output shaft. Such a device is called a *multiple-plate* or *disc* clutch. Figure 10.5 shows a part section of such a device. The intermediate plates are splined to the outer, driven housing while the friction plates are splined to the driving shaft.

Brakes

The performance of a disc brake can be simply modelled in an entirely analogous fashion to that of a plate clutch. Figure 10.6 shows diagrammatically such a device in which each pad subtends an angle $\Delta\theta$ at the centre of the disc. If we consider only the retarding effects produced on one face of the disc and suppose again that the product of the interfacial pressure and relative sliding velocity is constant, then it follows that M, the contribution to the retarding torque, is given by

$$M = \frac{\mu\Delta\theta r_i}{4}\left(\frac{r_o}{r_i} + 1\right)P. \tag{10.11}$$

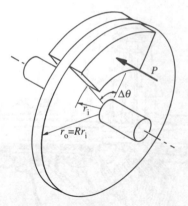

Fig. 10.6   Essential features of a disc brake. The rotating disc is decelerated by 'nipping' it between the two sector-shaped pads.

In this case, the effective radius of the brake, defined in the same way as in eqn (10.10), is given by

$$\frac{r_{\text{eff}}}{r_{\text{i}}} = \frac{\Delta\theta}{4}\left(1 + \frac{r_{\text{o}}}{r_{\text{i}}}\right). \tag{10.12}$$

When the pad is newly installed, before it has had the opportunity to run or bed-in, the application of the force P may lead to the disc and the pad making contact at a single point. The worst case (i.e. that producing the minimum retarding torque) would occur when this contact radius is the minimum possible value, that is, equal to the inner radius r_{i}. A possible design requirement might be to ensure that under these conditions of operation the performance of the brake is not significantly different from those when fully run-in. This requires that $r_{\text{eff}} \leq r_{\text{i}}$. It thus follows from eqn (10.12) that the angle subtended by the pad $\Delta\theta$ and the radii ratio $r_{\text{o}}/r_{\text{i}}$ must satisfy the inequality

$$\Delta\theta \leq \frac{4}{(1 + r_{\text{o}}/r_{\text{i}})}. \tag{10.13}$$

Band and shoe brakes

The band brake was one of the earliest mechanisms designed to slow or stop a rotating shaft. Despite its antiquity, modern designs are still much used in such applications as crane winding gear and vehicle automatic transmissions. In essence a band brake consists simply of a band or rope, with one end fixed, which is wrapped around the shaft whose speed is to be controlled; see Fig. 10.7(a). Its operation depends on the fact that when such a band

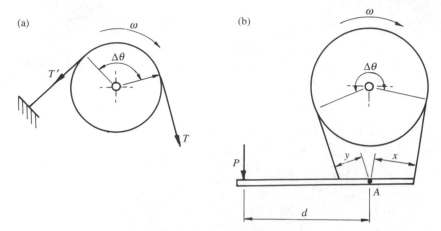

Fig. 10.7   (a) The band brake: rope or band wrapped around a rotating shaft. When the tension T is applied and the shaft rotates in the direction shown the tension T' is greater than T. (b) Lever-operated band brake; the lever of length d is pivoted at point A.

slips (or is just on the point of slipping) around a cylindrical shaft the ratio of the tensions on the tight and slack sides is equal to $\exp\mu\Delta\theta$, where μ is the local coefficient of friction and $\Delta\theta$ the angle of wrap subtended by the band at the centre of the rotation.

Suppose the shaft is of radius a and rotating in the direction shown in Fig. 10.7(a). When a tension T is applied to the band, the frictional amplification around the arc of contact, which subtends an angle $\Delta\theta$, leads to a retarding torque M given by

$$M = Ta\{\exp\mu\Delta\theta - 1\}. \tag{10.14}$$

Thus, if the product of $\mu\Delta\theta$ is such that $\exp\mu\Delta\theta \gg 1$ then modest applied tensions can produce large retarding torques. This is only the case if the direction of rotation of the shaft ω is as shown. The effect of the device will be much reduced if the direction of ω is reversed since then the retarding torque will be equal to $Ta\ \{1 - \exp(-\mu\Delta\theta)\}$.

In the arrangement shown in Fig. 10.7(b) the band tensions are applied by the force P acting on the lever of length d shown. Analysis of this geometry shows that the retarding torque M is given by

$$M = \frac{Pd\{\exp\mu\Delta\theta - 1\}}{y\exp\mu\Delta\theta - x} \tag{10.15}$$

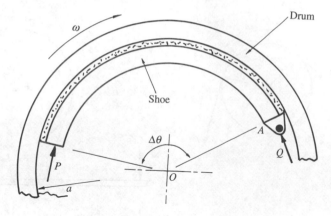

Fig. 10.8   Shoe brake: in the leading-shoe arrangement the direction of relative rotation is as shown; in a trailing shoe ω is reversed.

Thus if $x \geq y \exp \mu \Delta \theta$, so that the denominator becomes less than zero, the brake is effectively self-locking.

The essential features of a *shoe brake* are shown in Fig. 10.8. The shoe, which is of radius a and subtends an angle $\Delta \theta$ at the centre of the stationary back plate, is pivoted at the point A. The operating force, P is applied as shown. The shaft to be retarded is fixed to the brake drum which rotates about the centre O. If we make the idealization that the shoe is 'flexible' then although it is a structure loaded in compression its behaviour is analogous to a flexible belt in tension. The force Q at the pivot A is related to the applied force P by the equation

$$\frac{Q}{P} = \exp \mu \Delta \theta. \tag{10.16}$$

The retarding torque M experienced by the brake drum is equal to $(Q - P)a$, and is therefore given by the expression

$$\frac{M}{Pa} = \{\exp \mu \Delta \theta - 1\}. \tag{10.17}$$

In this arrangement, known as the *leading-shoe* configuration, the ratio M/Pa can be greater than unity; for example if $\mu = 0.4$ and $\mu = 120°$ then M/Pa is approximately 1.3. However, if the direction of relative rotation between the drum and the shoe is reversed, the *trailing-shoe* arrangement, the effectiveness of the brake is much reduced, since Q is now less than P and the retarding torque is related to Pa by

$$\frac{M}{Pa} = 1 - \frac{1}{\exp \mu \Delta \theta}. \qquad (10.18)$$

In the example above the retarding torque falls to about 40 per cent of its previous value.

Of course, real brake shoes are not inifinitely flexible, but more nearly rigid, and this makes the prediction of their braking performance more involved (Baker 1992). The pressure distribution along the are of contact will not be uniform and this usually means that the friction lining wears unevenly during service. This effect places a limitation on the angle $\Delta \theta$ that can be subtended by the shoes. If this is too great then the centre of the lining may be worn dangerously thin before any significant amount of material is lost from the ends of the arc. In practice, $\Delta \theta$ is usually of the order of 120°.

The performance of both band and shoe brakes is dependent on the factor $\mu \Delta \theta$. The angle $\Delta \theta$ is fixed during design while the coefficient of friction can vary with material composition, conditions of service, and so on. The *brake sensitivity s* is defined by the relation

$$\frac{dM}{M} = s \frac{d\mu}{\mu} \qquad (10.19)$$

and can be established by differentiating either of eqns (10.17) or (10.18). Distinguishing leading and trailing shoes by the suffices l and t, we then have

$$s_l = \frac{\mu \Delta \theta \exp \mu \Delta \theta}{\exp \mu \Delta \theta - 1} \quad \text{and} \quad s_t = \frac{\mu \Delta \theta \exp(-\mu \Delta \theta)}{1 - \exp(-\mu \Delta \theta)}. \qquad (10.20)$$

These relations are plotted in Fig. 10.9. Experience has shown that for a satisfactory performance the brake sensitivity should be no more than 2.

Friction materials

Early brakes and clutches often used organic materials such as wood and leather and although these were usually quiet and smooth in operation their surfaces gradually became carbonized, eventually causing an almost complete and perhaps disastrous loss of frictional traction. For many years most high-friction materials were based on some form of woven asbestos, impregnated, initially with bitumen, but more recently with various thermosetting resins and plastics. These are able to provide a consistent, reasonably high value of μ (typically about 0.4) together with good fade and wear resistance; fading is the tendency for μ to fall with increasing surface temperature. In recent years there have been considerable efforts to replace asbestos as the filler material, not least because of the safety aspects associated with its manufacture and use. One solution finding increasing

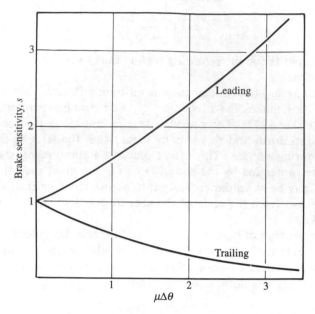

Fig. 10.9 Brake sensitivity s as a function of $\mu\Delta\theta$ for leading- and trailing-shoe brakes.

application is that of sintered metal facings, usually bronze, running against steel (see Table 10.1); this combination can also be used as a dry bearing.

10.3 Dry rubbing bearings

It is rarely feasible to run two loaded metal surfaces together for any length of time in the complete absence of a lubricant without producing severe damage to at least one of them. Successful dry bearings invariably consist of a combination of a metallic and a non-metallic component. The range of non-metallic materials (together with a lubricant-impregnated sintered bronze, see Section 10.4) commonly used in tribological situations is summarized in Table 10.1.

A typical dry rubbing bearing will consist of a metal shaft, nearly always steel, supported in a non-metal bush, and the maximum pressure tabulated above is the radial load divided by the nominal or projected bearing area in such a situation. The mechanical properties of both thermosets and thermoplastics can be improved by incorporating suitable amounts of inert material known as *fillers*. These may take the form of finely divided particles

Table 10.1 Dry rubbing bearings: materials and limiting properties.

Type of bearing material	Maximum static pressure (MPa)	Maximum service temperature (°C)	'Pressure–velocity' limit (MW/m^2)
Thermosets (phenolics, epoxies, etc. + fillers (textiles, etc.)	30–50	175	0.5–1.5
Thermoplastics (nylon, acetal, etc.) + fillers	15–20	150	0.17
Thermoplastics — unfilled	10	100	0.05
PTFE + fillers	2–7	250	0.15
Carbon graphites			
metal filled	3–5	350	0.10
epoxy filled	1–3	500	0.17
Sintered bronze (+ PTFE and Pb)	35–70	250	2.5

(e.g. wood or cellulose powders) or short fibres (usually of glass or carbon) or of a sheet or ribbon of a woven textile. However, although they improve the modulus of the material, such additions have little effect on the maximum temperature which can be tolerated in the bearing. Because plastics start to thermally soften or degrade at quite low temperatures this thermal limitation is much more severe than for conventionally lubricated metallic bearings.

Table 10.1 shows that the exceptions to this general rule are carbon-graphite materials. These are produced by the isostatic compaction of mixtures of crystalline and amorphous carbon and are often subsequently impregnated with either epoxy resins or low-melting-point metals to improve their thermal, mechanical, or electrical properties. As well as their use as electrical brushes, these materials are widely used in tribological situations, such as bearings and seals, in which their combination of self-lubrication, high thermal conductivity, good temperature stability, and chemical inactivity are particularly attractive. More details can be found in Paxton (1979).

The relevance of the maximum tolerable 'pressure–velocity' term, also indicated in Table 10.1, can be established as follows. Frictional heat is generated within the bearing at a rate $\dot{H}$ which will be related to the load W and sliding speed U by the coefficient of friction:

$$\dot{H} = \mu \times W \times U. \tag{10.21}$$

However, the rate at which this energy can be transferred or lost from the bearing and its associated components will depend predominantly on their surface area, say the product $L \times R$, where the bearing is of length L and radius R. The mean steady state temperature rise $\Delta\Theta$ must depend in some way on the ratio of $\dot{H}$ to the product $L \times R$, that is,

$$\Delta\Theta = f\left\{\frac{\dot{H}}{LR}\right\} = f(pU). \qquad (10.22)$$

This simple qualitative explanation thus justifies the often-quoted maximum or limiting 'pressure–velocity' (or 'PV') value beyond which plastic or dry bearings are unable to operate because of thermal limitations.

This product of nominal pressure and sliding speed can also be associated with the life of a dry journal bearing where this is limited by the development of an unacceptable degree of dimensional wear. If we assume that wear is governed by an 'Archard' relation, eqn (5.1), then the wear volume per unit sliding distance w is proportional to W, the radial load on the bearing. The total wear volume per unit time is thus dependent on the product of W and sliding velocity U. In a journal bearing, what really counts is the rate of increase of the radial clearance between the shaft and the bearing; the bearing will have reached the end of its useful life when this gap reaches some critical value. The radial dimensional wear loss is equal to the worn volume divided by the circumferential area of the bearing, so that it follow that:

$$\text{bearing life is proportional to } \frac{W \times U}{L \times R}, \qquad (10.21)$$

that is, once again to the product pU.

Limiting values of pressure and sliding velocity for a number of dry bearing materials can also be read off Fig. 10.10. At low pressures and high velocities the cut-off, or maximum permitted velocity, is set by the greatest temperature that the material can sustain. Conversely, at low values of sliding velocity, the maximum value of pressure p is set by the static strength or hardness of the material.

The dry bearing material combinations whose performance is illustrated in Fig. 10.10 will often only give satisfactory component lives when a stable transfer film is formed on the metallic counterface; this is particularly true of those involving PTFE additions. A running-in period is often required before steady state conditions are achieved and wear rates in the range of 10^{-6}–$10^{-3}\,\text{mm}^3\,\text{N}^{-1}\,\text{m}^{-1}$ are established. In the case of thermo-plastics, the essential wear mechanism is then one of adhesion as the transfer layer is continuously lost and replenished. On the other hand, thermosetting plastics do not form transfer layers but wear instead by abrasion or processes of surface fatigue: at high temperatures or sliding speeds abrasion may be exacerbated by direct thermal degradation of the mechanical properties of the plastic.

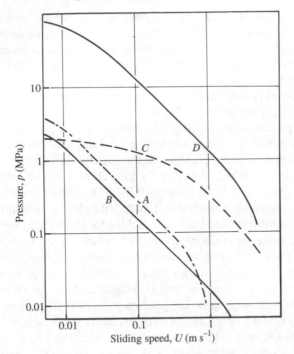

Fig. 10.10 Performance chart for dry rubbing bearing materials. At low speeds pressure is limited by material strength; at low pressures and high velocities by maximum temperature. *A*: thermoplastics; *B*: PTFE; *C*: thermoset-impregnated carbon-graphite (with additional PTFE); *D*: porous bronze material (from Neale 1973.

The wear of amorphous polymers (such as polystyrene, polyvinyl chloride, or polymethylmethacrelate) is much influenced by the surface temperature. If this is below their glass transition temperature Θ_g (see Section 4.4.3) so that their structure exhibits a degree of crystallinity, wear is mainly by fatigue and abrasion. On the other hand, if the interfacial temperature is above Θ_g then they form transfer films and wear by adhesion.

10.4   Marginally lubricated porous metal bearings

10.4.1   *Principles of operation*

Porous metal bearings can often be used in situations in which the operating conditions are more severe (in terms of specific loading and relative sliding speed) than could be accomodated by dry rubbing bearings, but in which both rolling contact and full hydrodynamic bearings are ruled out on the grounds of complexity and cost. A porous metal journey bearing consists of a sleeve of pressed and sintered material which has been processed to give a component with a density of about 80–90 per cent of the theoretical maximum. The porosity arises from incomplete solidification of the compressed particles of metal powder and, to a large extent, the pores between the metal particles are interconnecting. The most commonly used metal is bronze (typically 90 per cent copper, 10 per cent tin) although iron–copper mixtures can be useful in high-load/low-speed applications. Dry porous bearings can be made by incorporating graphite in the original mix or by impregnating with PTFE after sintering. However, in most cases, prior to use as a bearing, the sleeve is filled with a lubricating oil by a process of vacuum impregnation. The oil contained in the pores is intended to last a finite time, although this might be as long as the life of the machine to which they have been fitted. The lubricant wets the surface of the shaft which, because of its rotational motion, sets up a load-carrying circumferential pressure gradient in the oil film. Some of the oil drawn into the pressurized convergent wedge passes with the pores of the material and is recirculated to the unloaded region under the action of local pressure gradients. Even if there is an insufficient quantity of lubricant to maintain a full hydrodynamic wedge it is possible that the lands between adjacent pores act as miniature hydrodynamic pad bearings and so provide an additional element of load support.

Figure 10.11 represents the geometry of a fluid film formed between porous member, which is of thickness B, and an impermeable counterface. Within a porous matrix the mean volumetric flow rate per unite area q is given by Darcy's equation

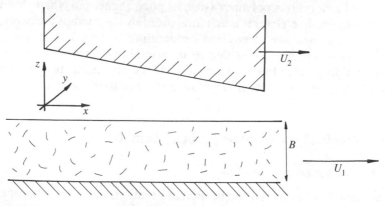

Fig. 10.11 Pad bearing with a porous slider of thickness B. The entraining velocity $\bar{U}$ is equal to $\frac{1}{2}(U_1 + U_2)$.

$$q = -\frac{\Phi}{\eta}\frac{\partial p}{\partial z},\qquad(10.23)$$

where Φ is a factor representing the material permeability, η is the viscosity of the, lubricant, and $\partial p/\partial z$ is the local pressure gradient. The minus sign indicates that volumetric flow is always down the pressure gradient. Φ has the dimensions of area and for sintered bearing materials is usually in the rage 10×10^{-9} to $200 \times 10^{-9}\,\text{mm}^2$.

Within the porous material, away from the material surface, continuity of flow must also be maintained, which implies that q must satisfy the relation

$$\frac{\partial q}{\partial x} + \frac{\partial q}{\partial y} + \frac{\partial q}{\partial z} = 0,\qquad(10.24)$$

that is,

$$\frac{\partial^2 p}{\partial x^2} + \frac{\partial^2 p}{\partial y^2} + \frac{\partial^2 p}{\partial z^2} = 0$$

which is Laplace's equation.

In addition, within the film itself, we must satisfy Reynolds' equation:

$$\frac{\partial}{\partial x}\left\{h^3\frac{\partial p}{\partial x}\right\} + \frac{\partial}{\partial y}\left\{h^3\frac{\partial p}{\partial y}\right\} = 12\eta\left\{\bar{U}\frac{\partial h}{\partial x} + \bar{V}\frac{\partial h}{\partial y} + (\mathcal{W}_2 - \mathcal{W}_1)\right\}.$$

$$(7.57\ \text{bis})$$

This equation can be simplified by setting $\bar{V}$, the entraining velocity in the y-direction, and velocity $\mathcal{W}_2$ equal to zero. The velocity $\mathcal{W}_1$ represents the velocity of flow of lubricant into the film at $z = 0$; it can be replaced (using eqn (10.23) since $\mathcal{W}_1 = q$) by

$$\mathcal{W}_1 = -\frac{\Phi}{\eta}\left\{\frac{\partial p}{\partial z}\right\}_{z=0}\qquad(10.25)$$

so that

$$\frac{\partial}{\partial x}\left\{h^3\frac{\partial p}{\partial x}\right\} + \frac{\partial}{\partial y}\left\{h^3\frac{\partial p}{\partial y}\right\} = 12\eta\left\{\bar{U}\frac{\partial h}{\partial x} + \frac{\Phi}{\eta}\left\{\frac{\partial p}{\partial z}\right\}_{z=0}\right\}.\qquad(10.26)$$

The full solution to the problem of designing a bearing operating on these principles requires the solution of eqns (10.24) and (10.26) subject to the appropriate boundary conditions for the particular case of interest. An approximate solution can be obtained for the case of a *short* or *narrow* journal bearing, of axial length L, in a manner similar to that for the case outlined in Section 7.7.1. If the radial clearance is c, then the pressure is given by the equation

$$p = \frac{3\eta\omega\varepsilon\sin\theta}{c^2\{(1+\varepsilon\cos\theta)^3 + 12B\Phi/c^3\}}\left(\frac{L^2}{4} - y^2\right), \tag{10.27}$$

where ε is the eccentricity ratio and θ the circumferential coordinate as in Fig. 7.15. The term $12B\Phi/c^3$ represents the loss of lubricant from the film into the porous material; if this has the value zero than the standard narrow bearing solution is recovered. The non-dimensional group $\Psi = B\Phi/c^3$ is thus a suitable descriptive parameter for porous bearings. The shaft has to rotate at a certain speed before it can carry sufficient oil around with it to compensate for leakage into the porous shell while still providing a full fluid film at the point $\theta = \pi$. The load W carried by the bearing will be dependent both on ε and Ψ. In terms of the Sommerfeld number S, defined in Chapter 7 by

$$S = \frac{\eta\omega LD}{W}\left\{\frac{R}{c}\right\}^2, \tag{10.28}$$

we could write

$$S = f(\varepsilon, \Psi). \tag{10.29}$$

The value of S which just causes ε to become unity (and therefore at which solid contact just occurs) is clearly a critical value, say S_{crit}: only if S is greater than this is the shaft carried on an oil film. For such a full oil film the frictional resistance, expressed as $\mu(R/x)$ where μ is the coefficient of friction, depends on the value of S. Friction is a minimum for a value of S only a little larger than S_{crit} and this operating point can be used as a design criterion for a porous journal. Figure 10.12 illustrates the variation of $\mu(R/c)$ with S for various values of the parameter Ψ.

More complete solutions confirm that Ψ is the appropriate design parameter to use provided $B < R/2$. Any increase in B above this value has little effect on the operation of the bearing (all the pressure flow being near the inner surface), so that if B is larger than $R/2$ the group Ψ can be taken as being equal to $\Phi R/2c^3$. Bearings of a finite length (i.e. $L/D > 0$) have a greater specific load-carrying capacity than infinitely short bearings, so that for a given value of Ψ the Sommerfeld number S increases with the L/D ratio. This effect is illustrated in Fig. 10.13.

10.4.2 *Practical considerations*

Thrust loads in shafts can be carried either by a separate thrust washer or the use of a flanged journal bearing; both these arrangements are illustrated in Fig. 10.14 which also indicates two possible means of providing lubricant replenishment. In practice the design parameter Ψ is likely to be in the range $10^{-3} < \Psi < 10^{-2}$. Figure 10.13 can then be used to estimate

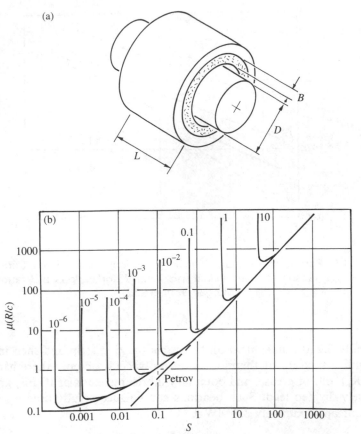

Fig. 10.12 Narrow porous metal journal bearing: (a) general arrangement in which the the porous sleeve is of thickness B; and (b) variation of coefficient of friction versus Sommerfeld number S with Ψ as parameter (after Morgan in Cameron 1966).

the Sommerfeld number and thus the load-carrying capacity of the bearing. The porous annulus is usually an interference fit in its housing and care must be taken to ensure that if, after fitting, it is necessary to enlarge the bore then this is not carried out not by a conventional reaming tool, which will tend to smear over the pores and so restrict oil flow at the surface. This problem can sometimes be overcome by using a burnishing tool of the opposite hand to the direction of rotation of the assembled shaft.

Quite complex commercial composite porous metal bearings are now available. These are manufactured by sintering a thin layer of bronze powder

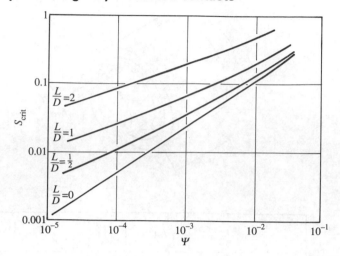

Fig. 10.13 Porous metal journal bearing: critical value of the Sommerfeld number S_{crit} versus the permeability variable Ψ for various L/D ratios (from Morgan *et al.* 1962).

(typically 0.2–0.3 mm thick) on to a steel backing strip and then infiltrating with a mixture of thermoplastics (polyacetal or polyetheretherketone [PEEK]), oil, or grease, and boundary lubricants (such as PTFE, graphite, or finely divided lead. Such bearings can be successfully used up to 'PV' limits of approximately 2.5 MW m^{-2}.

10.5   Tribology of sealing

Nearly all bearings function in association with some form of sealing device. This serves to prevent both the loss of lubricant from the bearing and the entry of contamination (often either hard abrasive particles or water) from the external environment into the bearing. Seals fall within two basic classifications: those for static duties and those for dynamic conditions. Static seals, or *gaskets*, are designed to produce leak-tight joints between stationary solid surfaces. We shall not consider these in any further detail; guidance on their design can be found in many books on machine design. Since an important requirement of any *dynamic seal* is also the minimization of any fluid leakage across the sealing surfaces which are in relative motion, consideration of their design and performance falls strictly within the scope of the tribologist.

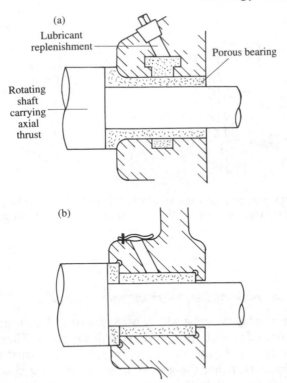

(a)

Lubricant replenishment

Porous bearing

Rotating shaft carrying axial thrust

(b)

Fig. 10.14 Porous metal bearing carrying both thrust and radial loads: (a) flanged journal; and (b) separate thrust washer. In each case there is provision for the replenishment of the lubricant.

At low temperatures, pressures, and sliding velocities, elastomeric seals in the form of either 'O' rings or 'U' or lip seals are widely used. These designs are illustrated in Fig. 10.15. Both rely on interference effects between the seal and the moving counterfaces for at least a component of their functional operation.

When conditions become more severe the more robust solution of a packed gland seal (sometimes known as a *stuffing box*, see Section 10.5.2) may be called for. If control of leakage is of great importance then the more sophisticated solution of a *mechanical seal* may be appropriate. Figure 10.16 gives some general guidance on the acceptable limits of operation, in terms of the sealed pressures p and sliding velocities U, for these types of seals.

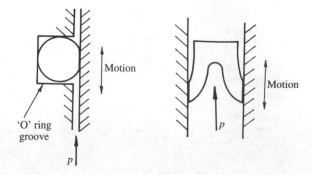

Fig. 10.15 'O' ring and 'U'-type lip seals often used for light-duty reciprocating and rotating sliding seals. p is the sealed service pressure.

10.5.1 *Elastomeric seals: 'O' rings and lip seals*

'O' rings are probably the most widely used type of hydraulic and pneumatic seal in general use for both static and dynamic purposes. They are inexpensive, easy to install, and generally reliable in service. For satisfactory performance it is important that the groove in which the ring sits should be of the correct dimensions, that the sealing surfaces have the appropriate surface finishes, and that the material of which the ring is made does not react adversely with the sealed fluid. Some practical elastomeric materials used for 'O' ring manufacture are shown in Table 10.2. The depth of the seal groove should be such as to impose a compression of about 10–20 per cent of its cross-section on the ring; over-compression can cause damage and permanent set to the ring or even cause it to extrude out of the groove, while too little squeeze may allow leakage. For dynamic situations the surface finish of the sliding surface should be no more than 0.4 microns R_a. Simple 'O' rings can be used for reciprocating seals or slow-speed rotation though rings of other more specialized sections are available.

The most commonly used form of elastomeric seal on a circular shaft is known as a *lip seal* and the general construction is illustrated in Fig. 10.17. The actual seal is made between the moulded elastomeric component and the surface of the rotating or reciprocating shaft, the two components being pushed into contact by a combination of the service pressure p and the circumferential force in the retaining or *garter* spring. Despite first appearances, the rubbing contact between the elastomer and the shaft is not truly dry; in a successful seal, a very thin film of the sealed fluid, often no more than a micron in thickness, is formed at this interface as a result both of

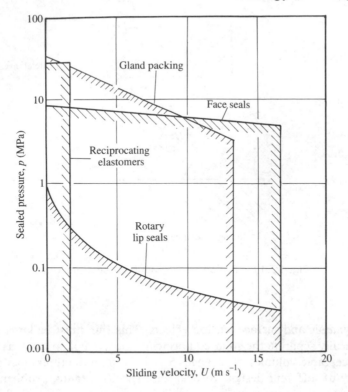

Fig. 10.16 Limits of sealed pressure *p* and sliding velocity *U* for various types of seal (from Neale 1973).

Table 10.2 Application and temperature range of elastomeric seal materials.

Elastomer type	Properties	Operating temperature range (°C)
Natural rubber	Suitable for animal and vegetable oil	−60 to +80
Neoprene	Resistant to outdoor weathering and mineral oils	−30 to +150
Butadiene styrene, SBR	Suitable for synthetic, animal, and vegetable oils	−60 to +90
Butyl rubber	Resistant to phosphate ester and gas permeation	−40 to +90
Silicone rubbers	Resistant to high and low temperatures but poor mechanical strength	−60 to 200

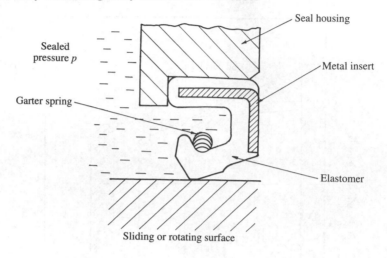

Sealed pressure p

Garter spring

Seal housing

Metal insert

Elastomer

Sliding or rotating surface

Fig. 10.17 Typical elastomeric lip seal design.

hydrodynamic and surface tension effects. This film must be large enough to physically separate the solid components and yet not so thick as to lead to unacceptable volumes of leakage. Specialist manufacturers can provide a variety of different designs to deal with specific sealing problems.

10.5.2 *Packed gland seals and mechanical or face seals*

The elastomeric lip seal is essentially a low-pressure seal and finds the majority of its applications in sealing fluids, often hydrocarbons, at gauge pressures of rarely much more than 0.1 or 0.2 MPa (i.e. 1–2 bars). A major advantage of the lip seal is its compactness when compared with the competing designs of gland seals and mechanical or face seals. However, these more bulky devices are capable of handling a much greater variety of service fluids under more demanding operating conditions. The construction of a typical packed gland seal or 'stuffing box' is illustrated in Fig. 10.18(a). The actual seal is made between the rotating or reciprocating shaft and the compressed packing. Most modern packings are of substantially square cross-section fabricated from a variety of materials. These can range from simply cotton fibres to composite constructions with metal wire reinforcement. They may be impregnated with suitable lubricants such as graphite or PTFE or rely on the service fluid to reduce the local frictional forces. Their major disadvantage is that they require periodic maintenance, principally adjustment of the compression load, to retain the best compromise of low friction without excessive leakage. They are widely used when

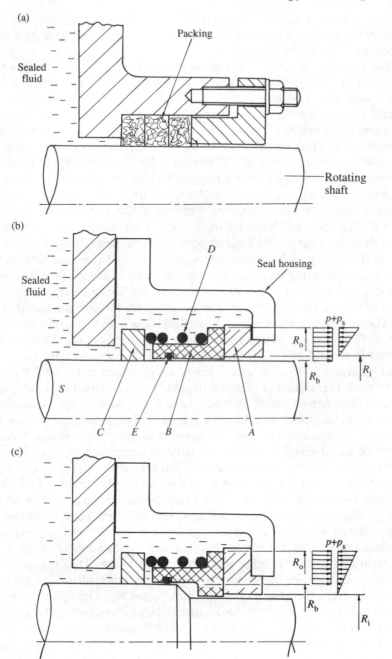

Fig. 10.18 (a) Packed gland seal. (b) Unbalanced and (c) balanced designs for mechanical seals. The indicated pressure distributions are those experienced by the rotating sealing component *B*.

a certain amount of fluid loss can be tolerated, for example, for many aqueous liquids or when sealing steam, and the extra expense of a mechanical seal cannot be justified.

The essential features of a *mechanical* or *radial face* seal are shown in Figs 10.18(b) and (c). The rotating shaft S passes from a region in which it is surrounded by the service fluid at gauge pressure p to atmosphere. The actual dynamic sealing takes place at the interface between the stationary seal ring A (which is effectively part of the casing of the seal housing or pressure-containing vessel) and the annular component B which rotates with the shaft S. Component B is held against A by a combination of the service pressure p and a spring pressure p_s generated by the spring D, the other end of which bears against the collar C which rotates with the shaft S. There must of course also be a static seal between B and S represented by the 'O' ring E. The main requirements of the device are to provide a seal with as little leakage of the service fluid as possible while maintaining a satisfactory life of the two wearing components A and B. One of these (usually, but not always, the rotating member) is conventionally made of carbon-graphite material (often either epoxy or metal filled, see Table 10.1) and operating lives of several thousand hours can be achieved. Despite the fact that in a satisfactory design there is very little fluid flow across the face of the seal the surfaces are not actually running totally dry over the whole of their apparent contact area. It is generally agreed that a significant part of the load is carried by hydrodynamic pressures generated in the very thin fluid film which is maintained between the sealing face. This film is very much thinner than 'conventional' hydrodynamic films, perhaps only a fraction of a micron in thickness, and the mechanisms leading to its formation are still a matter of some speculation. It seems clear that they involve the small pressure and thermal distortions of the seal components themselves.

Assuming that there is such a fluid film, then the pressure within it must drop from the service pressure p to zero across the annular face, which extends from radius R_i to R_o. The pressure distributions of Figs 10.18(b) and (c) illustrate the pressure loading on the seal component B; its rear face experiences the sum of the service pressure p and the spring pressure p_s, while the face pressure is shown (for simplicity) falling away linearly. Figure 10.18(b) illustrates a so-called *unbalanced* seal design. The pressurized side of component B will experience a uniform pressure of magnitude $p + p_s$ extending from the inner radius R_b to the outer R_o. The force represented by the difference between the uniform and linear pressure distributions must be carried by a combination of hydrodynamic load support and occasional dry contacts on the seal faces. For all but very modest values of sealed pressure p this difference is likely to be too large to be handled without excessive wear. When large sealed pressures are encountered the likelihood of truly dry contact (and therefore of reduced seal life) can be reduced by increasing the relative size of the seal faces. This produces a *balanced* seal, illustrated in Fig. 10.18(c). Balanced seals are characterized by the stepped shaft that they require. The seal *balance factor B_f* is defined by the relation

$$B_f = \frac{R_o^2 - R_b^2}{R_o^2 - R_i^2} \tag{10.30}$$

and experience shows that for success this should often be in the range $0.6 < B_f < 0.7$.

Mechanical seals for relatively light duty, for example, in low-pressure water pumps in domestic appliances, may be only marginally more complicated than the illustrative designs of Fig. 10.18. However, where service conditions are much more severe (either in terms of the pressures to be sealed, or the operating temperatures or speeds) or where there is possible contamination of the fluid by aggressive chemicals or particles, the final design may be much more complex. It may involve perhaps several stages of sealing separated by a recirculating filtered fluid or much more sophisticated secondary sealing arrangements. In some designs the hydrodynamic lift generated on the sealing face is augmented by incorporating grooves or pockets into which the fluid is pumped by the relative motion of the faces, rather like a thrust bearing; see Figs 10.19 and 7.11(c).

The primary function of a seal is to prevent the loss of the pressurized fluid, but often an almost equally important requirement is the prevention of the entry of potentially damaging abrasive contaminants from the external environment. The life of the sealing faces when they are operating within their permitted pressure–velocity envelope is likely to be limited by abrasive wear, and consequently there is a natural tendency to use the

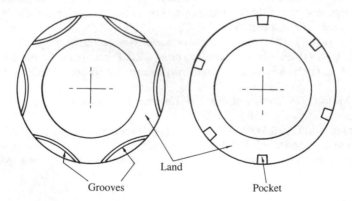

Land

Grooves

Pocket

Fig. 10.19 Grooves and pockets in the sealing faces of mechanical radial face seals leading to augmented hydrodynamic lift and hence improved seal life.

hardest material possible for their construction. If the recommended 'PV' limit of the material is exceeded, or if the seal were to run dry for any length of time, then the surface temperatures are likely to become excessively high, producing a characteristic surface cracking known as *heat checking*. Once the surface topography bas become broken up in this way, leakage and wear rates can increase dramatically and the seal will have failed prematurely. Further details of the technology of fluid sealing can be found in Angus (1965), Warring (1967), Fern and Nau (1976), Mayer (1972), Austin and Nau (1974), and the appropriate ESDU guides.

10.6 Further reading

Angus, George & Co. (1965) *Fluid sealing*. George Angus & Co, Wallsend.

Austin, R. M. and Nau, B. S. (1974) *The seal user's handbook*. BHRA Fluid Engineering, Cranfield, UK.

Baker, A. K. (1992) *Industrial brake and clutch design*. Penech Press, London.

Braithwaite, E. R. (1964) *Solid lubricants and surfaces*. Pergamon Press, Oxford.

Cameron, A. (1966) *The principles of lubrication*. Longman, London.

ESDU Guide 80012, *Dynamic sealing, of fluids I: a guide to the selection of rotary seals*; ESDU Guide 83031, *Dynamic sealing of fluids II: a guide to the selection of reciprocating seals*.

Faires, V. M. (1965) *Design of machine elements* (4th edition). Macmillan, New York.

Fern, A. G. and Nau, B. S. (1976) *Seals*. Engineering Design Guides No 15, Oxford University Press.

Ferodo Ltd. (1968) *Friction material for engineers*. Ferodo Ltd, Chapel-en-le-Frith, Derbyshire.

Mayer, E. (1972) *Mechanical seals*. Ilife Books, London.

Morgan, V. T., Cameron, A., and Stainsby, A. E. (1962) Critical conditions for hydrodynamic lubrication of porous metal bearings. *Proceedings of the Institution of Mechanical Engineers*, **176**, 761–70.

Neale, M. J. (ed.) (1973) *Tribology handbook*. Newnes-Butterworth, London.

Paxton, R. R. (1979) *Manufactured carbon*. CRC Press, Boca Raton, Florida.

Shigley, J. E. (1977) *Mechanical engineering design* (3rd edition). McGraw-Hill, New York.

Spotts, M. F. (1978) *Design of machine elements* (5th edition). Prentice-Hall, New York.

Summers-Smith, D. (ed.) (1992) *Mechanical seal practice for improved performance*. Mechanical Engineering Publications, London.

Warring, R. H. (1967) *Seals and packings*. Trade and Technical Press, Morden.

11
Rolling contacts and rolling-element bearings

11.1 The rolling contact of elastic solids

Everyday experience tells us that rolling contacts can exhibit a far lower resistance to motion than those involving simple sliding. However, to achieve this ease of movement the rolling component, that is, the wheel, must roll on a comparatively hard and smooth counterface so that the stresses in both surfaces are entirely elastic. For a given load and geometry, increasing the rigidity of the rolling components will reduce the resistance to motion still further. However it is important to realize that this will always be at the 'cost' of increasing the surface stresses (and those near the surface) at the point of contact. This effect can be quantified by examining the expressions for μ_R, the resistance of a cylinder rolling over an elastic 'road-way', and p_0 the peak contact pressure in an elastic contact. An expression for the first of these can be obtained from an argument involving elastic hysteresis (see Section 11.3.1) and is equal to

$$\mu_R = \frac{4\alpha}{3\pi} \left\{ \frac{W}{\pi R E} \right\}^{1/2}, \tag{11.1}$$

where α is a material loss factor associated with the hysteresis of the elastic solid. The second, the value of the peak contact pressure, can be taken from the Hertz analysis; see Section 3.3.4. From eqn (3.48) we can write that

$$p_0 = \left\{ \frac{W E^*}{\pi R L} \right\}^{1/2}. \tag{11.2}$$

Comparison of eqns (11.1) and (11.2) shows immediately that while increasing the rigidity of the contact, that is, increasing the contact modulus E^*, has the benefit of reducing the rolling resistance μ_R, it does so at the penalty of a greater contact stress p_0. The advantage, from both points of view, of increasing the effective radius R is also apparent from the two equations since R appears in both denominators.

Table 11.1 Comparison of pneumatic tyre/roadway and steel wheel/rail contact: typical values of rolling resistance and peak pressures

	Pneumatic tyre	Steel wheel/rail
Typical coefficient of rolling resistance μ_R	1×10^{-2}	2×10^{-4}
Typical maximum contact pressure p_0 (MPa)	75	1000

Two familiar, and important, examples of rolling contacts are those of pneumatically tyred road wheels and steel-rimmed railway wheels. The loads on each are quite similar but the properties of the solids involved are very different. The rolling resistance of a tyred road wheel is very much greater than that of a steel wheel on a steel rail because of the very much lower value of the contact modulus E^* of a rubber tyre on a concrete road, as well as its much greater value of α when compared to those of a steel wheel on rail (see Table 11.1). On the other hand, the magnitude of the contact stresses, which are potentially damaging to both road and rail surfaces, are very much lower for a pneumatic tyre than for the steel-rimmed wheel.

The smooth operation of virtually all but the very simplest machines relies, at least to some extent, on the inclusion in their design of rolling-element bearings. In such a device a number of third bodies or components — balls or rollers — are placed between the two moving elements so that the kinematics of sliding can be transformed into those of rolling, leading to a consequential reduction in frictional effort. Rolling-element bearings are simple, elegant devices and have a long history (see Section 1.2 and Nisbet (1974) and Harris (1984)). However, it is in relatively recent times that improvements in the quality of steels and the accuracy of manufacturing techniques have made possible their widespread use.

Although rolling friction is generally much less than sliding friction it is nevertheless finite and its origins have been the study of much investigation. As a result of the differences in the elastic deformation of the contacting surfaces, there is always likely to be an element of *microslip* within the contact zone and this makes a contribution to the frictional resistance to motion. However, in steel components rolling on one another the magnitude of this element of Reynolds' slip, and the friction associated with it, is quite small and insufficient to account for the whole of the measured rolling resistance.

A second mechanism, known as Heathcote slip, which contributes to the rolling resistance of ball bearings, was recognized around 1920. If the geometry of the rolling components is *conformal*, exemplified by that of a sphere rolling in a groove, then within the contact zone there will be two lines along which there is zero surface slip; on either side of them there must be regions in which the relative sliding between the sphere and the track is in opposite directions, and the contribution to rolling resistance because of friction in these areas can be significant (see Section 11.1.5 below). In addition to the energy dissipation associated with both these elements of surface slip, the small but finite hysteresis loss (present even in bard steels) can also be an important factor. Finally, in real bearings the contact is usually lubricated by elasto-hydrodynamic action and energy loss within the viscous lubricant film will also play a part.

11.1.1 *Microslip and creep*

Two solid bodies in contact are said to roll together if there is a difference in the components of their angular velocities measured along an axis which is parallel to their common tangent plane. Consider two cylinders touching along a common generator, as shown in Fig. 11.1(a). If the angular velocity ω_1 is not equal to ω_2, both in magnitude and sense (i.e. clockwise or counter-clockwise) the two cylinders are in rolling contact. The *angular velocity of roll* has a magnitude which is equal to the difference between ω_1 and ω_2. In this figure the *Oxy*-plane is the common tangent plane; if there is any difference in the components of the linear velocities of the two points in contact at O within this plane then, as well as rolling on one another, the surfaces also have a relative *sliding velocity*. In Fig. 11.1(a) this sliding velocity is equal in magnitude to $U_1 - U_2$ and thus equal to $|R_1\omega_1 - R_2\omega_2|$.

The term *rolling velocity* $\bar{U}$ is used in an analogous way to that of entraining velocity in lubricated contacts and must be equally carefully defined — especially if the point of contact between the solid surfaces is not at rest — in other words if one or both of the centres of the cylinders has a linear motion; see eqn (7.68). If, however, bodies 1 and 2 are rotating about *fixed* centres then the rolling velocity is $\bar{U} = \frac{1}{2}(U_1 + U_2)$.

In the case of a more general three-dimensional contact, such as that of two spheres in contact, Fig. 11.1(b), or of a sphere on a plane, there can also be a difference in the components of the angular velocities of the two bodies in the direction of the common normal, that is, at right angles to the common tangent plane. This represents a *spinning* motion and in Fig. 11.1(a) is of magnitude $|\omega_{1z} - \omega_{2z}|$.

Returning to the effectively two-dimensional case of Fig. 11.1(a) we can see that when the cylinders are pressed together by a normal load the contact will spread over a finite area which can be determined for this non-conformal contact by the Hertz theory; see Section 3.3.4. Once we take into account the fact that the contact spreads into a strip which has a finite

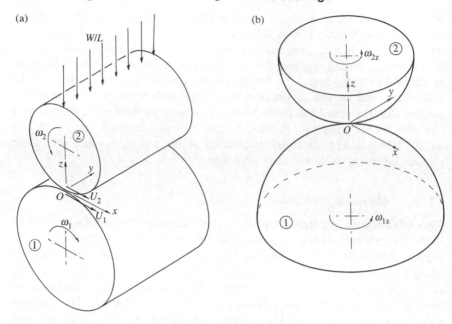

Fig. 11.1   (a) Two cylinders 1 and 2 in contact with their axes parallel: there can be rolling and sliding at the point of contact. (b) In the three-dimensional case body 2 can also spin relative to body 1.

width, rather than occurring along a line, the specification of 'sliding' is not quite so straightforward. This is because we can now imagine that in the presence of friction, or *traction*, between the surfaces some pairs of points in contact at the interface may 'slip' relative to one another while others 'stick'. A difference between the tangential surface strains in the two bodies in the region of 'stick' can lead to a small degree of overall relative movement which is known as *creep*.

This effect can be illustrated as follows. First, think about a rigid cylinder of radius R rolling over a rigid plane. In the time it takes to make one revolution, the centre of the roller will have moved forward a distance equal to $2\pi R$. Now suppose that the cylinder is deformable while the plane remains infinitely stiff. If, when carrying the normal load, the tangential strain in the surface of the roller is tensile then its outer layer will be stretched; consequently the cylinder will behave as if it had a slightly enlarged radius. In the time it takes to make one revolution its centre will have progressed a distance greater than $2\pi R$; the fractional increase in this distance is known as the *creep ratio* ξ. Since this movement has taken place in the same time interval as one revolution of the rigid case it follows that the effective velocity of the roller has been increased by an incremental term, the *creep velocity* δU.

If now we think about the actual velocity of a *particular element* of either body then we also need to consider how this will be influenced by the local state of strain in the deformed region. Suppose that when the body deforms the position of a particular particle of material is modified from its unloaded value, changing its position coordinate from (say) x to x'. The actual velocity U at any instant will be given by

$$U = \frac{dx'}{dt}. \tag{11.3}$$

Considering points at the interface, the difference between x' and x is the component of surface deformation parallel to the Ox-axis, that is, w_x. This surface deformation will in general be a function of both position and time. Thus

$$x' = x + w_x(x, t)$$

Hence,

$$U = \frac{dx'}{dt} = \frac{dx}{dt} + \frac{\partial w_x}{\partial x}\frac{dx}{dt} + \frac{\partial w_x}{\partial t} \tag{11.4}$$

but since $dx/dt = \bar{U} + \delta U$, neglecting second-order quantities, we can write that

$$U = \bar{U} + \delta U + \bar{U}\frac{\partial w_x}{\partial x} + \frac{\partial w_x}{\partial t}. \tag{11.5}$$

If the strain field does not change with time, in other words if we are dealing with a steady rolling process, then the final term, $\partial w_x/\partial t$, must be zero. In such a situation, at any point on the interface, the difference in the velocities U_1 and U_2 represents the velocity of *microslip* $\dot{s}$ between the solids 1 and 2; thus

$$\dot{s} = U_1 - U_2 = (\delta U_1 - \delta U_2) + \bar{U}\left\{\frac{\partial w_{x1}}{\partial x} - \frac{\partial w_{x2}}{\partial x}\right\}. \tag{11.6}$$

This equation can be written in a convenient non-dimensional form by dividing throughout by $\bar{U}$, thus

$$\frac{\dot{s}}{\bar{U}} = \xi + \left\{\frac{\partial w_{x1}}{\partial x} - \frac{\partial w_{x2}}{\partial x}\right\}, \tag{11.7}$$

where

$$\xi = \frac{(\delta U_1 - \delta U_2)}{\bar{U}} \tag{11.8}$$

is the creep ratio between the two surfaces.

'Slipping' and 'sticking' regions of the contact can be distinguished in two ways. Within any sticking region $\dot{s}$ must, by definition, be zero. But in addition, the surface traction stress $\tau(x)$ can be equal to (but certainly no more than) μ times the local normal pressure. Within a slipping region, where $\dot{s}$ is not equal to zero, the traction stress *must* equal μ times the local interfacial pressure. In addition, it must be acting in the direction *opposite* to that of slipping. To summarize:

in *sticking* regions

$$\text{microslip } \dot{s} = 0 \quad \text{and} \quad \tau(x) \leqslant \mu \times p(x); \qquad (11.9)$$

while in *slipping* regions

$$\dot{s} \neq 0; \qquad |\tau(x)| = \mu \times p(x) \qquad (11.10a)$$

and

$$\frac{\tau(x)}{|\tau(x)|} = -\frac{\dot{s}(x)}{|\dot{s}(x)|}. \qquad (11.10b)$$

One of the principal difficulties in the analysis of practical situations in which there are both sticking and slipping regions is the determination of the boundaries of these two zones of the contact.

11.1.2 *Free rolling of cylinders*

When rolling occurs without gross sliding the motion is sometimes referred to as *pure rolling*. However, as we have seen, this term is potentially ambiguous as even in rolling contacts in which there is no overall sliding there may still be elements of microslip at the interface. A more useful distinction is between those situations in which there is no tangential force, or traction, transmitted at the interface and those in which such a force is present. The term *free rolling* is usually used for conditions in which there is no net traction, while *tractive rolling* is reserved for those situations in which such tangential forces exist. Tractive rolling is typified by the driving or braking wheels of a vehicle, where the interfacial tangential or frictional force acts to change the velocity of the mass in motion.

Once again, to start the analysis, we consider the case of two cylinders loaded together by a normal force, say of intensity W/L per unit length. As in the case of static or sliding contacts our aim is to establish a realistic picture of how the load is transferred across the contact, and to obtain sensible estimates of the subsurface stresses that are generated, so that engineering components can be designed to provide adequate performance over their working lives. Initially we shall suppose that no torque is transmitted between the cylinders, so that they are rotating *freely*, and we further suppose that frictional effects are sufficiently large to entirely suppress slip at the interface. It is worth noting that if the two bodies were to be identical both in geometry and material constants then the situation would always be symmetrical about the interface and we should expect no tendency to slip anyway, whatever the value of μ. In free rolling, slip at the surface depends on the dissimilarity of the two rollers.

If there is no slip at the interface, so that in eqn (11.7) $\dot{s} = 0$, it follows that

$$\xi + \frac{\partial w_{x1}}{\partial x} - \frac{\partial w_{x2}}{\partial x} = 0. \tag{11.11}$$

The local displacement gradients, or surface tangential strains $\partial w_{x1}/\partial x$ and $\partial w_{x2}/\partial x$ can be obtained from the pressure and traction stress distributions. Referring back to eqns (3.30) and (3.71) we can write, supposing that the contact strip extends symmetrically from $x = -a$ to $x = +a$, that

$$\frac{\partial w_{x1}}{\partial x} = -\frac{(1 - 2v_1)(1 + v_1)}{E_1}p(x) - \frac{2(1 - v_1^2)}{\pi E_2}\int_{-a}^{+a}\frac{\tau(s)}{x - s}\,ds \tag{11.12}$$

and

$$\frac{\partial w_{x2}}{\partial x} = -\frac{(1 - 2v_2)(1 + v_2)}{E_2}p(x) + \frac{2(1 - v_2^2)}{\pi E_2}\int_{-a}^{+a}\frac{\tau(s)}{x - s}\,ds. \tag{11.13}$$

The traction term in eqn (11.13) must be equal in magnitude, but opposite in sign, to that in eqn (11.12). Substituting eqns (11.12) and (11.13) into (11.11) then leads to

$$\pi\beta p(x) + \int_{-a}^{+a}\frac{\tau(s)}{x - s}\,ds = \frac{\pi E^*}{2}\xi \tag{11.14}$$

where

$$\beta = \frac{1}{2}\left\{\frac{(1 - 2v_1)(1 + v_1)/E_1 - (1 - 2v_2)(1 + v_2)/E_2}{(1 - v_1^2)/E_1 + (1 - v_2^2)/E_2}\right\}. \tag{11.15}$$

If the two cylinders are of the same material then the parameter β is equal to zero; if the two solids are different then $|\beta| > 0$. The values of β for a number of fairly common materials chosen as body 1 are given in Table 11.2; in each case body 2 is steel.

Within the contact zone the *vertical* components of the surface deformation of the two solids (w_{z1} and w_{z2} respectively) must satisfy the compatibility condition expressed by eqn (3.8), that is, that

Table 11.2 Values of the parameter β for various materials: body 2 is steel

Body 1	β
Rubber	~0
Perspex/plexiglass	0.19
Glass	0.21
Duralumin	0.12
Cast iron	−0.24

$$w_{z1} + w_{z2} = \Delta - \frac{x^2}{2R}. \tag{3.8 bis}$$

where Δ is the relative distance of approach of distant points in the two bodies. This equation can be differentiated to introduce the displacement gradients, so that

$$\frac{\partial w_{z1}}{\partial x} + \frac{\partial w_{z2}}{\partial x} = -\frac{x}{R}, \tag{11.16}$$

then, substituting the expressions for surface displacement gradients from eqns (3.31) and (3.71) and simplifying, we can write

$$-\pi\beta\,\tau(x) + \int_{-a}^{+a} \frac{p(s)}{x-s}\,ds = \frac{\pi E^* x}{2R}. \tag{11.17}$$

Equations (11.14) and (11.17) form a pair of coupled integral equations for the distribution of normal pressure $p(x)$ and traction stress $\tau(x)$ in cylindrical rolling contact. The problem of finding their solution is much simplified by making the reasonable assumption that in such non-conformal rolling contacts the pressure distribution remains that given by the static Hertzian semi-ellipse, that is, that

$$p(x) = p_0\{1 - x^2/a^2\}^{1/2} \quad \text{where} \quad p_0 = \frac{2W}{L\pi a}. \tag{3.41 bis}$$

When this is substituted into eqn (11.17) the solution for the traction stress $\tau(x)$ becomes

$$\tau(x) = \frac{2\beta p_0}{x}\left\{\frac{x}{(a^2-x^2)^{1/2}} + \frac{1}{2a}(a^2-x^2)^{1/2}\log\left|\frac{a+x}{a-x}\right|\right\}$$
$$+ \frac{E^*\xi}{2}\frac{x}{(a^2-x^2)^{1/2}}. \tag{11.18}$$

At the *leading* edge of the contact, where $x = -a$, strain must be continuous as material flows into the contact strip. The traction at this point,

$\tau(-a)$, must therefore be zero. This condition enables the value of ξ, the creep ratio, to be established from eqn (11.18); it has the value given by

$$\xi = \frac{4\beta p_0}{\pi E^*} = \frac{2\beta a}{\pi R}. \tag{11.19}$$

Equation (11.18) can then, in turn, be simplified to

$$\frac{\tau(x)}{p_0} = \frac{\beta}{\pi} \{1 - x^2/a^2\}^{1/2} \ln \left| \frac{a+x}{a-x} \right|. \tag{11.20}$$

Examination of Table 11.2 indicates that β is never in excess of about 0.3 and the distribution of the traction stress $\tau(x)$ for this value is plotted from eqn (11.18) as the solid curve in Fig. 11.2. Note that the traction is never more than of the order of 0.1 of the normal pressure. The traction $\tau(x)$ acts *outwards* on the more compliant surface and *inward* on the more rigid one.

The other extreme case occurs when friction forces are so small that they effectively have no effect on the degree of interfacial slip. If this is the case then it is reasonable to assume that the tangential surface *deformations* are just those arising from the Hertzian pressure distribution, that is, are those given by substituting the Hertzian equation (3.41) into eqn (3.30); thus in each surface

$$\frac{\partial w_x}{\partial x} = -\frac{(1-2\nu)(1+\nu)}{E} p_0 \{1 - x^2/a^2\}^{1/2}. \tag{11.21}$$

Equations (11.7) then becomes

$$\frac{\dot{s}}{\bar{U}} = \xi - \left\{ \frac{(1-2\nu_1)(1+\nu_1)}{E_1} + \frac{(1-2\nu_2)(1+\nu_2)}{E_2} \right\} p_0 \{1 - x^2/a^2\}^{1/2}$$

which can be simplified to

$$\frac{\dot{s}}{\bar{U}} = \xi - \frac{\beta a}{4R} \{1 - x^2/a^2\}^{1/2}. \tag{11.22}$$

Equation (11.22) shows that if ξ has a positive value then the degree of relative slip $\dot{s}/\bar{U}$ will be greater than zero at the ends of the contact but may be less than zero, that is, in the opposite direction, close to the centre of the contact where x/a has small values. The frictional traction, equal in magnitude to $\mu \times p(x)$, although small, will change direction at two points equidistant from, but on either side of, the origin. The location of these points, and hence the value of the creep ratio are ξ, are determined by the fact that there is no overall tangential force. Setting $\int_{-a}^{+a} \tau(x) \, dx$ equal to zero shows that these points of reversal occur at the two positions $x/a = \pm 0.404$.

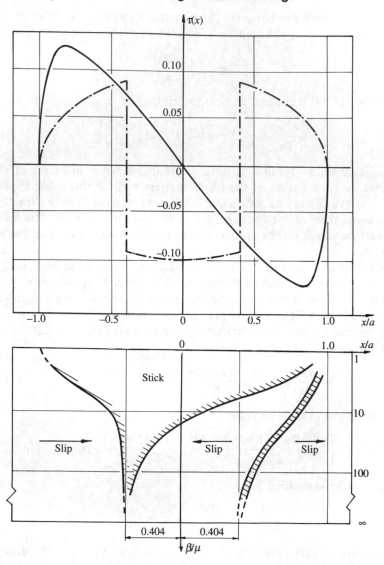

Fig. 11.2 Free rolling contact between dissimilar cylinders. (a) Distribution of traction stress plotted as $\tau(x)/p_0$ across the contact zone: solid line, no slip, $\beta/\mu = 0.3$; chained line, complete slip, $\mu = 0.1$. (b) Stick and slip regions within the contact zone as functions of ratio β/μ (from Bentall and Johnson 1967).

The true state of affairs must lie somewhere between these two extremes of high and low friction. Using a numerical analysis it can be shown that there are indeed two 'stick' regions and that these are separated by three regions in which there is slip in alternate directions — see Chapter 3 of Johnson (1985). The relative sizes of these portions of the contact depend on the value of the group β/μ and the variation is indicated in Fig. 11.2(b).

Once there are regions of slip at the interface there must be a loss of mechanical energy through friction which will make a contribution to the torque required to maintain the cylinders in rotation. This *rolling resistance* will be low when μ is small (since then, although microslip will occur, the friction forces are small), and low again when μ is large (since then microslip is suppressed). The maximum rolling resistance occurs at some intermediate state of affairs when the ratio β/μ has a numerical value of approximately 2.

11.1.3   *Rolling with traction*

As might be anticipated the situation becomes even more complex when either driving or braking tractions are applied. Consider first two elastically similar cylinders, loaded together by a normal force of intensity W/L per unit length and rotating so that a torque is transmitted across the interface. Referring to Fig. 11.3 we suppose that cylinder 1 is the *driving* roller and cylinder 2 the *driven*.

By analogy with the stationary situation (Section 3.4.2) we might expect the contact area to be divided into regions of stick and slip. In the static case there was a central sticking region, extending from $-b < x < +b$, with side slip zones disposed symmetrically. As a first attempt at analysing the rolling situation we can examine this static stress distribution and see whether the conditions summarized in eqns (11.9) and (11.10), which refer to the rolling case, are satisfied. The shear stress distribution in the static case was found by superposing two traction distributions $\tau'(x)$ and $\tau''(x)$, where

$$\tau'(x) = \mu p_0 \{1 - x^2/a^2\}^{1/2} \tag{3.78 bis}$$

$$\tau''(x) = -\frac{b}{a}\mu p_0 \{1 - x^2/b^2\}^{1/2}. \tag{3.80 bis}$$

The associated displacement gradients obtained by differentiating eqns (3.79) and (3.81) are

$$\frac{\partial w_x'}{\partial x} = -\frac{2(1 - v^2)}{Ea}\mu p_0 x \tag{11.23}$$

and

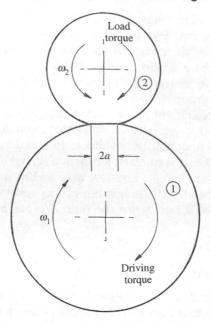

Fig. 11.3   Rolling contacts; the lower cylinder 1 drives the upper cylinder 2 through frictional effects at the interface which is of width $2a$. In free rolling there is no load torque applied to the driven cylinder.

$$\frac{\partial w_x''}{\partial x} = \frac{b}{a}\frac{(1-v^2)}{Eb}\mu p_0 x. \tag{11.24}$$

Since the tractions acting on each surface are equal but opposite in sign it follows that so are the tangential strains, that is, that

$$\frac{\partial w_{x1}}{\partial x} = \left\{\frac{\partial w_x'}{\partial x} + \frac{\partial w_x''}{\partial x}\right\} = -\frac{\partial w_{x2}}{\partial x}.$$

Thus, eqn (11.7) becomes

$$\frac{\dot{s}}{\bar{U}} = \xi + \left\{\frac{\partial w_{x1}}{\partial x} - \frac{\partial w_{x2}}{\partial x}\right\} = \xi + 2\frac{\partial w_{x1}}{\partial x}. \tag{11.25}$$

Within the slip regions, $a < |x| < b$, it is clear that the condition expressed in eqn (11.10a) is satisfied. However, comparison of eqns (11.23) and (3.78) indicates that, while in the slip zone towards the *trailing* edge of the contact (i.e. at $x > 0$) the shear stress is opposed to the slip direction, it will be in the same sense in the slipping region that occurs towards the *leading* edge, that is, at $x < -b$. This is clearly in violation of eqn (11.10b); the friction

force and the direction of slip must be in opposing senses. This observation
leads to the idea of displacing the sticking zone, from its symmetrical static
position at the centre of the contact, towards the leading edge, as indicated
in Fig. 11.4(a). The shear stress component $\tau''(x)$ has here moved by distance d so that slipping at the leading edge is just suppressed. The tangential

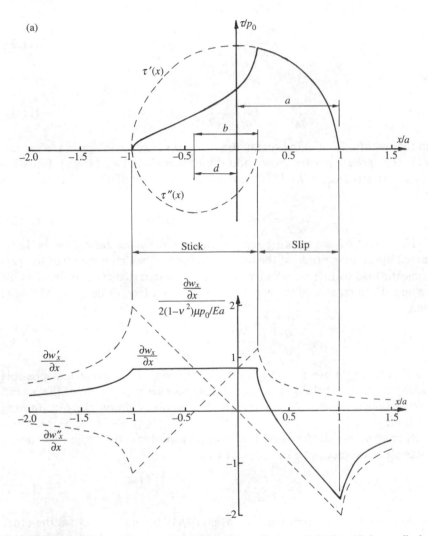

Fig. 11.4   Tractive rolling contact of similar cylinders; 1 is the driving cylinder
and 2 the driven. The forces shown are those experienced by 1. (a) Tangential
tractions: the net surface traction $\tau(x)$ is made up of the sum of $\tau'(x)$ and $\tau''(x)$;
(b) surface tangential strains, plotted non-dimensionally. Both are drawn for the
case $b/a = 0.6$ or $F/\mu W = 0.64$.

traction $\tau(x)$ is now made up of the superposition of $\tau'(x)$ and $\tau''(x)$ where the first is given by eqn (3.78) and the second by

$$\tau''(x) = \mu p_0 \frac{b}{a} \left\{ 1 - \frac{(x+d)^2}{b^2} \right\}^{1/2}. \tag{11.26}$$

The corresponding displacement gradients are thus

$$\frac{\partial w_x'}{\partial x} = 2 \frac{(1-v^2)}{Ea} \mu p_0 x \tag{11.27}$$

and

$$\frac{\partial w_x''}{\partial x} = -2 \frac{b}{a} \frac{(1-v^2)}{Eb} \mu p_0 (x+d) \tag{11.28}$$

so that within the sticking region $\partial w_x / \partial x = \partial w_x' / \partial x + \partial w_x'' / \partial x$ is a constant. These surface strains are shown non-dimensionally in Fig. 11.4(b). Equation (11.25) is satisfied, with $\dot{s} = 0$, providing the value of the creep ratio ξ is given by

$$\xi = -4(1-v^2)\mu p_0 d / Ea. \tag{11.29}$$

The extent of the sticking region, measured by the ratio b/a, is determined by the magnitude of the tangential force F, which is required to overcome the load torque on cylinder 2, when compared to its maximum possible value μW. In an analogous way to the static case (eqn (3.86)), we can write that

$$\frac{b}{a} = 1 - \frac{d}{a} = \left\{ 1 - \frac{F}{\mu W} \right\}^{1/2}. \tag{11.30}$$

As F is made larger (i.e. a greater proportion of its maximum possible value μW) so the sticking region becomes a smaller proportion of the overall contact length. When F reaches μW the whole junction slips in limiting friction.

By making use of the usual Hertzian relationships, the creep ratio under these circumstances can be expressed as

$$\xi = -\frac{\mu a}{R} \left\{ 1 - \left(1 - \frac{F}{\mu W} \right)^{1/2} \right\}. \tag{11.31}$$

Under conditions of high friction, when $F/\mu W$ is small, we are at the other extreme condition and slip at the trailing edge vanishes; the distribution of traction stress approaches a limit given by

$$\tau'(x) = \frac{p_0}{2} \frac{a+x}{(a^2 - x^2)^{1/2}} \frac{F}{W} \tag{11.32}$$

with the corresponding creep ratio

$$\xi = \frac{aF}{2RW}.$$ (11.33)

In deriving these results we have made the idealization that the two cylinders are elastically similar. When this is not the case, additional tractions will be generated (see, for example, Kalker (1990), However, if the elastic constants of bodies 1 and 2 are not widely different, reasonably realistic numerical predictions can be made by replacing $(1 - v^2)/E$ by $1/2E^*$ in the equations of this section.

11.1.4   Rolling contact of spheres

The three-dimensional problem which is analogous to that of two cylinders rolling together is that of two spheres in contact. Once again we can simplify this problem by supposing that both have the same elastic constants, and that the normal pressure and size of the circular contact area can be evaluated from the appropriate Hertz analysis. Suppose the total normal load is W, the maximum Hertz pressure p_0, and the contact radius a. By analogy with the argument of the previous section we should expect a region of stick at the leading edge, and a region of slipping contact towards the trailing edge. The analysis can be facilitated by thinking of the circular contact spot to be split into a series of narrow strips each parallel to the rolling direction, as indicated in Fig. 11.5

If we assume negligible interaction between adjacent strips then we can apply the argument of Section 11.1.3 to each. For a typical strip, located at a distance y from the Ox-axis, the contact length will be $2a'$, where

$$a' = a\left(1 - \frac{y^2}{a^2}\right)^{1/2}$$ (11.34)

and the pressure $p'(x)$ on this strip will be distributed as

$$p'(x) = p_0'\left(1 - \frac{x^2}{a'^2}\right)^{1/2} \quad \text{where} \quad \frac{p_0'}{p_0} = \frac{a'}{a} = \left(1 - \frac{y^2}{a^2}\right)^{1/2}.$$ (11.35)

Treating this narrow strip as the contact between two cylindrical bodies we should expect a traction distribution as indicated in Fig. 11.5. The sticking length is $2b'$ and the centre of this zone of the contact is located by distance d'. The creep ratio ξ is given by eqn (11.29) as

$$\xi = -\frac{4(1 - v^2)}{Ea'}\mu p_0'd',$$

that is

(a)

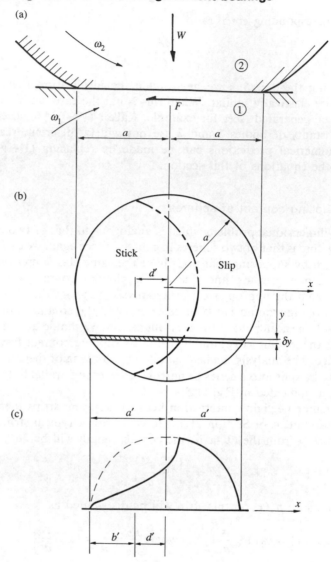

(b)

(c)

Fig. 11.5 Rolling contact between similar spheres. (a) At the centre of the contact spot has radius a. (b) In plan view the boundary of the sticking region is shown by the chained line. (c) Along a strip distance y from the centre line the distribution of traction is as shown.

$$\xi = -\frac{4(1-v^2)}{Ea}\mu p_0 d'. \tag{11.36}$$

Since ξ must have the *same* value for all such strips within the contact, it follows that d' is similarly constant. This means that the curve separating the stick and slip regions is just the reflection of the rim of the contact in the line $x = -d'$, and so gives rise to a characteristic 'lemon-shaped' sticking zone. The numerical value of d' depends on how large the total applied traction F is compared to its maximum possible value μW.

If W' is the load on our typical strip, then

$$W' = \left\{ \int_{-a'}^{+a'} p' \, dx \right\} \delta y = \frac{\pi a' p_0'}{2} \delta y \tag{11.37}$$

so that, substituting from eqn (11.35),

$$W' = \frac{\pi a p_0}{2} \left(1 - \frac{y^2}{a^2} \right) \delta y. \tag{11.38}$$

But from eqn (11.30) if F' is the traction experienced by the strip in question,

$$\frac{F'}{\mu W'} = 1 - \left(1 - \frac{d'}{a'} \right)^2. \tag{11.39}$$

The *total* traction F is made up by accounting for all such strips, thus

$$F = \frac{\mu \pi a p_0}{2} \int_{-a}^{+a} \left(1 - \frac{y^2}{a^2} \right) \left\{ 1 - \left(1 - \frac{d'}{a'} \right)^2 \right\} dy. \tag{11.40}$$

where a' is obtained from eqn (11.34).

For a Hertzian contact between spheres we have, from eqn (3.53),

$$p_0 = \frac{3W}{2\pi a^2},$$

so that

$$\frac{F}{\mu W} = \frac{3}{4a} \int_{-a}^{+a} \left(1 - \frac{y^2}{a^2} \right) \left\{ 1 - \left(1 - \frac{d'}{a'} \right)^2 \right\} dy. \tag{11.41}$$

This equation can be integrated to give

$$\frac{F}{\mu W} = 1 + \frac{3d'}{2a} \arcsin \sqrt{1 - (d'/a)^2} - \frac{1}{2} \{ 2 + (d'/a)^2 \} \sqrt{1 - (d'/a)^2}. \tag{11.42}$$

This relation between the extent of the slipping zone measured by d'/a and the ratio $F/\mu W$ is plotted in Fig. 11.6. As the ratio of the applied

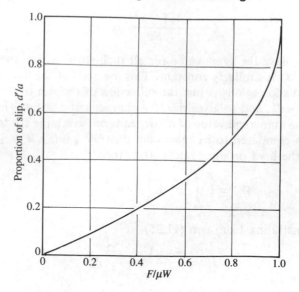

Fig. 11.6 Growth of the slipping zone between similar spheres measured by the ratio d'/a in tractive contact as a function of the traction coefficient $F/\mu W$.

traction to the maximum traction that the junction can tolerate without gross movement increases, so the relative extent of the slipping zone grows. The ratio $F/\mu W$ is known as the *traction coefficient*.

11.1.5 *Conformal contacts; Heathcote slip*

The case of two cylinders in contact is an example of non-conformal contact and is characteristic of many tribological pairs including roller bearings. However, in ball bearings, the balls roll along a closely conforming groove, and the idealizations of non-conforming contacts are no longer strictly tenable. This difficulty has already been alluded to in Section 11.1. If we think, initially at least, of the ball and the conforming track as both being rigid but fitting closely together, as shown in Fig. 11.7, it is clear that the velocity U_p of some point P within the contact on the freely rolling ball is dependent on its height z above the base of the track, so that

$$U_P = U_C - (R - z)\omega, \tag{11.43}$$

where U_C is the velocity of C, the centre of the ball whose radius is R and rate of rotation is ω. Provided the angle subtended by the contact arc is reasonably small, we can replace z by $y^2/2R$, so that

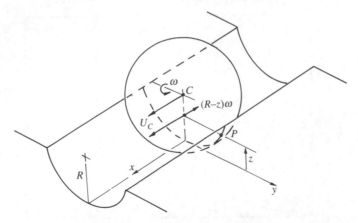

Fig. 11.7 A ball of radius R rolling in a conforming groove illustrating Heathcote slip. U_C is the velocity of the centre of the ball which rotates at ω. If there is to be no slip at P then $U_C = (R - z)\omega$.

$$U_P = U_C - \left\{ R - \frac{y^2}{2R} \right\} \omega. \tag{11.44}$$

Clearly, U_C and R have unique values, so there can only be one pair of symmetrically placed points, where the y-coordinate has a particular magnitude, which will allow U_P to be zero. At every other point on the contact there must be some slip between the ball and its track. As there is no net traction, this slip must be in opposite directions on either side of the no-slip points.

This form of slip was first described by Heathcote (1921) and is generally known by his name. He deduced, neglecting any effects of elastic deformation, that the rotation of the ball will be resisted by a frictional moment M given by

$$M = 0.08\mu b^2 W / R \tag{11.45}$$

where $2b$ is the transverse width of the contact. However, more often than not, the elastic compliance of the surfaces is *not* negligible; in these circumstances the situation can be analysed by a form of the strip theory already developed. The resulting retarding moment M depends on the conformity Γ of the contact which is defined by the expression

$$\Gamma = \frac{b^2 E^*}{4\mu p_0 R}, \tag{11.46}$$

the contact half-width b and the peak pressure p_0 are given by the Hertz theory. E^* is the contact modulus and μ the coefficient of friction at the

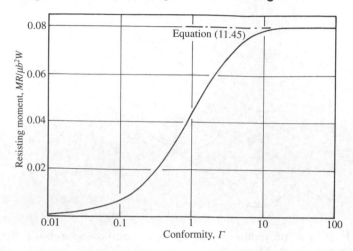

Fig. 11.8 Resisting moment M on a ball in a conforming groove. Equation (11.45) is shown as the chained line.

interface. The variation of M with Γ is shown in Fig. 11.8. When conformity of the surfaces is close (or the coefficient of friction small) Γ is large, and the moment M approaches the result for complete slip given by eqn (11.45).

11.2 Rolling-element bearings

11.2.1 *Nomenclature*

The purpose of a ball or roller bearing is to provide close relative positioning of two loaded components yet allow rotational motion between them. There are many detailed variations in design but bearings can generally be classified according to whether they support a radial load, a thrust load, or a combination of both. The most widespread type of radial ball bearing is the deep groove ball bearing (or 'Conrad' bearing) in which the depth of the tracks in the races is approximately one-quarter of the ball diameter. As well as carrying a radial load, bearings of this design will also sustain modest thrust loads. The essential parts of such a bearing are shown in Fig. 11.9(a); this figure also illustrates various other designs of rolling-element bearings.

The inner ring or *race* is carried by the rotating shaft and the outer race, mounted in the machine casing or bearing pedestal, is often stationary. The

(a)

(b)

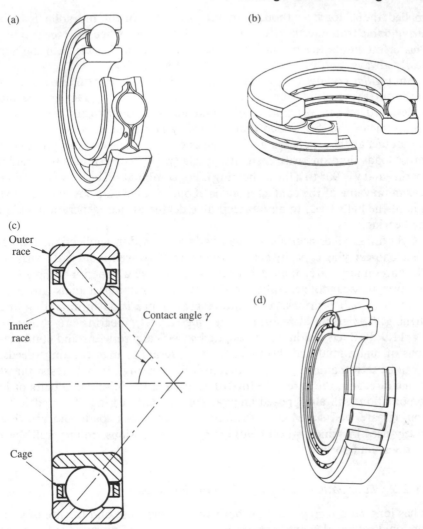

(c)

Outer race

Inner race

Contact angle γ

Cage

(d)

Fig. 11.9 Component parts of (a) a typical deep-groove ball bearing; (b) ball thrust bearing; (c) angular contact ball bearing; and (d) tapered roller bearing.

cage prevents contact between the rolling elements and thus reduces friction, wear, and noise from the regions where severe sliding would occur.

The simplest form of ball *thrust* bearing is shown in Fig. 11.9(b). A single row of balls set in a cage runs in two similar grooves formed in the stationary and rotating races. These grooves are usually shallower than those in a radial bearing. Thrust bearings are not suited to carry any significant applied radial loads, although an intrinsic component of radial load is unavoidable from inertia effects in the balls and cage when the race rotates. This effect limits the maximum speed of operation. Thrust bearings are nearly always used in conjunction with radial bearings.

An *angular contact* ball bearing, Fig. 11.9(c), is made with one shoulder of the outer ring counter-bored almost to the bottom of the race groove and is designed to support both radial and unidirectional thrust loads. The magnitude of the thrust load that can safely be sustained is dependent on the contact angle γ. A bearing with a large contact angle can sustain heavier thrust loads than one with a smaller angle ($\gamma = 0°$ corresponds to a radial bearing and $\gamma = 90°$ to a thrust bearing). For combination bearings a feasible maximum value of the contact angle is about 40°; beyond this the excessive spin of the balls leads to an unacceptable degree of heat generation within the bearing.

Cylindrical roller bearings are generally only used in applications requiring a load-carrying capacity beyond the range of conventional ball bearings. They are much stiffer than ball bearings and have a superior fatigue life; however, they cannot generally sustain any significant thrust loads, are more costly than ball bearings, and require greater precision in mounting and aligning. The essential features of a *tapered roller* bearing are shown in Fig. 11.9(d). Such bearings are designed specifically to withstand combinations of high radial and thrust loads at moderate to high rotating speeds. It can be seen that a single-row tapered roller bearing can only resist thrust in one direction; they are therefore often used either in opposed pairs or in combination with an opposed angular contact ball bearing. Tapered roller bearings are well suited to withstand repeated shock loads and are thus widely used in vehicle wheel bearings and transmissions, rolling mill shaft bearings, and so on.

11.2.2 *Lubrication of rolling-element bearings*

It has long been recognized that both the performance and the life of conventional rolling-element bearings is much reduced if no suitable lubricant is provided. This is despite the fact that for many years it was thought that the very high surface pressures and non-conforming contact between the balls or rollers and the bearing races prevented the formation of any sort of fluid film at all. It is now accepted that such films do exist and indeed are vital in preventing premature surface failure by scuffing or adhesive plastic flow. Their formation depends on both the elasticity of the surfaces and the pressure–viscosity characteristic of the lubricant; in other words on elasto-hydrodynamic effects (see Section 8.4). Only a very small quantity of lubricant is actually required to provide satisfactory surface protection, and in many applications rolling bearings can be packed at installation with grease which contains only small amounts of oil and then permanently sealed; such bearings will often operate satisfactorily for almost indefinite lengths of time. In other applications, especially where the lubricant is also required to act as a coolant, a more generous supply may be necessary.

The elasto-hydrodynamic fluid film in rolling bearings is very thin, but if it is to prevent asperity contacts between the rolling surfaces it is thought that it must still be of a magnitude comparable with their mean surface roughness. The extent by which the film thickness exceeds the surface roughnesses is conventionally indicated by the film parameter Λ, the lamda ratio defined as the ratio of the minimum film thickness to the mean component surface roughness (see Section 9.3). Conventionally, for safety Λ should be greater than 3. The final polishing process in the manufacture of the balls or rollers leaves them with an R_a surface roughness value of typically 0.05 micron, while the races are lapped to about $Ra = 0.175$ micron. This means that a typical elasto-hydrodynamic film thickness should be in excess of approximately 0.55 micron. The methods of Section 8.4 can be used to estimate elasto-hydrodynamic film thicknesses, and thus we need to know the load W_{max} on the most heavily loaded ball in the bearing. Although one or both of the races is likely to be rotating, it is usually (though certainly not always) at a speed insufficient to cause significant centrifugal or gyroscopic forces, so that a quasi-static analysis is in many cases satisfactory. However, establishing accurately even this static load distribution is not straightforward (since it depends on the initial clearance in the bearing) and is really only feasible by finite element or similar numerical methods. This procedure is rarely justified: experience and experiment suggest that the following two simplifying idealizations are usually satisfactory. Firstly, it is assumed that rolling elements above the centre of the bearing carry no load, and secondly, that the radial load on any element is proportional to $\cos\theta$, where θ is the angle between the individual contact load W_i and the line of the total applied load W, as shown in Fig. 11.10. Generalizing the case illustrated here to N balls or rollers, we can write that

$$W_i = W_{max} \cos \theta_i = W_{max} \cos \left\{ \frac{2\pi i}{N} \right\} \quad \text{where } |i| < N/4$$

and

$$W_i = 0 \quad \text{for } |i| \geqslant N/4, \tag{11.47}$$

where W_{max} is the load on the element currently at the lowest point and so carrying the greatest load. The vertical component of the load at each contact point is equal to

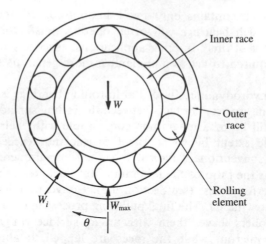

Fig. 11.10 Schematic diagram of a rolling element bearing. The total radial load on the bearing is W and W_i is the radial load on one of the balls or rollers.

$$W_i \cos\left\{\frac{2\pi i}{N}\right\}, \tag{11.48}$$

so that vertical equilibrium leads to the relation

$$W = W_{\max} + 2W_{\max} \cos^2\left\{\frac{2\pi}{N}\right\} + 2W_{\max}\cos^2\left\{\frac{4\pi}{N}\right\} + \cdots. \tag{11.49}$$

To a good approximation, for ball bearings with practical values of N, we can often write that

$$W_{\max} \approx \frac{4W}{N}. \tag{11.50}$$

11.2.3 *Rolling-element bearing life*

Even if a ball bearing is correctly specified, fitted, lubricated, and protected from abrasive or corrosive contamination it will eventually fail to operate satisfactorily, usually because of the formation of pits or spalls on the surfaces of the bearing races. These are a result of fatigue induced by the repeated cyclical stresses generated as the balls rotate. Fatigue damage often initiates within the bulk of the material at an inclusion or structural defect in the steel and then propagates to the free surface. The presence of such a crack, visible on the surface, is usually taken as the criterion of bearing failure. Since there will always be a variation in the magnitude and distribution of such defects there will be a corresponding spread in the material

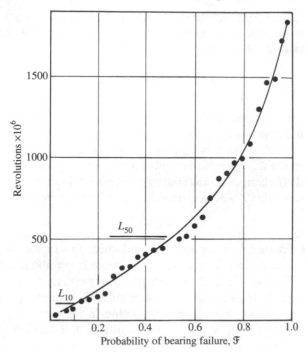

Fig. 11.11   Ball bearing fatigue life distribution. $\mathcal{F}$ is the probability of failure. L_{10} and L_{50} are the lives after which 10% and 50% of a large batch of bearings might be expected to have failed under these operating conditions (from Harris 1984).

resistance to fatigue; thus in a group of bearings, all subjected to identical working histories, there is likely to be a variation in the time to failure. This means that the specification of bearing life must be treated statistically.

The predominant factor in determining the fatigue life of ball bearings is the load that they carry. A typical distribution of the fatigue life for a population of identical bearings operated under nominally identical conditions is given in Fig. 11.11, in which $\mathcal{F}$ is the probability of bearing failure. The figure shows that the number of revolutions that a bearing can complete with 100% probability of survival is zero (i.e. $\mathcal{F} = 0$), and equally well that no bearing can be expected to have an infinite life. Bearing manufacturers conventionally express the rated fatigue life of a bearing by the L_{10} value, that is, the period, expressed either as number of revolutions or hours of running, after which it might be expected that 10% of a large batch of identical bearings, running at the particular load conditions concerned, would have failed.

It has been suggested that the fatigue life of a homogeneous group of ball bearings are dispersed according to Weibull statistics (see, for example, Harris (1984)), in which case

$$\ln \frac{1}{1 - \mathfrak{F}} = \left\{ \frac{L_s}{L_{10}} \right\}^e \ln \frac{1}{0.9} \qquad (11.51)$$

where $\mathfrak{F}$ is now the probability of failure at the particular life L_s, and e is the *dispersion exponent*, or *Weibull slope*. For ball bearings it has been found that $e \approx 10/9$. In this equation it is assumed that the only difference between the L_{10} operating conditions and those giving life L_s is one of applied load. If changes in materials or effective lubrication are made then the anticipated life L_s' will be related to L_s by the expression

$$L_s' = C_1 \times C_2 \times L_s \qquad (11.52)$$

where C_1 is a material factor and C_2 is a lubrication factor. Reductions in the impurity and inclusion content of steels can bring about an increase in the material factor, while C_2 is principally dependent on the lubrication film parameter Λ. If the numerical value of Λ falls to much below 2 then there may be as much as a six-fold reduction in the factor C_2 (for further details see Harris (1984) and Hamrock and Dowson (1981)).

11.3 Pneumatic tyres

11.3.1 *Elastic hysteresis in rolling*

As rolling proceeds so elements of material in each surface undergo a cycle of stress and a corresponding history of strain. As the ball or roller moves forward so it deforms the material ahead of it, doing mechanical work. The material behind the contact patch recovers elastically so pushing the rolling body forward. As the material moves through the Hertzian contact region it undergoes a cycle of reversed shear and compression, as illustrated in Fig. 11.12.

If we were dealing with a perfectly elastic material there would be no net energy expenditure and so no contribution to rolling friction from this source. However, no solid is perfectly elastic and during such a cycle of loading and unloading some energy is dissipated through 'internal friction' or elastic hysteresis. This loss is sometimes expressed as a fraction α of the maximum strain energy stored in the material during the deformation cycle. For metals operating within the elastic regime α is very small (less than 1%) but for polymers and elastomers it can be very much greater (up to more than 90% for some natural rubbers). Referring to Fig. 11.12, which shows the contact between a translating flat surface and a cylindrical roller, the

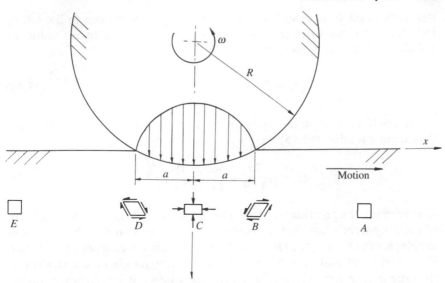

Fig. 11.12 Elastic deformation in rolling contact. The lower plane surface moves from left to right; it is loaded against the upper roller which rotates about its fixed centre. An element of the material originally undeformed at *A* undergoes shear at points such as *B* before moving into a region of pure compression at *C*. As it emerges from the deformation zone there is an element of reverse shear, at *D*, until it regains its original shape at *E*.

strain energy of material elements increases as they move towards the centre plane ($x = 0$) due to the work of compression done by the contact pressure working on the front half of the contact area. As the element moves out of the contact zone it does work against the contact pressure. Neglecting any effects of interfacial friction we can evaluate the work done by the pressure on the front half of the contact. During an increment of time dt the amount of work done d(work) will be given by

$$\text{d(work)} = \frac{U}{R} \times \int_0^a p(x)x\,dx \times dt. \tag{11.53}$$

The pressure distribution $p(x)$ can be obtained from the Hertz equations (3.41), so that eqn (11.53) becomes

$$\frac{\text{d(work)}}{dt} = \frac{2}{3\pi} Wa\omega \tag{11.54}$$

where W is the total normal load on the contact, a is the contact semi-width, and ω the rate of rotation of the cylinder. If α is the small proportion of this work which is dissipated by hysteresis, and so contributes to the rolling resistance, we can write that M_R, the moment required to maintain the cylinder in motion, is given by

$$M_R = \frac{\alpha}{\omega} \frac{d(work)}{dt} = \frac{2}{3\pi} \alpha W a. \tag{11.55}$$

The coefficient of rolling resistance μ_R can be defined as M_R/WR and thus from equation (11.55) we have

$$\mu_R = \alpha \frac{2a}{3\pi R} = \frac{4\alpha}{3\pi} \left\{ \frac{W}{\pi R E^*} \right\}^{1/2}, \tag{11.56}$$

so expressing the resistance of imperfectly elastic bodies to rolling in terms of their hysteresis loss factor. However, there are two drawbacks to this simple theory. Firstly, the factor α is not generally a true material constant: for a variety of materials it depends on the magnitude of the strain (i.e. the ratio a/R in eqn (11.56)). Secondly, the hysteresis loss factor in rolling cannot readily be identified with that found in a simple tension or compression test because the nature of the stress cycles in the two situations are so different. In rolling contact, as the element proceeds from A to E (in Fig. 11.12) the total strain energy is little changed although the principal axes are continuously rotating. The hysteresis loss in such circumstances is not easily predicted from uniaxial data in which no rotation of the principal axes occurs.

11.3.2   *Vehicle tyres*

Vehicle tyres are complex structures. They consist of a rubber carcass or casing reinforced by steel or polymeric fibres, and a tread which rolls, or perhaps on occasion slides, on the roadway surface. The limiting coefficient of friction between dry rubber and concrete can be close to, or even in excess of, unity. If it is known that the tyre will never operate under wet conditions there is no virtue in cutting or moulding a pattern on to the tread as this will only reduce the true area of contact. However, if such 'slicks' are run at speed in the presence of only a very thin film of water, then a hydro-dynamic film may be built up between the solid surfaces and the coefficient of friction falls disastrously (by as much as an order of magnitude); this is the phenomenon of *aquaplaning*. Moulding a pattern of grooves on the tyre surface allows the water drawn into the contact gap to be squeezed out and drained away from the 'footprint' of the tyre, so restoring, at least to some extent, the conditions of dry contact. In a good design, the effective value of μ can be brought back to about 0.4.

A *freely* rolling pneumatic tyre produces a coefficient of rolling friction of between 0.01 and 0.03. There is a relatively small amount of interfacial slip or creep and hysteresis losses in the walls of the tyre account for the greater part of the energy dissipation. Since the usual aim is to reduce such energy losses, it makes sense to use a rubber with an intrinsically low value of α for the tyre carcass. The situation is rather different under tractive rolling. When the vehicle is being either accelerated or braked the amount of interfacial slip increases; the coefficient of friction between the tyre surface and the roadway is very important in determining such things as the minimum stopping distance or the maximum rate of acceleration. In wet conditions, the contribution to the frictional resistance made by hysteresis losses, within and around the tyre footprint, are particularly significant. To maximize these, and so keep the stopping distance to a minimum, the rubber of the tread should have a hysteresis (or α value) greater than that of the tyre walls.

When a vehicle turns a corner there will always be some side slip within the area of contact because the plane of the wheel will be slightly skewed to the direction of motion of the vehicle. The angle α_s between these two is called the *slip angle*. This results in additional friction forces and moments. The total traction force F generally acts a little way behind the wheel centre, as illustrated in Fig. 11.13(a). This force system is equivalent to that of Fig. 11.13(b), where the forces bave been transferred to the wheel centre. The transverse force component F_y, which brings about the change in direction of the vehicle motion, is known as the *cornering force*, and the moment M_z represents the moment which must be applied by the steering mechanism (and hence by the driver at the steering wheel) to overcome the natural tendency of the wheels to continue in the same direction. It is thus the *self-aligning torque*. The relation between the cornering force F_y and the self-aligning torque M_z is illustrated in Fig 11.13(c). At modest rates of cornering, and thus lower values of the cornering force, the torque M_z increases with F_y and the steering retains a good 'feel'; however, at higher cornering speeds, as the point A in the curve is approached, there is little change in M_z as the rate of comering increases and the driver may be dangerously unaware of how close to the limit of adhesion the vehicle actually is.

The behaviour of a vehicle tyre under these circumstances is extremely complex — not least because of its very non-uniform construction which does not lend itself to the sort of analytical treatment that can be developed for continuous, isotropic solids. Nevertheless, some simple, predominantly one-dimensional, models have been proposed which can account for the main features of the observed behaviour. A description of these is beyond the scope of this discussion and the interested reader is referred to the relevant sections of the references below.

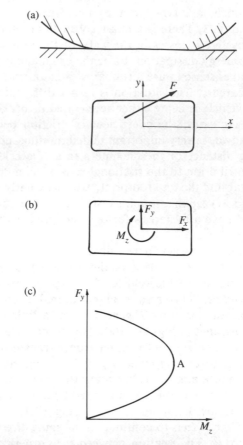

Fig. 11.13 Traction and frictional forces at a tyre–road contact: (a) the total force F acts behind the centre of the contact patch; (b) its effect is equal to the sum of F_x, the traction force, F_y, the cornering force, and M_z, the self-aligning torque; (c) variation of M_z with cornering force F_y.

11.4   Repeated loading; shakedown in rolling

Shakedown in repeated loading is the process whereby plastic deformation in the first few cycles of load leads to a steady cyclic state which lies within the elastic limit. The maximum load for which shakedown occurs is called the *shakedown limit*. In two-dimensional or line contacts there are two processes that can contribute to this phenomenon. Firstly, protective residual stresses introduced by the initial plastic flow can act to inhibit further plasticity in the steady state, and secondly, strain hardening of the

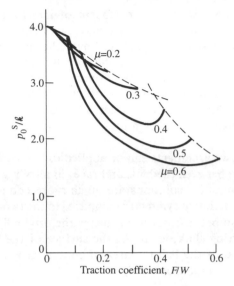

Fig. 11.14 Shakedown map for partial slip in a rolling contact involving elastic–perfectly-plastic material of flow shear stress k. The shakedown maximum Hertz pressure is p_0^S (from Johnson 1990).

material can raise its elastic limit. The application of these ideas to a sliding line contact has been discussed in Chapter 3.

When rolling takes place at a cylindrical contact in the presence of a tangential force F which is less than the value of limiting friction μW (as for example in the driving wheel of a vehicle or in some other forms of traction drive) we have seen that 'microslip' occurs over part of the contact area, while there is no slip, but rather 'stick', over the remainder. The shakedown limits for this form of rolling contact can be plotted on axes of the maximum Hertz pressure p_0^S versus the value of the traction coefficient, that is, the ratio F/W, producing a 'shakedown map' as in Fig. 11.14. Provided that the coefficient of friction between the two solid surfaces is below some critical value (of about 0.3) both first yield and subsequent shakedown in repeated loading are governed by subsurface stresses and the results are not very different from those for complete slip; these conditions lie on the dotted line in Fig. 11.14 and are taken from curve D of Fig. 3.23. The fact that there is little difference between the shakedown pressures in sliding and rolling contact under these conditions is not surprising since the *subsurface* stresses are not likely to be much influenced by the distribution of *surface* tractions.

On the other hand, when the coefficient of interfacial friction is high (in particular greater than about 0.3) shakedown behaviour is significantly influenced by the existence of partial slip at the interface. Figure 11.14 illustrates the damaging effect, that is, the reduction in the shakedown limit, that can be brought about even when the overall traction coefficient at which the contact is operating is relatively modest. For example, if $F/W = 0.3$ and $\mu = 0.2$ then shakedown will occur at values of p_0 below about $3k$. However, if μ increases to 0.6 then the value of p_0^S is reduced to rather less than $2k$.

Similar qualitative arguments can be applied to loaded point contacts (as occur for example between the balls and races in a rolling-element bearing) but here the quantitative solutions are much more complex. This is both because of the reduction in symmetry compared to the two-dimensional case of rollers, and also because in this geometry the originally flat surface will not remain so during shakedown. As the surface of the half-space begins to conform to that of the ball the area of contact grows and so the stress levels inevitably fall (Johnson 1990).

11.5 Further reading

Bentall, R. H. and Johnson, K. L. (1967) Slip in the rolling contact of two dissimilar elastic rollers. *International Journal of Mechanical Sciences*, **9**, 389–404.

Bidwell, J. B. (ed.) (1962) *Rolling contact phenomena*, Elsevier, New York.

Hamrock, B. J. and Dowson, D. (1981) *Ball bearing lubrication*. John Wiley, New York.

Harris, T. A. (1984) *Rolling bearing analysis*, John Wiley, New York.

Heathcote, H. L. (1921) The ball bearing: in the making, on test and in service. *Proceedings of the Institute of Automobile Engineers*, **15**, 569.

Hills, D. A., Nowell, D., and Sackfield, A. (1993) *Mechanics of elastic contacts*. Butterworth-Heinemann, Oxford.

Johnson, K. L. (1985) *Contact mechanics*. Cambridge University Press.

Johnson, K. L. (1990) A graphical approach to shakedown in rolling contact. In *Applied stress analysis* (ed. T. H. Hyde and E. Ollerton) Elsevier, Amsterdam, pp. 263–74.

Kalker, J. J. (1990) *Three-dimensional elastic bodies in rolling contact*. Kluwer Academic Publishers, Dordrecht.

Moore, D. F. (1975) *The friction of pneumatic tyres*. Elsevier, Amsterdam.

Nisbet, T. S. (1974) *Rolling bearings*. Engineering design Guides, Oxford University Press.

Peck, H. (1971) *Ball and parallel roller bearings: design application*. Pitman Publishing, London.

Weibull, W. (1949). A statistical representation of fatigue failures in solids. *Acta Polytechnica*, Mechanical Engineering Series 1, R.S.A.E.E No 9.49.

Problems

(Unless otherwise stated material properties can be taken from Appendix 4)

Chapter 2

2.1   A cylindrical turned surface is produced with a tool having a tip radius R. The feed rate is f per revolution. Show that if $R \gg f$ then the resulting surface will have the roughness parameters R_a and R_q measured along a generator of the cylinder of approximately

$$R_a \approx \frac{f^2}{36R} \quad \text{and} \quad R_q \approx \frac{f^2}{26.8R}.$$

Hint: If $R \gg f$ then the circular arcs of the surface profile can be adequately described by parabolae.

2.2   A rough surface has the profile illustrated in Fig. 1. Sketch the form of the probability density function and evaluate the R_q in terms of the dimension a. What are the values of the skewness and kurtosis of this surface?

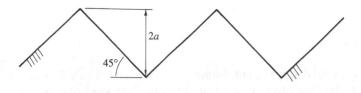

Fig. 1

2.3   On a particular rough surface the statistical distribution of asperity heights $\phi(z)$ is linear, so that

$$\phi(z) = \frac{2}{a^2}(a - z).$$

Evaluate the standard deviation σ, the skewness Sk, and kurtosis K for this surface.

2.4 Find the values of R_a, R_q, Sk, and K for a surface whose profile is given by $z(x) = a\cos(\pi x/a)$.

Chapter 3

3.1 Describe how the contact pressure between two metal cylinders, having their axes parallel, varies as they are pressed together by a gradually increasing load up to a value which results in appreciable plastic deformation.

3.2 The pivot shown in Fig. 2 has a cylindrical tip with radius 5 mm. It rests on a flat surface and supports a mass of 200 kg. Both surfaces are of hardened steel and the allowable compressive stress at the point is 1600 MPa. What length is required for the pivot? Estimate the stress at a depth of 0.5 mm below the contact line.

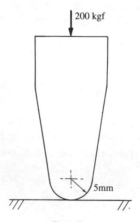

200 kgf

5mm

Fig. 2

3.3 Two steel discs, each of diameter 75 mm, and width 20 mm, and hardness HV 400, are pressed into contact with their axes parallel. It is found that a load of 40 kN can be applied before the discs show evidence of plastic deformation. If one of the discs is exchanged for a disc of light alloy of hardness HV 150, diameter 125 mm, and width 20 mm, calculate the maximum load which can now be applied without plastic deformation.

3.4 A pair of steel gear wheels have pitch circle radii of 20 mm and 60 mm respectively; the face width is 9 mm and the pressure angle 20°. If the power transmitted is 4 kW at a pinion (smaller wheel) speed of 1200 rpm, find:

(a) the normal contact force;

(b) the width of the band of contact when contact occurs at the pitch point; and

(c) the greatest maximum shear stress under these conditions.

3.5 A roller bearing, which has inner and outer race radii R_1 and R_2 respectively, is subjected to a load W caused by the weight of the rotor. The inner race is stationary and the outer race rotates. If the design is changed so that the outer race is stationary, to what value can the load be increased without the maximum stress at the most highly loaded point on the outer race exceeding the maximum stress at the most heavily loaded point on the inner race in the original design? Assume that stresses remain in the elastic range, and that the fraction of the load carried by a single roller remains the same.

3.6 Show that the area of contact between two equal cylinders pressed into contact with their axes at right angles is approximately circular.

The cylindrical wheels of a railway vehicle, each having a mass of 250 kg and radius 300 mm, support the vehicle through a soft spring. The wheel load is 50 kN. The rail head has a radius of curvature of 300 mm and may be taken to be rigidly supported by the track.

A certain length of the track has small corrugations on the rail surface of 50 mm pitch measured along the track. At what speed of the vehicle might resonant vertical oscillations of the wheel be expected.

3.7 A lathe attachment is to be designed to induce beneficial residual stresses by cold rolling as indicated in Fig. 3. The hardened steel roller has diameter 50 mm and a tip radius of 6 mm. For a steel shaft of diameter $d = 75$ mm determine the minimum roller force required to produce yielding at the surface where work hardening has raised the yield strength to 1400 MPa in compression.

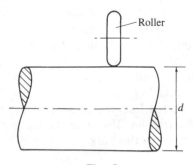

Fig. 3

3.8 The contact of two nominally flat rough surfaces is idealized by a model which comprises a rigid flat surface in contact with a deformable rough surface which has spherically capped asperities of uniform radius R

and random height z. The statistical distribution of asperity heights $\phi(z)$ is linear, that is,

$$\phi(z) = \frac{2}{h_0^2}(h_0 - z).$$

Assuming the deformation of the asperities to be perfectly elastic show that the total area of contact is proportional to $(\text{load})^{6/7}$, and find an expression for the average real contact pressure.

3.9 (a) A smooth surface is brought into contact with a surface having a random roughness. Outline the steps in the argument which shows that the real area of contact is approximately proportional to the load, independently of the mode of deformation of each of the asperities.

(b) The electrical conductance of two metallic solids which touch at a small circular area of radius a is $2ak$ where k is the electrical conductivity of the material. Show that the conductance of the two surfaces in (a) is proportional to the load pressing them together.

(c) The two surfaces in (a) are in sliding contact. The shear stress at each individual point of contact depends on the normal pressure at that point. Show that the total force of sliding friction will be proportional to the total normal load independently of the relation between the shear stress and normal pressure.

3.10 A steel disc of radius 40 mm and a brass disc of radius 60 mm rotate about parallel axes but in opposite senses with angular speeds of 200 rpm and 400 rpm respectively. The discs are 20 mm wide and pressed together with a load of 5 kN.

(a) Estimate the width of the contact zone between them.

(b) If the coefficient of dry friction between steel and brass is 0.3 estimate the maximum surface temperature above the bulk.

3.11 A vehicle of mass 1000 kg has two disc brakes. Each disc is nipped between two square pads 50 mm × 50 mm at an effective radius equal to one third of the radius of the road wheels. When the vehicle is travelling at 20 m s^{-1} the brakes are applied to produce a retardation of 10 m s^{-2}. Using the data given below and in Fig. 3.30 show that, shortly after the brakes are applied, the heat conducted into the pads is a very small fraction of the total heat generated by friction. Estimate the mean 'flash temperature' of the surface of the pad above the bulk.

		Discs	Pads
Conductivity	(W/m deg C)	54	0.8
Diffusivity	(m^2/s)	15×10^{-6}	3.2×10^{-3}

3.12   A railway locomotive has conical wheels of mean rolling radius 0.5 m which run on rails with a mean transverse radius of curvature of 0.5 m. The load carried by each wheel is 100 kN. Find the shape and dimensions of the contact patch.

The locomotive is hauling a train at a steady speed of $10\,\mathrm{m\,s^{-1}}$ when one of the wheels slips in such a way that its angular velocity doubles while the forward velocity of its centre remains unchanged. Estimate the mean temperature rise in the contact region if the coefficient of friction between the wheel and the rail is 0.2.

3.13   Why does temperature play such an important role in the tribological behaviour of surfaces? Estimate the mean 'flash temperature' of the surface's of a set of bicycle brakes when the rider executes an emergency stop from a speed of $10\,\mathrm{m\,s^{-1}}$. Numerical data on material properties can be taken from problem 3.11.

Chapter 6

(Viscosity–temperature data for mineral oils can be found in Fig. 1.8.)
6.1   A weightless flat circular plate of outer diameter D and inner diameter d moves on a larger flat horizontal surface separated from it by a film of lubricant of uniform thickness h and viscosity η. Derive expressions for
(a) the force to move the plate linearly with velocity U; and
(b) the torque to rotate the plate about its axis with angular velocity Ω.
6.2   A weightless conical pivot of diameter D and semi-angle α rotates in a conical seat of the same angle at speed Ω, separated from it by a uniform oil film of thickness h and viscosity η. Derive an expression for the driving torque.
6.3   A shaft carries a 21 kN thrust load and rotates at 600 rpm supported by a circular hydrostatic pad bearing with an outside diameter of 75 mm and a recess diameter of 40 mm. A pump with constant delivery of $20 \times 10^{-6}\,\mathrm{m^3\,s^{-1}}$ is available. The oil temperature may vary between 43°C and 65°C. Select a SAE oil such that the film thickness will never be less than 0.04 mm. What will be the oil pressure in the recess? What is the torque absorbed by the bearing.
6.4   A square hydrostatic pad bearing with four pockets (see Fig. 6.10) of side 800 mm is to support a load of 112 kN. The land width is 100 mm. Calculate the pressure required and the flow capacity of the pump to maintain a film thickness of 0.1 mm with SAE 20 oil at 40°C. What is the power requirement for each pad if the supply pressure is twice the recess pressure?
6.5   Air at 20°C is used in a circular hydrostatic bearing to support a centrifuge weighing 4 kg and rotating at 75,000 rpm. If the air inlet pressure

is 70 kPa, the ratio of recess to outside diameter is 0.6, and the film thickness is 25 micron, what should be the bearing dimensions and the air flow rate in litre/min? Assume the air to be incompressible and use the viscosity data from Table 1.9. What is the torque for rotation?

6.6   A hydrostatic bearing pad consists of a long rectangular recess of width w and length L with lands of width d. The recess pressure is p_r. Show that if the contribution of the ends is neglected, the load W and leakage rate Q are, respectively

$$W = p_r L(w + d) \quad \text{and} \quad Q = p_r L h^3/6\eta d.$$

Suggest approximate formulae for a square recess ($w \times w$) surrounded by a land of width d.

6.7   A hydrostatic thrust bearing of a turbine generator is designed for a load of 24 kN. The outside diameter is 0.2 m and the diameter of the recess is 0.1 m. The film thickness is to be 0.1 mm.

(a) Determine the recess pressure and the volume flow required if the oil used is of viscosity 30 cP.

(b) Assuming that $p_r/p_s = 0.5$ determine the dimensions of a suitable capillary compensator.

(c) When the generator is running at 750 rpm find the power loss due to viscous friction and also that due to pumping, and estimate the temperature rise of the oil which has density 900 kg m^{-3} and specific heat 1.88 kJ kg^{-1}K^{-1}.

(d) Determine the stiffness of the bearing.

6.8   A hydrostatic bearing pad has an upthrust Cp_r with $C = 0.01$ m^2 and a leakage rate with a particular oil of $\alpha h^3 p_r$, where p_r is the recess pressure and α is a constant. It is supplied through a capillary restrictor from a pump giving a fixed pressure of 5 MPa. The bearing is designed to support a load of 20 kN with a film thickness of 40 μm. What is the stiffness? How much greater could the stiffness be with a different pump and restrictor?

6.9   A hydrostatic pad bearing consists of a circular recess of radius a surrounded by an annular and of outer radius b. The depth of the recess is large compared to the film thickness h over the face of the land. Oil is supplied to the recess at gauge pressure p_1. Show that the flow rate Q and load W supported by the pad can be expressed as

$$Q = \frac{Q^* h^3 p_1}{\eta} \quad \text{and} \quad W = W^* \pi b^2 p_1$$

where

$$Q^* = \frac{\pi}{6\ln(b/a)} \quad \text{and} \quad W^* = \frac{1 - a^2/b^2}{2\ln(b/a)}.$$

Why does this arrangement, of itself, not constitute a satisfactory design?

Three such pads located at the corners of an equilateral triangle of side c support a load. The recess of each pad is supplied with oil from a source at constant gauge pressure p_0 through a restrictor such that the flow of oil to each pad is $Q = k(p_0 - p_1)$. Find an expression for the couple required to tilt the assembly about a median line of the triangle through a small angle $\Delta\theta$.

6.10   A vertical machine component is located by two circular hydrostatic pad bearings, as described in problem 6.9, operating in opposition to one another. The recess of each pad is supplied from a source at constant gauge pressure p_0 through a restrictor, such that the flow of oil to each pad is equal to $k(p_0 - p_1)$ where k is a constant. With no load applied to the central member the gap over each land is h_0 and the recess pressure is $0.5p_0$. Obtain an expression for the magnitude of the horizontal force in terms of W^*p_0 (where the form of W^* is given in problem 6.9) which must be applied to reduce the smaller gap to $0.5h_0$. When this load is applied by what factor has the oil flow changed compared to that in the unloaded condition?

6.11   A hydrostatic bearing pad consists of a square recess of side $2a$ surrounded by a land of uniform width $b - a$. The depth of the recess is large compared with the film thickness h over the face of the land. Oil of viscosity η is supplied to the recess at gauge pressure p. The flow rate Q and the load W supported by the pad can be expressed as

$$Q = Q^* \frac{h^3 p}{\eta} \quad \text{and} \quad W = W^* 4b^2 p$$

where Q^* and W^* are factors which depend on the ratio a/b.

(a) By making sensible approximations obtain your own estimates of Q^* and W^*.

(b) It bas been found that satisfactory performance is obtained from bearings which are designed using the relationships

$$Q^* = \frac{3a/b + 1}{6(1 - a/b)} \quad \text{and} \quad W^* = \frac{(1 + a/b)^2}{4}.$$

Use these expressions to show how the pumping power absorbed by the bearing depends on the values of Q^* and W^* in the particular case in which the dimension b, the load W, and the film thickness h are all fixed. Show that for this power to be a minimum the bearing land should be approximately one-third of the overall bearing width.

(c) For a bearing of this design what are the differences between your estimates of Q^* and W^* from part (a) and the values from the expressions given in (b)? Comment briefly on the difference.

6.12 To maximize the stiffness of the bearing of problem 6.9 the restrictor is chosen so that the recess pressure is one-half of the supply pressure. In this case show that the specific pumping power H_p', that is, the pumping power per unit load, can be written as

$$H_p' = \frac{2h^3 W Q^*}{\pi^2 \eta b^4 W^{*2}}.$$

Now if a relative sliding velocity of magnitude U is imposed between the two elements of the bearing show that the *total* specific power H_t' can be written as

$$H_t' = \frac{2h^3 W Q^*}{\pi \eta b^4 W^{*2}} + \frac{\eta U^2 \gamma \pi b^2}{hW} \qquad \text{where} \quad \gamma = 1 - (a/b)^2,$$

and where the additional term represents the viscous power loss. Show that H_t' will be minimized when the pumping loss is one-third of the viscous loss, and that this occurs when

$$h^4 = \frac{\pi^3 \gamma \eta^2}{6 Q^*} \left\{ \frac{U b^3 W^*}{W} \right\}^2.$$

Chapter 7

(Viscosity–temperature data for mineral oils can be found in Fig. 1.8.)

7.1 The errors in the measurement of journal and bearing bore diameters by a micrometer can be of the same magnitude as the radial clearance itself. Where the friction torque at the bearing under no-load conditions can be readily measured the clearance c can be directly calculated on the basis of the Petrov equation. Do this for a bearing with nominal diameter 50 mm, length 40 mm, and speed 3600 rpm, in which the friction torque is 0.74 Nm and SAE 10 W oil is used at 65°C.

7.2 To confirm that the effects of gravity and inertia may be safely neglected consider an oil film of unit area and thickness 25 microns. Calculate the pressure required on one edge (a) to lift it and (b) to accelerate it at $300 \, \mathrm{m \, s^{-2}}$. Compare these with the typical pressures of 700 kPa–3.5 MPa developed in lubricating films.

7.3 The pad of an infinitely wide fixed-inclination pad bearing has the form of a step of height Δh whose segments of length $B/2$ are both parallel to the counterface. Find the relation between the load supported and the minimum film thickness when the sliding speed is U and the lubricant has constant viscosity η. Sketch the relations between:
(a) load per unit width W/L and the minimum film thickness h_0;
(b) $W h_0^2 / \eta U L B^2$ and $\Delta h / h_0$.

How does the maximum load capacity compare with that of a Michell pivoted pad bearing? Compare the behaviour as the load is increased with that of the Michell bearing, and explain which is the more desirable in practice.

7.4 Calculate the pressure distribution under a plane slider bearing with the inlet film thickness h_i, outlet film thickness h_o, and length B. Assume pressure is zero at both inlet and outlet. Find W/L, the load carried per unit length. Evaluate the dimensionless load $Wh_o^2/\eta ULB^2$ for the cases $h_i = 2h_o$, $h_i = 2.2h_o$, and $h_i = 2.5h_o$.

7.5 A tilting-pad slider bearing is to be designed to the following application:

Load per unit width $10\,\mathrm{kN\,m^{-1}}$
Speed $2\,\mathrm{m\,s^{-1}}$
Lubricant viscosity $0.01\,\mathrm{Pa\,s}$
Minimum film thickness $30\,\mu\mathrm{m}$.

Use the one-dimensional results of Section 7.3 and design for the maximum load-carrying capacity. What should be the length of the pad and where should it be pivoted? What is the nominal pressure? What will be the coefficient of sliding friction and the heat dissipated per unit width?

7.6 A $75\,\mathrm{mm} \times 75\,\mathrm{mm}$ full journal bearing supports a load of $4\,\mathrm{kN}$, $c/R = 0.0015$, and the speed of rotation is $150\,\mathrm{rpm}$. The temperature of the SAE 20 oil is maintained at $60°\mathrm{C}$. Estimate the frictional torque and power.

7.7 Repeat problem 7.6 except that the oil is SAE 50 and the speed of rotation is $600\,\mathrm{rpm}$.

7.8 The load on a $100\,\mathrm{mm}$ diameter full bearing of length $0.1\,\mathrm{m}$ is $9\,\mathrm{kN}$ and the speed of rotation is $320\,\mathrm{rpm}$; $c/R = 0.0011$ and the operating temperature is $100°\mathrm{C}$. The minimum clearance is to be $0.022\,\mathrm{mm}$. Select an oil that will satisfy these conditions. For this lubricant determine the frictional power loss, the maximum pressure, and the amount of end leakage.

7.9 A lightly loaded bearing has $125\,\mathrm{mm}$ diameter, and is $125\,\mathrm{mm}$ long. The radial clearance is $0.04\,\mathrm{mm}$, the load is $7\,\mathrm{kN}$, and the speed is $500\,\mathrm{rpm}$. For a coefficient of friction of 0.01 what is the average viscosity of the lubricant? If the temperature is $90°\mathrm{C}$ what SAE grade would be recommended?

7.10 A $360°$ journal bearing with diameter $50\,\mathrm{mm}$ and length $50\,\mathrm{mm}$ consumes $86\,\mathrm{W}$ in friction at an operating speed of $1432\,\mathrm{rpm}$. SAE 10 W oil is used and the film temperature is $80°\mathrm{C}$. The radial clearance is $0.025\,\mathrm{mm}$. Determine the load and the minimum film thickness.

7.11 A $180°$ partial journal bearing has an L/D ratio of 1, a $100\,\mathrm{mm}$ diameter, a radial clearance of $0.08\,\mathrm{mm}$, an operating speed of $600\,\mathrm{rpm}$ and an applied load of $7\,\mathrm{kN}$. If the minimum film thickness is to be no less than $0.03\,\mathrm{mm}$, find the required viscosity and the frictional torque.

7.12 A journal bearing is to support 100 N from a shaft rotating at 6000 rpm using air of viscosity 0.018 cP as the lubricant. The allowable unit loading is 35 kPa, the length-to-diameter ratio is 1.2:1, and the clearance ratio is 0.0008. Determine the dimensions, minimum film thickness, and torque required for rotation.

7.13 A bearing 50 mm in diameter by 25 mm long with a diametral clearance of 0.05 mm is loaded with 8 kN on a shaft rotating at 2400 rpm. If the minimum film thickness is to be 7.5 microns and the oil enters with a temperature of 50°C find the effective temperature of the oil film and suggest an appropriate SAE grade of lubricant. Repeat the calculation for inlet temperatures of 30°C and 70°C and for a load of 12 kN.

7.14 A full journal bearing is to support a load of 4.5 kN at a rotational speed of 500 rpm. It is suggested that a 'square' bearing with $L = D = 75$ mm could be used with a radial clearance of 0.06 mm lubricated with SAE 30 oil supplied at 50°C. Estimate the lubricant temperatures associated with this design.

7.15 A 50 mm × 50 mm bearing sustains a load of 25 kN; $c/R = 0.001$, $\mathfrak{N} = 40$ rps, and $\eta = 0.03$ Pa s. Determine the minimum film thickness and the frictional power loss for (a) a full 360° bearing, and for partial bearings subtending angles of (b) 180° and (c) 60°.

7.16 A simple viscous oil pump is shown in Fig. 4. It consists of a circular cylinder of radius R and length L rotating at speed Ω in a fixed housing. The clearance gap h_2 subtends an angle α at the centre of the cylinder; over the remainder of the circumference the clearance is h_1. Both h_1 and h_2 are small compared with R. Derive an expression relating the pressure rise across the pump to the net volumetric flow rate Q. The fluid can be taken to be incompressible and of constant viscosity η.

In a particular design the ratio $h_1:h_2$ is fixed at 5 and the flow rate is $\Omega R L h_2$. Determine the value of the angle α which will maximize the

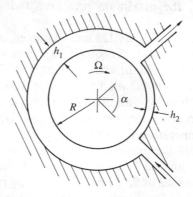

Fig. 4

pressure rise. Explain how a theoretical estimate of the efficiency of the pump might be made.

7.17 A full journal bearing of diameter D and length L has $L \gg D$. Show that if the Sommerfeld conditions are assumed the coefficient of friction of the bearing, defined as (cM^*/RW^*), has a minimum when the Sommerfeld number S is 0.132. Estimate the corresponding value of S if the bearing is square.

7.18 Explain why the film thickness in a journal bearing can be taken as $h = c + e\cos\theta$, where c is the radial clearance and e is the displacement of the centre of the shaft.

A lightly loaded 180° journal bearing runs with a very small eccentricity ratio ε such that $\varepsilon^2 \ll 1$. Show that Reynolds' equation is satisfied by a pressure distribution of the form

$$p = K\varepsilon(\sin\theta - \theta\cos\theta^*),$$

where θ^* is a constant whose significance should be explained.

Find the resultant force on the bearing expressed as the Sommerfeld number S when the inlet is at $\theta = 0$ and the outlet at $\theta = \pi$. Why is the Reynolds exit condition not applicable here? Sketch the bearing in its running configuration marking the positions of the load line, maximum pressure, and minimum film thickness.

7.19 The table below gives the calculated behaviour of a 'square' 180° journal bearing assuming that the oil film pressure is zero at entry A and is terminated by cavitation at point B, see Fig. 5. The minimum film thickness occurs at E located by the angle φ (degrees).

ε	0.1	0.2	0.3	0.4	0.5	0.6	0.7	0.8	0.9	0.95
φ	78	69	62	56	50	45	39	32	24	18
$\dfrac{W(c/R)^2}{\eta U}$	0.23	0.48	0.77	1.15	1.67	2.49	3.90	6.87	16.5	37.0
$\dfrac{M(c/R)^2}{\eta Uc}$	3.21	3.40	3.68	4.13	4.77	5.67	7.09	9.54	15.2	23.3

W is the load per unit length, M the frictional torque, c the radial clearance, R the radius, U the peripheral speed, and η the fluid viscosity.

(a) Explain what ε represents.

(b) Explain how the minimum film thickness for a given operating condition (load, speed, and oil viscosity) will vary with the choice of radial clearance for the bearing.

(c) A bearing is to be designed for a shaft of radius 15 mm carrying a load of 1 kN. The bearing should be 30 mm wide (i.e. *long*) and should use oil of viscosity 0.03 Pa s. Shaft speeds of 400–2000 rpm are expected.

What clearance do you recommend, and what will then be the minimum film thickness at maximum and minimum speeds? At what rate will heat be generated in the bearing?

(d) What will happen at starting and stopping?

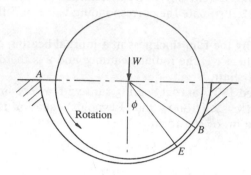

Fig. 5

7.20 A rigid sphere is constrained to move through an oil bath with velocity V towards a rigid plane. Show that, when the distance between the sphere and the plane is h_0, the resisting force is

$$6\pi\eta V \frac{R^2}{h_0}.$$

7.21 A light shaft is supported in two plain circular bearings of diameter 50 mm and length 12 mm. The bearings each have a radial clearance of 0.05 mm and are supplied with oil of viscosity 20 cP. The shaft is subject to a force of 900 N which rotates at a speed of 1500 rpm in the same direction as the shaft. Determine the motion of the journal centre when the shaft rotates at 1500 rpm, 3000 rpm, and 4500 rpm.

For each case, indicate whether the force vector lags or leads the line joining the bearing and journal centres.

7.22 The stiffness k of a journal bearing can be defined either as the load W divided by the deflection of the journal centre e, or as W divided by the component of e in the direction of the load. For both definitions derive k as a function of the eccentricity ratio ε for a short bearing. For the case when ε is small do the two stiffnesses increase or decrease with increasing ε?

7.23 A 125 mm diameter circular pad is supported on a flat platen by a film of SAE 30 oil at 70°C which is 0.025 mm thick. It is loaded by a force of 25 kN. In what period of time does the thickness decrease to (a) 1/10, (b) 1/100, and (c) 1/1000, of its original value?

Chapter 8

8.1   An air-lubricated journal bearing has the following specification: journal diameter 50 mm, radial clearance 25 microns, journal speed 22000 rpm, bearing length 50 mm, mean temperature of the air film 100°C, and entry and exit air pressure 106 kPa. Estimate the maximum steady safe load if the minimum safe film thickness is to be 10 microns; also estimate the associated power loss.

8.2   Use the half-Sommerfeld solution for the pressure distribution between a pair of rigid rollers to deduce the limiting film thickness at which the maximum pressure becomes infinite when the viscosity of the lubricant (a) obeys the Barus law $\eta = \eta_0 \exp \alpha p$, and (b) obeys the relation $\eta = \eta_0 (1 + Cp)^{16}$.

8.3   A roller bearing which supports a horizontal rotating shaft has an inner race of outside diameter 80 mm and contains 15 cylindrical rollers whose diameter and length are both 20 mm. The inner race rotates at 2000 rpm while the outer race is stationary. The bearing is made of steel and is lubricated by a mineral oil whose viscosity at atmospheric pressure and temperature is $\eta_0 = 7 \times 10^{-3}$ Pa s and pressure–viscosity index is $\alpha = 2 \times 10^{-8}$ Pa^{-1}. The bearing carries a vertical load of 20 kN. Identify the position in the bearing of the minimum lubricant film thickness between the rollers and the races and make estimates of its value (a) neglecting elastic deformation of the bearing elements and the influence of pressure on viscosity, and (b) taking both these effects into account.

8.4   In the elasto-hydrodynamic lubrication of highly elastic materials such as rubber the pressures in the lubricant film are suiticiently high to deform the material but not high enough to alter the lubricant viscosity. A long rubber cylinder of radius R slides over a rigid plane on a film of water. Write down the equations governing the geometry of the film and the development of the hydrodynamic pressure. What are the most appropriate boundary conditions? Investigate the dependence of the minimum film thickness on the load and the sliding speed.

In a series of experiments in which only the speed is varied the film thickness is found to vary as the 0.63 power of the speed. The experiments are then repeated, using a rubber of double the elastic modulus but with no other changes. By what factor is the film thickness changed?

A clamp is now released so that the cylinder is free to rotate. Suggest how the film thickness for a given translational speed might alter. Note that the surface deformation w_z of a semi-infinite elastic solid of modulus E and Poisson's ratio v under a pressure $p(s)$ is given by eqn (3.29).

8.5   Outline the principal experimental evidence for elasto-hydrodynamic action of rollers. By using Crook's approximation for the film shape:

$$h = h_0 + \frac{a^2}{R} \frac{2\sqrt{2}}{3} \left\{ \frac{|x|}{a} - 1 \right\}^{3/2}$$

show that h_0 is proportional to

$$(\alpha \eta_0 \bar{U})^{3/4} \quad R^{3/8} \left\{ \frac{E^*}{W} \right\}^{1/8}$$

where the symbols have their usual meaning.

8.6   Explain how the pressure distribution for lubrication by a fluid whose viscosity is a function of pressure may be determined from that for an isoviscous fluid used in the same configuration.

Fig. 6 shows the geometry of a pad bearing. The inlet region, of slope θ (θ is small), is of length a which is a small proportion of the overall bearing size. Over the region of constant film thickness $\bar{h}$ the pressure in the film is uniform. The lower, plane surface moves from left to right with constant velocity U. The contact is lubricated with a fluid whose viscosity η varies with pressure p according to $\eta = \eta_0 \exp(\alpha p)$ where η_0 and α are constants.

By considering conditions within the entry zone estimate the limiting film thickness $\bar{h}$ assuming that both surfaces are rigid and that the pressure reaches its maximum value at the end of the inlet region.

Show that the result may be written in the form

$$\bar{h}(\bar{h} + a\theta)^2 = 3\alpha \eta_0 U a^2 \theta.$$

If the group $\alpha \eta_0 U / a\theta^2$ is large then show that

$$\bar{h} \approx 1.44 a^{2/3} (\theta \alpha \eta_0 U)^{1/3}.$$

What happens at the downstream end of the parallel section of the pad?

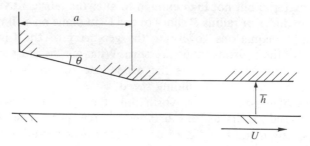

Fig. 6

8.7   Use the approximate pressure distribution $p = Kx/h^2$, where K is a constant, to obtain an expression for the lubrication between a rigid cylinder of radius R and a rigid plane with a lubricant whose viscosity η varies with pressure p according to

$$\eta = \eta_0 \exp(\alpha p),$$

and show that this predicts a minimum film thickness h_0 given by

$$h_0 = H_k(\alpha \eta_0 \bar{U})^{2/3} R^{1/3}$$

where $\bar{U}$ is the entraining velocity and H_k is a constant whose value should be found. Explain why this theory fails to explain all aspects of the actual physical situation.

Chapter 10

10.1 A rectangular box of weight W rests against an inclined plane along its edge A, which is horizontal, and is supported by a force P applied to a lever BCD which is hinged to the ground at D as shown in Fig. 7. $BC = 4CD$. The angle of friction for the surfaces at contact at A and C is $30°$. Show that slipping will start at A or C depending on whether the block is raised or lowered from this position, and determine the corresponding values of P.

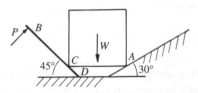

Fig. 7

10.2 Four equal thin rods, each of length a, are freely jointed together to form a rhombus. They are then hung with one diagonal vertical over a rough circular peg of radius $r(<a)$. If equilibrium is possible in the configuration in which the length of the horizontal diagonal is one-half that of the vertical diagonal, what is the minimum coefficient of friction at the peg?

10.3 A thrust rod carries pins of diameter $a/5$ with centres A and B distance a apart. The pins bear loosely in holes in blocks and these are guided by a pair of plane surfaces at right angles, the lines of movement of A and B meeting at O. A force F in the direction AO acts on the block at A and overcomes the force nF in the direction BO on the block at end B. At the instant when the angle BAO is $60°$ determine the value of n if the angle of friction between the block and the planes is $20°$,

(a) assuming frictionless pins, and

(b) assuming the same friction angle at the pins.

10.4 A light square peg A is a loose fit in a hole in a fixed plate B, as shown in Fig. 8. The coefficient of friction between the peg and the plate is μ. If a force P is applied to the peg as shown, prove that it will not slip provided that

$$\tan \theta > \frac{a}{\mu(c + \mu b)}$$

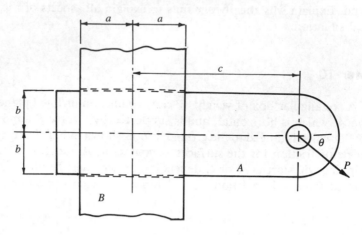

Fig. 8

10.5 A pulley having a diameter of 100 mm and a mass of 2.5 kg is supported loosely on a shaft having a diameter of 25 mm. Determine the torque required to cause it to rotate steadily. The coefficient of kinetic friction between the shaft and the pulley is 0.4. Also calculate the angle which the normal force at the point of contact makes with the horizontal. The shaft itself cannot rotate.

10.6 A single-disc clutch having both faces effective is to be used in an application where 45 kW are to be developed at 600 rpm. If the outside diameter is 230 mm and the inner diameter is 200 mm, what will be the actuating force and the average face pressure for a clutch material having a coefficient of friction of 0.3 for the case of assumed uniform face pressure? Repeat the calculation for the case in which the product of local face pressure and slipping velocity is constant.

10.7 A multiple-disc clutch baving four pairs of mating friction surfaces operates at a speed of 945 rpm. The maximum permissible magnitude of the axial operating force is 4 kN. The coefficient of friction is 0.15. The inside and outside radii of the friction surfaces are 25 mm and 50 mm respectively. If conditions are such that uniform wear prevails, determine (a) the torque capacity per pair of friction surfaces, (b) the total power capacity, (c) the maximum face pressure, and (d) the minimum face pressure.

Answers to problems

2.2 $a^2/3$, Sk $= 0$, and $K = 1.8$.
2.3 $\sigma = a/3\sqrt{2}$, Sk $= 0.566$, and $K = 2.4$.
2.4 $R_a = 2a/\pi$, $R_q = a/\sqrt{2}$, Sk $= 0$, $K = 1.8$.

3.2 5.6 mm; 210 MPa.
3.3 12.5 kN.
3.4 1.7 kN, 0.21 mm, 350 MPa.
3.5 WR_2/R_1.
3.6 16.4 m s^{-1}.
3.7 241 N.
3.8 $p = 0.345(h_0^2 WE^{*6}/NR^4)^{1/7}$.
3.10 0.62 mm, 132°C.
3.11 130°C approximately.
3.12 1112°C.

6.1 $\dfrac{\pi U \eta}{4h}(D^2 - d^2)$; $\dfrac{\pi \Omega \eta}{32h}(D^4 - d^4)$.

6.2 $\dfrac{\pi \Omega \eta D^4}{32h}(1 + \cot \alpha)$.

6.3 SAE 30; 8.35 MPa; 0.088 Nm at 43°C; 0.035 Nm at 65°C.
6.4 244 kPa, 0.7 l min^{-1}; 5.7 W.
6.5 17 mm, 10 mm, 3.6 l min^{-1}, 0.637 N mm.

6.6 $p_r(w + d)^2$; $\dfrac{p_r h^3}{6\eta d}(2w + d)$.

6.7 1.41 MPa; 0.37 mm radius by 10 mm long; 272 W; 6.2°C; 360 MN m^{-1}.
6.8 0.9 MN m^{-1}; 1.5 MN mm^{-1}.

6.9 $W^*Q^* \dfrac{3\pi c^2 b^2}{2} \dfrac{p_0 h^2}{\eta k} \left\{ 1 + \dfrac{Q^* h^3}{\eta k} \right\}^{-2} \Delta \theta$.

6.10 0.660 $W^* p_0$; falls to 0.88 of previous value.
6.11 Power depends on Q^*/W^{*2}.

7.1 $c = 26$ microns.

7.3 $Wh_0^2/\eta ULB^2 = 3\Delta h/2h_0[(1 + \Delta h/h_0)^3 + 1]$.

7.4 0.159; 0.160; 0.158.

7.5 53 mm, 31 mm from leading edge, 189 kPa, 0.0026; 52.7 W m^{-1}.

7.6 0.33 Nm; 5.03 W.

7.7 2.22 Nm; 139 W.

7.8 SAE 70; 44 W; approximately 3 MPa; 0.17 l min^{-1}.

7.9 0.017 Pa s; SAE 40.

7.10 15.9 kN; 3.8 microns.

7.11 0.023 Pa s; 1.12 Nm.

7.12 59 × 49 mm dia.; 8 microns; 0.004 N m.

7.13 ~77°C; SAE 30; SAE 20; SAE 70.

7.14 Approximately 60°C.

7.15 12 microns, 474 W; 355 W; 207 W.

7.16 $\Delta p = \dfrac{12\eta WR^2}{h_2^2} \dfrac{\alpha(2\pi - \alpha)}{2\pi + 124\alpha}$; $\alpha = 29.6°$; efficiency $\approx \dfrac{Q\Delta p}{M\Omega}$.

7.17 0.35.

7.18 $S = 0.212/\varepsilon$.

7.19 12.7 and 7.3 microns; 1.7 and 29 W.

7.21 $\varepsilon \approx 0.25$; line of centres lags load, half-speed whirl, leads.

7.22 $\pi\eta R\omega L^3/4c^3\{1 + 2.31\varepsilon^2 + \ldots\}$ increases;
$\pi^2\eta R\omega L^3/16c^3\varepsilon\{1 + 1.81\varepsilon^2 + \ldots\}$ decreases.

7.23 ~5 s, 8.4 min, 14 hours.

8.1 214 N, 27 W.

8.2 $\frac{3}{2}(\alpha\eta_0 \bar{U})^{2/3}R^{1/3}$; $\frac{3}{2}[c(n - 1)\eta_0 \bar{U}]^{2/3}R^{1/3}$, with $n = 16$.

8.3 (a) 0.0057 micron; (b) 0.14 micron.

8.4 0.774.

8.7 $H_k = \left[\dfrac{3\sqrt{3}}{2\sqrt{2}}\right]^{2/3}$.

10.1 0.19 W, 0.09 W.

10.2 $5(0.1 - r/a)$.

10.3 0.840; 0.726.

10.5 0.455 Nm, 68.2°.

10.6 11.1 kN; 1.09 MPa.

10.7 22.5 Nm; 8.9 kW, 1.02 MPa; 0.51 MPa.

Appendix 1

Skewed cylinders in contact

The lower cylinder is of radius R_1 and the upper of R_2 as shown in Fig. A1.1. They touch at the point O. Ox_1y_1 is a set of Cartesian axes with Oy_1 along a generator of the lower cylinder, and Ox_2y_2 a second similar set with Oy_2 along a generator of the upper cylinder.

The two Oy-axes are inclined to each other at an angle θ. Close to the origin we can approximate the circular section of each cylinder by a

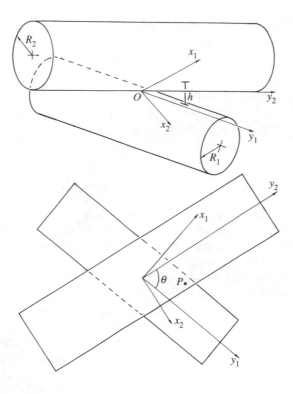

Fig. A1.1 The geometry of two circular cylinders in skewed contact.

parabolic profile and so write that the separation h of the two solid surfaces at the point P, which has coordinates (x_1, y_1) in the first coordinate system or (x_2, y_2) in the second, is given by

$$h \approx \frac{x_1^2}{2R_1} + \frac{x_2^2}{2R_2}. \tag{A1.1}$$

It is possible to choose a common set of axes (Oxy), in which Ox is inclined to the Ox_1-direction by the angle α, such that

$$h \approx Ax^2 + By^2 \tag{A1.2}$$

provided that A and B satisfy the conditions

$$B - A = \frac{1}{2} \left\{ \left(\frac{1}{R_1} \right)^2 + \left(\frac{1}{R_2} \right)^2 + \frac{2}{R_1 R_2} \cos 2\theta \right\}^{1/2}, \tag{A1.3}$$

$$B + A = \frac{1}{2} \left\{ \frac{1}{R_1} + \frac{1}{R_2} \right\} \tag{A1.4}$$

and α is given by the solution of

$$\frac{R_2}{R_1} \sin 2\alpha = \sin 2(\theta - \alpha). \tag{A1.5}$$

Equation (A1.2) can be written as

$$h = \frac{x^2}{2R'} + \frac{y^2}{2R''}, \tag{A1.6}$$

by defining

$$R' = 1/2A \quad \text{and} \quad R'' = 1/2B; \tag{A1.7}$$

R' and R'' are known as the principal radii of *relative* curvature.

It is evident from equation (A1.6) that contours of constant gap h between the undeformed surfaces will be *ellipses*, the lengths of whose axes are in the ratio $(R'/R'')^{1/2}$.

When a normal load W is applied, the point of contact spreads into an area, whose shape is not known with certainty in advance. However, it can be shown that the resultant contact patch remains elliptical, say with semi-axes a and b, and is such that the eccentricity, that is, the ratio b/a, is *independent* of the load and depends only on the ratio of R'/R''. For mildly elliptical contacts, that is, $A/B < 5$, the ratio b/a is given by

$$\frac{b}{a} \approx \left(\frac{A}{B} \right)^{2/3} \tag{A1.8}$$

We can then define an 'equivalent' radius R_e as

$$R_e = (R'' \times R')^{1/2} = \tfrac{1}{2}(AB)^{-1/2}. \qquad (A1.9)$$

To estimate the contact area or Hertz stress the circular contact equations can be used with R_e replacing R. For further details of this analysis the reader is referred to Chapter 4 of Johnson, K. L. (1985) *Contact mechanics*, Cambridge University Press.

If the two cylinders make contact with their axes *parallel* so that $\theta = 0$, it follows from eqns (A1.3), (A1.4), and (A1.7) that

$$\frac{1}{R'} = \frac{1}{R_1} + \frac{1}{R_2}, \qquad (A1.10)$$

that is, eqn (3.7) is recovered. On the other hand, if the axes of the two cylinders are *perpendicular* to one another, $\theta = 90°$, then $R' = R_1$ and $R'' = R_2$. It follows that

$$h = \frac{x^2}{2R_2} + \frac{y^2}{2R_1}. \qquad (A1.11)$$

Consequently, for the particular case of a pair of *equal* cylinders crossing at an angle of 90°, contours of constant surface separation close to the contact point will be circles.

Appendix 2
Engineering Sciences Data Unit Guides

ESDU Guides are available from: ESDU International Ltd., 27 Corsham Street, London N1 6UA.

65007 (1965)	General guide to the choice of journal bearing type.
67033 (1967)	General guide to the choice of thrust bearing type.
78026 (1986)	Equilibrium temperatures in self contained bearing assemblies. Part I: outline method of estimation.
78027 (1979)	Equilibrium temperatures in self contained bearing assemblies. Part II: first approximation to temperature rise.
78028 (1978)	Equilibrium temperatures in self contained bearing assemblies. Part III: estimation of thermal resistance of an assembly.
78029 (1978)	Equilibrium temperatures in self contained bearing assemblies. Part IV: heat transfer coefficient and joint conductance.
78035 (1978)	Contact phenomena I: stresses, deflections and contact dimensions for normally loaded unlubricated elastic contacts.
79002 (1979)	Equilibrium temperatures in self contained bearing assemblies. Part V: examples of the complete method.
80012 (1980)	Dynamic sealing of fluids I: a guide to the selection of rotary seals.
81005 (1981)	Designing with rolling bearings. Part I: design considerations in rolling bearing selection with particular reference to single row radial and cylindrical roller bearings.
81037 (1981)	Design with rolling bearings. Part II: selection of single row, angular contact ball, tapered roller and spherical roller bearings.
82014 (1982)	Design with rolling bearings. Part II: special types.
82029 (1982)	Calculation methods for steadily loaded fixed inclined pad thrust bearings.

83004 (1983)	Calculation methods for steadily loaded, off-set pivot, tilting pad thrust bearings.
83031 (1983)	Dynamic sealing of fluids II: a guide to the selection of reciprocating seals.
84017 (1984)	Contact phenomena II: stress fields and failure criteria in concentrated contacts under combined normal and tangential loading.
84031 (1991)	Calculation methods for steadily loaded axial groove hydrodynamic journal bearings.
85007 (1985)	Contact phenomena III: calculation of individual stress components in concentrated elastic contacts under combined normal and tangential loading.
85027 (1985)	Film thicknesses in lubricated Hertzian contacts (EHL). Part I: Two-dimensional contacts (line contacts).
85028 (1985)	Calculation methods for steadily loaded axial groove hydrodynamic journal bearings: superlaminar operation.
86008 (1986)	Calculation methods for steadily loaded axial groove hydrodynamic journal bearings: low viscosity process fluid lubrication.
86040 (1986)	Selection of surface treatments and coatings for combating wear to load bearing surfaces.
87007 (1987)	Design and materials selection for dry rubbing bearings.
88018 (1988)	Selection of alloys for hydrodynamic bearings.
89044 (1989)	Friction in bearings.
89045 (1989)	Film thicknesses in lubricated Hertzian contacts. Part II: point contacts.
90027 (1990)	Calculation methods for steadily loaded central circumferential groove hydrodynamic journal bearings.
92026 (1992)	Calculation methods for externally pressurised (hydrostatic) journal bearings with capillary restriction control.

Appendix 3
Conversion factors

1	To convert units of column 1 to those of 2 multiply by	To convert units of column 2 to those of 1 multiply by	2
Length			
m	3.28	0.304 8*	ft
mm	0.039 4	25.4*	in
μ m, micron	39.37	0.025 4*	μ in
Mass			
kg	0.453 592 37*	2.205	lb
Velocity			
m m s^{-1}	0.197	5.08	ft min^{-1}
Area			
m^2	10.76	0.092 0	ft^2
mm^2	0.001 55	645.2	in^2
Volume			
litre	0.035 31	28.32	ft^3
m^3	35.31	0.028 32	ft^3
cm^3, cc	0.061 0	16.39	in^3
mm^3	6.1×10^{-5}	1.639×10^4	in^3
litre	0.219 98	4.546	gal (imp)
litre	0.26.4 2	3.785	gal (US)
in^3	0.113 61	277.3	gal (imp)
Flow rate			
litre s^{-1}	2.12	0.472	ft min^{-1}
m^3 s^{-1}	13.2×10^3	75.77×10^{-6}	gal (imp) min^{-1}
Density			
kg m^{-3}	0.062 4	16.02	lb ft^{-3}
Force			
kgf	9.81	0.1019	N
N	0.225	4.448	lbf
kN	0.100 4	9.964	tonf
Torque or moment			
N m	8.85	0.113	lbf in

1	To convert units of column 1 to those of 2 multiply by	To convert units of column 2 to those of 1 multiply by	2
Pressure or stress			
N m^{-2}, Pa	0.000 145	6895	lbf in^{-2}, psi
N mm^{-2}, MPa	0.145	6.895	ksi
kPa	9.67×10^{-6}	103.4	atmosphere
Pa	1×10^{-5}	1×10^{5}*	bar
Energy			
kJ	0.947 1	1.056	Btu
J	0.238 9	4.187	cal
Power or heat flow			
kW	1.34	0.746	hp
W	0.056 9	17.6	Btu min^{-1}
Rotational speed			
rad s^{-1}	9.55	0.1047	rpm
rad s^{-1}	0.159	6.28	rps
Viscosity			
cP, $\times 10^{-3}$ Pa s	0.145	6.895	μ reyn
cS	0.001 55	645.2	in^{2} s^{-1}
Temperature interval			
K	1.8	0.555 6	deg F
Temperature			
°C	temp $\times$ 1.8 + 32	(temp $-$ 32) $\times \frac{5}{9}$	°F
K	temp $-$ 273	temp $+$ 273	°C
Specific energy			
kJ kg^{-1}	0.429 9	2.326	Btu lb^{-1}
KJ kg^{-1} K	0.238 9	4.187	Btu lb^{-1} deg F
Thermal conductivity			
W m^{-1} K^{-1}	346.7	0.002 89	Btu ft^{-1} min^{-1} deg F^{-1}

*Exact conversion factors. All others are approximate.

Appendix 4
Characteristic physical properties of materials of tribological interest

Lubricants

Property	Units	'Typical' mineral oil	Water	Air
ρ, density	$kg\,m^{-3}$	880	1000	1.23
η, viscosity	$Pa\,s \times 10^{-3}$ (at 20°C)	70–1000 (see Fig. 1.8)	1	18×10^{-3} (Table 1.9)
K, thermal conductivity	$W\,m^{-1}\,K^{-1}$	0.13	0.58	0.024
c, specific heat	$KJ\,kg^{-1}\,K^{-1}$	1.88	4.2	1.01
κ, thermal diffusivity	$m^2\,s^{-1} \times 10^{-6}$	0.078	0.14	19
α, pressure-viscosity index	Pa^{-1}	2×10^{-8} (see Table 1.8)	–	–

Property	Units	Steel	Aluminium alloys	Brass	Cobalt-bonded tungsten carbide	Natural rubber	Alumina
E, elastic modulus	GPa	210	80	105	~450–650	–	390
ν, Poisson's ratio		0.30	0.33	0.35	0.22	0.5	0.27
$\mathcal{H}$, Vickers hardness,	GPa	~2 (low carbon) to ~9.5 (hardened)	~1–6	~1–5	~13–18	–	~18–22
ρ, density	$kg\,m^{-3} \times 10^3$	~7.8	2.75	8.4	~13–15	~0.8–0.9	3.9
K, thermal conductivity	$W\,m^{-1}\,K^{-1}$	54	150	120	~60–120	0.13	30
c, specific heat	$J\,kg^{-1}\,K^{-1}$	~460	~890	~350	~200	~2000	~800
κ, thermal diffusivity	$m^2\,s^{-1} \times 10^{-6}$	~15	~61	~40	~20–40	~0.068	~9.8

Appendix 5
Hertzian elastic contact

The contact modulus E^* is given by

$$\frac{1}{E^*} = \frac{1 - v_1^2}{E_1} + \frac{1 - v_2^2}{E_2}.$$

The reduced radius of curvature R is given by:

$$\frac{1}{R} = \frac{1}{R_1} + \frac{1}{R_2}.$$

(a) *Line contacts* (load W/L per unit length):

semi-contact width $a = \sqrt{\dfrac{4WR}{L\pi E^*}}$;

maximum contact pressure $p_0 = \dfrac{2W}{L\pi a} = \sqrt{\dfrac{E^*W}{RL\pi}}$;

mean contact pressure $p_m = \dfrac{\pi}{4} \times p_0$;

maximum shear stress: $\tau_1 = 0.30 p_0$ at $x = 0$, and depth $z = 0.78a$.

(b) *Circular point contacts* (load W):

radius of contact circle $a = \left\{ \dfrac{3WR}{4E^*} \right\}^{1/3}$;

maximum contact pressure $p_0 = \dfrac{3W}{2\pi a^2} = \left\{ \dfrac{6WE^{*2}}{R^2\pi^3} \right\}^{1/3}$;

mean contact pressure $p_m = \dfrac{2}{3} \times p_0$;

approach of distant points $\varDelta = \dfrac{a^2}{R} = \left\{ \dfrac{9W^2}{16RE^{*2}} \right\}^{1/3}$;

maximum shear stress: $\tau_1 = 0.31 p_0$ at $x = 0$, $z = 0.48a$;

maximum tensile stress: $\sigma_r = \frac{1}{3}(1 - 2v) p_0$ at $r = a$, $z = 0$.

Appendix 6
Surface temperatures in sliding

Stationary conditions

The mean surface temperature rise $(\bar{\Theta} - \Theta_0)$ due to a stationary heat source $2a \times 2a$ and intensity $\dot{h}$ per unit area into a material of thermal conductivity K and initial temperature Θ_0 is

$$\bar{\Theta} - \Theta_0 = 0.946 \frac{\dot{h}a}{K}.$$

Moving heat source

The Peclet number Pe is defined as $\text{Pe} = Ua/2\kappa$.

The maximum surface temperature rise under a source of length $2a$ and intensity $\dot{h}$ moving with velocity U, provided $\text{Pe} \gg 5$, is

$$\Theta_{\max} - \Theta_0 = \frac{2}{\sqrt{\pi}} \frac{\dot{h}a}{K} \text{Pe}^{-1/2}.$$

The mean surface temperature rise under a source of length $2a$ and intensity $\dot{h}$ moving with velocity U, provided $\text{Pe} \gg 5$, is

$$\bar{\Theta} - \Theta_0 = \frac{4}{3\sqrt{\pi}} \frac{\dot{h}a}{K} \text{Pe}^{-1/2}.$$

For $\text{Pe} \ll 1$ use stationary conditions, for intermediate values of Pe see Fig. 3.30.

Appendix 7
Hydrostatic lubrication

For an uncompensated pad bearing of area A supplied with fluid of viscosity η at a pressure p_r the load supported W and volume flow rate Q are given by

$$W = W^* \times A \times p_r \quad \text{and} \quad Q = Q^* \times \frac{p_r h^3}{\eta}$$

where W^* and Q^* are non-dimensional shape factors.

For a circular pad bearing of radius r_o with a pocket radius r_i,

$$W^* = \frac{1 - (r_i/r_o)^2}{2 \ln (r_o/r_i)} \quad \text{and} \quad Q^* = \frac{\pi}{6 \ln (r_o/r_i)}.$$

The width of the land, that is, $r_o - r_i$, is generally approximately 25% of the total bearing width.

If the fluid is supplied at pressure p_s through a restrictor or compensator which reduces its pressure to p_r, so that

$$Q = \frac{C}{\eta} (p_s - p_r),$$

where C is a geometric factor, the optimum bearing stiffness k_o and film thickness h_o are given by

$$k_o = \frac{3 W^* A p_s}{4 h_o} \quad \text{and} \quad h_o = \left\{ \frac{C}{Q^*} \right\}^{1/3}.$$

The recess pressure is then one-half of the supply pressure.

The temperature rise $\Delta \Theta$ is given by

$$\Delta \Theta = \frac{p_s (1 + S)}{\rho c},$$

where $S(> 3)$ is a numerical factor expressing the ratio between frictional and pumping losses and ρ and c are the density and specific heat of the fluid.

Appendix 8
Hydrodynamic lubrication

Reynolds' equation in two dimensions

$$\frac{\partial}{\partial x}\left\{\frac{h^3}{\eta}\frac{\partial p}{\partial x}\right\} + \frac{\partial}{\partial y}\left\{\frac{h^3}{\eta}\frac{\partial p}{\partial y}\right\} = 12\left\{\bar{U}\frac{\partial h}{\partial x} + \bar{V}\frac{\partial h}{\partial y} + (\mathcal{W}_2 - \mathcal{W}_1)\right\}$$

where $\bar{U}$ is the entraining velocity in the Ox-direction and $\bar{V}$ that in the Oy-direction.

Reynolds' equation in two dimensions for a fluid of constant viscosity

$$\frac{\partial}{\partial x}\left\{h^3\frac{\partial p}{\partial x}\right\} + \frac{\partial}{\partial y}\left\{h^3\frac{\partial p}{\partial y}\right\} = 12\eta\left\{\bar{U}\frac{\partial h}{\partial x} + \bar{V}\frac{\partial h}{\partial y} + (\mathcal{W}_2 - \mathcal{W}_1)\right\}.$$

$12\eta(\mathcal{W}_2 - \mathcal{W}_1)$ is the squeeze film term which can be written as $12\eta\,\mathrm{d}h/\mathrm{d}t$. Setting this and $\bar{V}$ to zero:

$$\frac{\partial}{\partial x}\left\{h^3\frac{\partial p}{\partial x}\right\} + \frac{\partial}{\partial y}\left\{h^3\frac{\partial p}{\partial y}\right\} = 12\eta\,\bar{U}\frac{\mathrm{d}h}{\mathrm{d}x}.$$

Reynolds' equation for steady conditions in one-dimensional flow

$$\frac{\mathrm{d}p}{\mathrm{d}x} = 12\eta\,\bar{U}\frac{h - \bar{h}}{h^3}.$$

Hydrodynamic lubrication of discs

If h_c is the minimum or central film thickness, then

$$W^* = \frac{Wh_c}{\eta RL\bar{U}} = 4.00 \qquad \text{for half-Sommerfeld boundary conditions;}$$

$$W^* = \frac{Wh_c}{\eta RL\bar{U}} = 4.89 \qquad \text{for Reynolds' boundary conditions.}$$

Appendix 9
Elasto-hydrodynamic lubrication

If α is the lubricant pressure-viscosity index, η_0 its viscosity at low pressures and q the reduced pressure defined by

$$q = \frac{1 - \exp(-\alpha p)}{\alpha}$$

then Reynolds' equation in one dimension becomes

$$\frac{\mathrm{d}q}{\mathrm{d}x} = 12\eta_0 \bar{U}\frac{h - \bar{h}}{h^3}.$$

Line contacts

With rigid surfaces

$$h_{\min} = 1.66(\alpha\eta_0\bar{U})^{2/3}R^{1/3}.$$

With elastic surfaces

$$h^* = 2.65\, g_1^{0.54}\, g_3^{0.06}$$

where

$$g_1 \equiv (\alpha^2 W^3/\eta_0 R^2 L^3\bar{U})^{1/2}$$
$$g_3 \equiv (W^2/2\eta_0 R\bar{U}E^*L^2)^{1/2}$$

and

$$h^* \equiv Wh_{\min}/\eta_0 RL\bar{U}.$$

Alternatively,

$$\frac{h_{\min}}{R} = 2.65\{2\alpha E^*\}^{0.54}\left\{\frac{\bar{U}\eta_0}{2E^*R}\right\}^{0.7}\left\{\frac{W}{2E^*RL}\right\}^{-0.13}.$$

Over the parallel section of the contact

$$\frac{\bar{h}}{R} = 3.11\{2\alpha E^*\}^{0.56}\left\{\frac{\bar{U}\eta_0}{2E^*R}\right\}^{0.69}\left\{\frac{W}{2E^*RL}\right\}^{-0.1}.$$

Circular point contacts

The minimum film thickness is given by

$$\frac{R}{h_{\min}} = 1.79\{2\alpha E^*\}^{0.49}\left\{\frac{\bar{U}\eta_0}{2E^*R}\right\}^{0.68}\left\{\frac{W}{2E^*R^2}\right\}^{-0.073}$$

while that on the centre line is given by

$$\frac{h_c}{R} = 1.90\{2\alpha E^*\}^{0.53}\left\{\frac{\bar{U}\eta_0}{2E^*R}\right\}^{0.67}\left\{\frac{W}{2E^*R^2}\right\}^{-0.067}.$$

Appendix 10
Greek alphabet

alpha	A, α
beta	B, β
gamma	Γ, γ
delta	Δ, δ
epsilon	E, ε
zeta	Z, ζ
eta	H, η
theta	$\Theta, \theta, \vartheta$
iota	I, ι
kappa	K, κ
lamda	Λ, λ
mu	M, μ
nu	N, ν
xi	Ξ, ξ
omicron	O, o
pi	Π, π
rho	P, ρ
sigma	Σ, σ
tau	T, τ
upsilon	Y, υ
phi	Φ, ϕ, φ
chi	X, χ
psi	Ψ, ψ
omega	Ω, ω

Author index

478 Author index

Subject index

Entries in *italics* refer to figures and in **bold** to tables.